Final Report of the
IGCP-Project No. 92
(Archaean Geochemistry)

Archaean Geochemistry

The Origin and Evolution of the
Archaean Continental Crust

Edited by
A. Kröner, G. N. Hanson and A. M. Goodwin

With 86 Figures

Springer-Verlag
Berlin Heidelberg New York Tokyo 1984

Professor A. KRÖNER

Institut für Geowissenschaften
Universität Mainz
Saarstraße 21
6500 Mainz, FRG

Professor G. N. HANSON

Department of Earth and
Space Sciences
State University of
New York at Stony Brook
Stony Brook, NY 11794, USA

Professor A. M. GOODWIN

Department of Geology
University of Toronto
Toronto, Ontario M5S 1A1, Canada

ISBN 3-540-13746-7 Springer-Verlag Berlin Heidelberg New York Tokyo
ISBN 0-387-13746-7 Springer-Verlag New York Heidelberg Berlin Tokyo

Library of Congress Cataloging in Publication Data. Main entry under title: Archaean
geochemistry. Includes bibliographies and index. 1. Geochemistry—Addresses, essays,
lectures. 2. Earth—Crust—Addresses, essays, lectures. 3. Geology, Stratigraphic—
Archaean—Addresses, essays, lectures. I. Kröner, A. (Alfred) II. Hanson, Gilbert N.
III. Goodwin, A. M. QE515.A73 1984   551.7′12   84-22139

© by Springer-Verlag Berlin Heidelberg 1984
Printed in Germany.

The use of registered names, trademarks etc. in this publication does not imply, even in
the absence of a specific statement, that such names are exempt from the relevant pro-
tective laws and regulations and therefore free for general use.

Typesetting: Fotosatz GmbH, Beerfelden.
Offsetprinting and bookbinding: Konrad Triltsch, Graphischer Betrieb, Würzburg
2132/3130-543210

Preface

Archaean Geochemistry 1972 – 1984

The realisation that the continental crust contains well-preserved relics which date as far back as 4/5 of the Earth's age has given a great impetus to the study of early Precambrian terrains. As late as the mid-sixties the Archaean still constituted the 'terra-incognita' of earth science. High metamorphic grades, poor outcrop, and not least a widely assumed obliteration of early crustal records by convective recycling and thermal reworking had combined to discourage research in this field. Many excellent local studies existed, notably around gold mining centres, but remained unrelated to a broader regional and theoretical understanding. This situation has changed as the consequence of two inter-related factors: (1) advances in isotopic methods and their application to Precambrian rocks, and (2) the recognition that some of the oldest terrains have retained a wealth of primary igneous and sedimentary textures and even geochemical characteristics.

These advances would not have been possible without the vital clues furnished first by field geologists. Detailed documentation of key Archaean terrains such as in West Greenland-Labrador, Zimbabwe, Transvaal-Swaziland, Ontario-Quebec, southern India, Western Australia and recently China and Brazil, coincident with NASA's lunar project dealing with rocks of only slightly older age, has focussed sophisticated laboratory studies on early crustal rocks. The outset of IGCP Project 92 "Archaean Geochemistry" dates back to the period immediately following the identification of early Archaean gneisses at Goodthaabfjord, SW Greenland. At the time few geochemical data were available for Archaean terrains, with the notable exception of the Canadian shield and Scotland. Furthermore, there existed little faith among geochemists in the primary significance of element distribution patterns in metamorphosed rocks, particularly regarding trace elements. An imbalance was evident between field and laboratory studies; on the one hand many geochemical and isotopic data were obtained on rocks collected from widely scattered localities or little-charted terrains, while on the other hand few analytical data were available for some of the best exposed and geologically well documented terrains.

This situation required international collaboration particularly since key Archaean terrains and geochemical/isotopic research centres are often in different parts of the world. The International

Geological Correlation Programme (IGCP) seemed to offer to facilitate collaboration, and the inception and development of Project 92 from Australia in the period 1972 – 1975 met with positive response from the international earth science community. Subsequent to its acceptance by the IGCP Board in 1975, project-sponsored meetings and field excursions (Leicester, 1975; Moscow 1976; Hyderabad 1977; Thunder Bay 1978; Perth 1980; Lake Baikal 1981; Salvador 1982; Beijing 1983) have served as foci for collaboration, and while it would be incorrect to credit the rapid advances in Archaean geochemistry during the decade to the IGCP alone, the project has played an important role in this development. In a number of instances the IGCP contributed directly to the acceptance of specific research proposals, for example in India and in Western Australia.

We wish to refer to a few examples of new findings in Archaean geochemical research, not as a comprehensive list, but reflecting our own bias, and some of these new results are presented in this volume. Some observations of interest were (1) definition of the komatiite suite, including peridotitic and high-Mg basaltic components, and detection of compositional gaps between these components; (2) definition of bimodal basalt-dacite suites vis-a-vis the comparative scarcity of andesites in many greenstone sequences; (3) comparisons between major and trace element data of least-altered komatiites and chondrites, establishing the primary significance of many Archaean rock compositions with implications to early mantle composition; (4) evidence for lateral heterogeneity of mafic-ultramafic volcanic suites with respect to Ti, Ni, Cr and REE; (5) variable ε_{ND} values indicated by Sm-Nd systematics and implying long-term LREE depletion of mantle sources, reservoir heterogeneity and, in some cases, contamination of mantle-derived magmas by older continental crust; (6) the dominantly tonalitic to trondhjemitic chemistry of the older "grey gneiss" components of Archaean terrains and the marked HREE depletion shown by many of these rocks, suggesting residual garnet or amphibole; (7) recognition of the role of fluids in the generation of some granulite terrains; (8) discovery of crustal components that may be as old as 4.2 Ga; (9) establishment of exploration guides such as low Pd/Ir ratios in rocks from which nickel sulfides have been segregated.

Hopefully, the accumulation of new data should serve to place constraints on conceptual interpretations of early crustal history. In contrast, the decade has seen a proliferation of models which fall into a number of categories, including plate tectonic models, modified plate tectonic models, rift or subsidence models, sialic basement models, two-stage mantle melting ensimatic models, etc. While geochemical and isotopic criteria proved successful in identifying the nature of source materials and fractionation processes, outstanding questions remain including, for example: (1)

was there a magma ocean on Earth as postulated for the early history of the Moon? (2) do crustal rocks older than 3.8 Ga really exist and where? (3) was the Earth affected by meteorite bombardment prior to that time and why has no evidence been observed to date? (4) what was the nature of plate motion and ocean-continent distribution patterns? (5) which tectonic processes governed the evolution of granite-greenstone and high-grade gneiss terrains? (6) what type of tectonic setting facilitated two-stage mantle melting, i.e. whether subduction or a temporally unique process? (7) what were the factors underlying the apparent major episodicity of mafic-ultramafic volcanic activity in the Archaean? (8) what was the nature of the Archaean lower continental crust, i.e. at the base of the voluminous granitoid batholiths whose roots are nowhere observed at the present surface? (9) what did the Archaean ocean crust look like, and where can we find its remnants?

The existence of such fundamental questions regarding early crustal origin and evolution hints at yet little understood principles. As these questions become progressively better defined, and the data base against which they can be tested is broadened, so does the promise for new breakthroughs in our understanding of the early Earth.

As in all multi-author volumes of this kind there are problems of standardization of language, style and presentation of data. The spelling follows the English usage, e.g. Archaean, not Archean, etc. The expressions "terrain" and "terrane" are used synonymously. All ages are abbreviated Ma (million years) and Ga (billion years) respectively, and all ages are quoted on the basis of the new decay constants. Cross-references were inserted later in a few instances by the editors and are printed in italics.

The editors express their sincere appreciation to the referees who have devoted considerable time to improve the manuscripts submitted.

This volume concludes the activities of IGCP Project No. 92, and it is hoped that the project and its results will stimulate further research into the fascinating subject of the evolution of the early Earth.

Canberra/Toronto/Stony Brook/Mainz   A. Y. Glikson
                                            A. M. Goodwin
                                            G. N. Hanson
                                            A. Kröner

Contents

Mantle Chemistry and Accretion History of the Earth
H. WÄNKE, G. DREIBUS and E. JAGOUTZ
(With 1 Figure)   1

Geochemical Characteristics of Archaean Ultramafic and
Mafic Volcanic Rocks: Implications for Mantle
Composition and Evolution
S.-S. SUN (With 7 Figures)   25

Archaean Sedimentary Rocks and Their Relation to the
Composition of the Archaean Continental Crust
S. M. MCLENNAN and S. R. TAYLOR (With 8 Figures)   47

Spatial and Temporal Variations of Archaean Metallogenic
Associations in Terms of Evolution of Granitoid-
Greenstone Terrains with Particular Emphasis on the
Western Australian Shield
D. I. GROVES and W. D. BATT (With 4 Figures)   73

Magma Mixing in Komatiitic Lavas from Munro Township,
Ontario
N. T. ARNDT and R. W. NESBITT (With 6 Figures)   99

Oxygen Isotope Compositions of Minerals and Rocks and
Chemical Alteration Patterns in Pillow Lavas from the
Barberton Greenstone Belt, South Africa
H. S. SMITH, J. R. O'NEIL and A. J. ERLANK
(With 7 Figures)   115

Petrology and Geochemistry of Layered Ultramafic to
Mafic Complexes from the Archaean Craton of
Karnataka, Southern India
C. SRIKANTAPPA, P. K. HÖRMANN and M. RAITH
(With 8 Figures)   138

Pressures, Temperatures and Metamorphic Fluids Across
an Unbroken Amphibolite Facies to Granulite Facies
Transition in Southern Karnataka, India
E. C. HANSEN, R. C. NEWTON and A. S. JANARDHAN
(With 5 Figures)   161

Origin of Archaean Charnockites from Southern India
    K. C. CONDIE and P. ALLEN (With 10 Figures)   182

Radiometric Ages (Rb-Sr, Sm-Nd, U-Pb) and REE
    Geochemistry of Archaean Granulite Gneisses from
    Eastern Hebei Province, China
    B.-M. JAHN and Z.-Q. ZHANG (With 11 Figures)   204

The Most Ancient Rocks in the USSR Territory by U-Pb
    Data on Accessory Zircons
    E. V. BIBIKOVA (With 8 Figures)   235

Age and Evolution of the Early Precambrian Continental
    Crust of the Ukrainian Shield
    N. P. SHCHERBAK, E. N. BARTNITSKY,
    E. V. BIBIKOVA and V. L. BOIKO (With 8 Figures)   251

Significance of Early Archaean Mafic-Ultramafic Xenolith
    Patterns
    A. Y. GLIKSON (With 3 Figures)   262

Subject Index   283

Mantle Chemistry and Accretion History of the Earth

H. Wänke, G. Dreibus and E. Jagoutz[1]

Contents

1 Introduction 2
2 Chemistry of the Earth 3
3 Discussion on the Overall Elemental Abundances in the Primitive Mantle 6
4 Accretion Sequence of the Earth 9
5 Consequences of an Inhomogeneous Accretion of the Earth 12
6 The Abundances of some Crucial Elements 14
7 Distribution of Volatiles in Earth's Mantle and Crust 19
References 21

Abstract

The chemical composition of the Earth's primitive mantle (present mantle + crust) yields important information about the accretion history of the Earth. For the upper mantle reliable data on its composition have been obtained from the study of primitive and unaltered ultramafic xenoliths (Jagoutz et al. 1979). Normalized to C 1 and Si the Earth's mantle is slightly enriched in refractory oxyphile elements and in magnesium. It might be that this enrichment is fictitious and only due to the normalization to Si and that the Earth's mantle is depleted in Si, which partly entered the Earth's core in metallic form. Alternatively, the depletion of Si may only be valid for the upper mantle and is compensated by a Si enrichment of the lower mantle.

For the elements V, Cr, and Mn the most plausible explanation for their depletion in the Earth's mantle is their partial removal into the core. Besides the high concentrations of moderately siderophile elements (Ni, Co, etc.) in the Earth's mantle, the similarity of their C 1 abundances with that of moderately volatile (F, Na, K, Rb, etc.) and partly even with some highly volatile elements (In) is striking.

We report on new data especially concerning halogens and other volatiles. The halogens (Cl, Br, I) are present in the Earth's mantle in extremely low concentrations, but relative to each other they appear in C 1 abundance ratios.

To account for the observed abundances an inhomogeneous accretion from two components is proposed. According to this model accretion began with the highly reduced component A, with all Fe and even part of Si as metal and Cr, V, and Mn in reduced state, but almost devoid of moderately volatiles and volatiles. The accretion continued with more and more oxidized matter (component B),

---

1 Max-Planck-Institut für Chemie, Saarstraße 23, 6500 Mainz, FRG

Archaean Geochemistry (ed. by A. Kröner et al.)
© Springer-Verlag Berlin Heidelberg 1984

containing all elements, including moderately and at least some volatile elements in C 1 abundances. The two component inhomogeneous accretion model is discussed in light of the abundances of a number of elements which are especially crucial to the model.

1 Introduction

A number of authors (e.g. Wänke et al. 1973; Ganapathy and Anders 1974) have tried to consider certain groups of elements related to each other by their condensation behaviour as building blocks of the planetary objects. However, only for the refractory elements do we have well-founded evidence that they appear in C 1 (carbonaceous chondrites type 1) abundance ratios relative to each other in all objects of the solar system studied so far (Wänke 1981).

Large-scale fractionations of chemical elements governed by volatility prior to or during the accretion of planets seem to be an undisputable necessity. If not in the solar nebula, these fractionations may have taken place on early-formed planetesimals heated by the decay of the now extinct radioisotope ^{26}Al (half-life 7×10^5 years) as proposed by Wänke et al. (1981). In the model of Ringwood (1970) these fractionations took place during planetary accretion. For some elements volatility is a strong function of the oxygen fugacity, and we must therefore expect respective changes in the condensation or volatilization behaviour. Heating experiments with samples of the C 2 meteorite Murchison showed that elements not volatile under highly oxidizing conditions (In, Zn) can be evaporat-

Table 1. Thermal volatilization studies of Murchison (C 2-chondrite)[a]

Depletion of volatiles (%) a) after heating under H_2O-steam	b) after heating under N_2
C 100	97
Na 0	0
S 100	23
Cl 100	41
K 0	0
Zn 0	41
As 50	7
Se 100	5
Br 100	27
In 0	10
Re 82	0
Os 70	0
W 34	0
Mo 70	0

[a] Temperature: 1000 °C; time 30 min. (Heating under inert atmosphere or vacuum leads to reducing conditions due to the carbon content of the meteorite.)

ed under more reducing conditions. Other elements show the opposite behaviour (Table 1).

2 Chemistry of the Earth

The mass of the Earth's mantle amounts to about two-thirds of the total mass of our planet. Except for the most incompatible elements for which the contribution of the crust (0.59% of the mass of the mantle) becomes important, the mantle determines the chemistry of the Earth for all oxyphile elements.

Violent volcanic eruptions of basaltic magmas carry more or less unaltered solid mantle material to the surface. The nodules of most interest to us are the spinel-lherzolites which represent upper mantle material from depths up to 70 km. These nodules frequently became severely contaminated by various processes before, during, or after their eruption. Therefore, we have to apply various criteria to distinguish between nodules which represent true mantle material and those which contain alteration products. In fact, unaltered or nearly unaltered samples have so far been found only among the spinel-lherzolites. Sheared garnet-lherzolites are often regarded as nearly unaltered mantle samples. However, all those samples studied in our laboratory are considerably contaminated by an interstitial alteration phase (Spettel and Jagoutz 1981).

The eruption process of ultramafic nodules is somehow related to the genesis of the magmas which carry them to the surface and, hence, it is no surprise that many of the nodules are depleted in those elements which are enriched in basalts relative to their mantle abundances.

Fortunately, we have clear cosmochemical constraints to distinguish those nodules which really represent unaltered and primitive or at least close to primitive mantle material if such material exists at all. We only have to remember that in the primitive mantle all the refractory oxyphile elements must be present in their C 1 abundance ratios (Wänke 1981). Here and in the following we will exclusively use data for C 1, compiled by Palme et al. (1981). The refractory elements Al, Ca, Sc, and Yb are all relatively compatible elements, and of none of them can we expect more than 10% of the original mantle concentration to have entered the crust. Hence, nodules with Ca/Al, Al/Sc, or Al/Yb ratios more than 10% of the C 1 value cannot represent primitive mantle material.

Among a great number of spinel-lherzolites studied by Jagoutz et al. (1979), five nodules were found which have not lost significant amounts of Al or Ca and can therefore be called "primitive" according to their major element chemistry. However, only one of them (SC 1) meets the above criteria with respect to trace element concentrations. Among the refractory oxyphile elements depletions are observed for Ba, La, and Ce in nodule SC 1 which otherwise has almost chondritic REE (rare earth elements) abundance patterns.

Measurements of Nd isotopes in SC 1 (spinel-lherzolite from San Carlos, Arizona, USA) proved its primitive, i.e. unfractionated, character with respect to the Sm/Nd system (Jagoutz et al. 1980). SC 1 clearly resided in a reservoir with chondritic Sm/Nd ratio from the accretion of the Earth until about 800 Ma ago and was only then subjected to a slight depletion of Nd and the LREE (light rare earth elements). On the other hand, it also became evident from the Nd data that the depletion of sample KH 1 (Kilbourne Hole, New Mexico, USA) occurred at the same time. In the case of the refractory elements Ti, La, Ba, U, etc. we can actually not expect to find relative C 1 abundances in ultramafic nodules, since for the most incompatible elements the portions that reside in the crust are in the order of 30% to 50%.

In the meantime it was recognized that the San Carlos nodule SC 1 is in fact contaminated with small amounts of a weathering residue widely present in the San Carlos area and generally known as "caliche". Leaching experiments as well as analysis of the contaminating phase showed that appreciable amounts of S, halogens, La and U measured in this nodule are derived from this phase. It also turned out that the nodule SC 1 is very inhomogeneous. In an aliquot of the sample used by Jagoutz et al. (1979) the unleachable K amounts to 127 ppm, while in another sample of SC 1 only 35 ppm K were found (Wänke 1981). It may be that these differences are due to variations of the modal composition. Leaching was done with hot water for the samples analysed for Cl and Br and with diluted HCl in the case of all other elements.

In the approach of Jagoutz et al. (1979) and Wänke (1981), the abundances of highly incompatible elements in the primitive mantle were estimated by the use of cosmochemical constraints. In this work we have calculated the composition of the primitive mantle (whole mantle plus crust) from analytical data of primitive nodules from the upper mantle to which we added the crustal contribution using data of Wedepohl (1975, 1981) and Hunt (1972). In other words, at least as a first step we assume that the primitive spinel-lherzolites represent the composition of the whole mantle. The lower mantle may differ in composition from that of the upper mantle for a number of elements. (We will return to this point in Arndt and Nesbitt and Smith et al., this Vol.) However, for major elements as well as for refractory and compatible trace elements drastic differences between the upper and lower mantle are unlikely. Any larger fractionation of these elements would lead to a considerable change in the ratios of compatible elements or in the ratios of non-mafic to mafic elements (Ca/Sc, Al/Yb, etc.). In SC 1 all these ratios agree within 10% with the C 1 values.

Furthermore, there are a number of lherzolite nodules from localities all over the world with very similar concentrations of oxyphile major and compatible refractory trace elements which differ from C 1 abundances only by a Si deficiency. If the upper mantle would indeed have (C 1 and Si) normalized abundances of Al, Ca, Yb, etc. of about 2.0 and 2.2, as claimed by Anderson (1982a, 1983), where do all the nodules come from that are only enriched in these elements by factors between 1.2 and 1.5 with all the "most primitive" ones falling into this category?

The composition of the primitive Earth's mantle (mantle + crust) obtained in this way is listed in Table 2 and illustrated in Fig. 1, together with a recently published estimate by Anderson (1983). There is also excellent agreement with the estimates of Sun (1982 and this Vol.) derived from his partial melting model using MORBs and Archaean komatiites as a data base.

Anderson (1983) used five components: ultramafic rocks 32.6%, crust 0.56%, MORB 6.7%, kimberlites 0.11%, orthopyroxene 59.8%. The propor-

Table 2. Composition of the primitive mantle

Element	mantle (primitive) (nodules)	Ref.[a]	Crust ×0.59%	Mantle + crust (This work) absol.	norm. to C1 + Si	(Anderson 1983)	Portion in crust %
Mg %	22.22	A	0.014	22.23	1.18	1.05	<0.1
Al	2.17	A	0.049	2.22	1.35	1.17	2.2
Si	21.31	A	0.166	21.48	1.00	1.00	0.77
Ca	2.50	B	0.029	2.53	1.40	1.17	1.1
Ti	0.132	A	0.0031	0.135	1.53	1.33	2.3
Fe	5.86	A	0.029	5.89	0.16	0.16	0.49
Li ppm	2.07	A	0.081	2.15	0.74	0.69	3.8
C	24	C	22.2	46.2	0.66×10^{-3}	–	48
F	16.3	A	3.1	19.4	0.18	0.25	16
Na	2745	A	144	2889	0.29	0.19	5.0
P	60	C	4.5	64.5	0.032	0.027	7.0
S*	8	C	5.2	13.2	1.1×10^{-4}	3.9×10^{-4}	(39)
Cl	0.50	C	11.3	11.8	8.7×10^{-3}	5.6×10^{-3}	96
K	127	C	104	231	0.22	0.14	45
Sc	16.9	A	0.126	17.0	1.43	1.21	0.74
V	81.3	A	0.79	82.1	0.73	0.66	1.0
Cr	3010	A	0.86	3011	0.56	0.42	<0.1
Mn	1016	A	5.0	1021	0.28	0.26	0.50
Co	105	B	0.15	105	0.10	0.096	0.14
Ni	2108	B	0.41	2108	0.097	0.087	<0.1
Cu	28.2	A	0.28	28.5	0.13	0.13	1.0
Zn	48	A	0.45	48.5	0.070	0.051	0.93
Ga	3.7	A	0.11	3.8	0.21	0.21	2.9
Ge	1.31	A	0.0078	1.32	0.021	0.017	0.59
As	0.14	F	0.012	0.152	0.041	–	7.9
Se* ppb	12.6	F	0.9	13.5	3.6×10^{-4}	5×10^{-4}	(6.7)
Br	4.6	C	41	45.6	9.0×10^{-3}	–	90
Rb	276	A	466	742	0.18	0.090	63
Sr ppm	26.0	C	1.73	27.7	1.60	0.90	6.3
Ag ppb	2.51	F	0.41	2.92	6.9×10^{-3}	6.8×10^{-3}	14
Cd	25.5	F	0.59	26.1	0.017	0.012	2.3
In	18.1	F	0.41	18.5	0.115	0.060	2.2
Sb	4.5	F	1.20	5.7	0.022	–	21
Te	19.9	F	0.012	19.9	4.2×10^{-3}	–	<0.1
I	4.2	D	9.1	13.3	0.012	–	68
Cs	1.44	F	7.7	9.14	0.024	0.05	84
Ba ppm	2.40	C	3.20	5.60	1.27	1.08	57
La*	0.35	C	0.171	0.52	1.06	1.11	33
Ce	1.41	C	0.32	1.73	1.35	1.05	18.5
Nd	1.28	C	0.15	1.43	1.50	1.03	10
Sm*	0.49	C	0.033	0.52	1.68	0.99	6.3
Eu*	0.18	C	0.0083	0.188	1.61	1.07	4.4
Gd*	0.69	C	0.048	0.74	1.80	–	6.5
Tb*	0.12	C	0.0060	0.126	1.69	1.16	4.8
Dy	0.73	C	0.036	0.766	1.50	–	4.7
Ho	0.17	A	0.011	0.181	1.58	–	6.1
Er	0.44	A	0.020	0.46	1.38	–	4.3
Yb	0.47	C	0.020	0.49	1.48	0.92	4.1
Lu	0.071	C	0.0034	0.074	1.47	1.14	4.6
Hf	0.26	A	0.0204	0.28	1.16	1.31	7.3

Table 2 (continued)

Element	mantle (primitive) (nodules)	Ref.[a]	Crust ×0.59%	Mantle + crust (This work) absol.	Mantle + crust (This work) norm. to C1 + Si	Mantle + crust (Anderson 1983)	Portion in crust %
Ta ppb	12.6	D	13	25.6	0.91	1.36	51
	37	C	13	50	1.78		26
W	16.4	A	7.7	24.1	0.13	–	32
Re	0.23	F	0.0060	0.236	3.2×10^{-3}	1.3×10^{-3}	2.5
Os	3.1	F	0.0060	3.106	3.2×10^{-3}	3.0×10^{-3}	0.20
Ir	2.8	F	0.0060	2.81	2.9×10^{-3}	3.0×10^{-3}	0.20
Au	0.50	A	0.024	0.524	1.9×10^{-3}	1.7×10^{-3}	4.6
U*	22.2	D	7.1	29.3*	1.78	1.16	(24)
	(47)	C	7.1	(54.1)	(3.28)		

Data for the crust from Wedepohl (1975, 1981), except in the case of C for which the value of Hunt (1972) is used.
Data for the mantle:
A = SC 1 Jagoutz et al. (1979);
B = average of spinel-lherzolites (Jagoutz et al. 1979);
C = SC 1* = SC 1 leached, i.e. SC 1 minus contaminating phase; new data, unpublished and Weckwerth (1983);
D = FR 1 (Jagoutz et al. 1979);
F = SC 1 new data (Nonaka 1982);
* = Abundance in primitive nodules may not be representative for the mantle as a whole; see text. In particular all normalized abundances of refractory elements which grossly deviate from the mean abundance value of 1.3 are suspicious. The largest deviation is observed in the case of U. For U we prefer a normalized abundance of 1.3 for mantle + crust, leading to 21 ppb.

tions of these components were calculated with the constraint to yield chondritic abundance ratios for refractory oxyphile elements. As can be seen from Table 2 the element pattern of Anderson's mantle is very similar to that found here which, in turn, is almost identical to previous estimates published by our group (Jagoutz et al. 1979, Wänke 1981). The high amount of orthopyroxene in Anderson's model calculation consequently leads to noticeably lower abundances for elements that are low in orthopyroxene. His C 1-normalized Yb/Sc ratio, for example, is only 0.76. For the same reason he also finds drastic differences between the upper and lower mantle. It is interesting to note that the abundance of his MORB component of 6.7% very well matches the total mass of the oceanic basalts generated and subducted over 4 Ga, assuming present rates of 20 km^3/a (Williams and von Herzen 1974) to be valid during the whole time. An irreversible loss of the subducted oceanic crust was previously postulated by Ringwood (1971).

3 Discussion on the Overall Elemental Abundances in the Primitive Mantle

Relative to Si and C 1 all oxyphile refractory elements and Mg – with a few exceptions that will be discussed below – have abundances between 1.2 and 1.5.

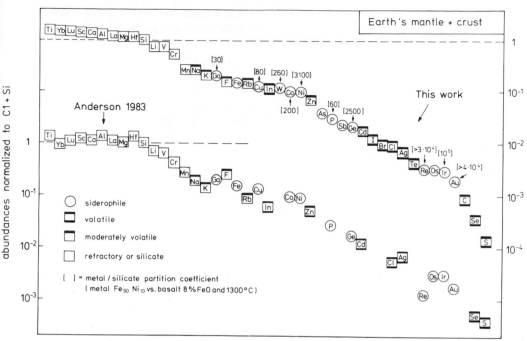

Fig. 1. Composition of the primitive mantle of the Earth (mantle + crust). This work: *upper curve*; Anderson (1983): *lower curve*. The metal/silicate partition coefficients ($D_{m/sil}$), as given in the figure, are derived from experiments with molten silicates (Rammensee and Wänke 1977; Schmitt and Wänke 1984; and unpublished data from our laboratory). In the actual metal segregation (core formation) process the amount of silicate melt will probably not exceed a value of 20%. Hence, in the case of incompatible elements with low solid silicate/liquid silicate partition coefficients like W, P, etc. the actual $D_{m/sil}$ will increase drastically (Newsom and Drake 1983). In the case of Ni, which is readily accepted by olivine (high $D_{ol/liq}$), the $D_{m/sil}$ is lowered. For elements like Ge or Ga with solid/liquid partition coefficients close to 1 $D_{m/sil}$ is independent of the amount of silicate melt. For the relevant conditions $D_{ol/liq}$ is about 7.4 (Irving 1978), which brings the metal/silicate partition coefficient from 3100 down to about 750 under mantle conditions at 1300 °C and low pressure

The depletion of V, Cr, and Mn first noted by Ringwood (1966) is striking, and as we shall see this is very informative. Ringwood and Kesson (1977) favoured a depletion mechanism based on the higher volatility of Cr and Mn compared to Si as observed in the case of CM, CO, and CV chondrites (Kallemeyn and Wasson 1981). Dreibus and Wänke (1979) presented evidence which makes the depletion of Cr, Mn, and V by volatility less likely. Meteoritic basalts from the eucrite parent body (EPB), formed under lower oxygen fugacity than terrestrial basalts, as well as those from the shergotty parent body (SPB), that formed under higher oxygen fugacity, indicate no depletion of Mn on these planetary objects. For both the EPB and SPB depletion of moderately volatile elements like Na or K, comparable or higher to that of the Earth, is beyond any doubt. In the case of Cr the volatility argument was never very convincing, and it fails completely in the case of V, which is in fact a refractory element, and its abundance should be compared to that of Al or Ca. The most likely explanation for the depletion of

Mn and that of Cr and V in the Earth's mantle is their removal into the Earth's core, either in reduced form as metals or sulphides or as oxides. In the case of Mn depletion by volatilization may also have played a minor role. Removal in reduced form was originally proposed by Ringwood (1966). Segregation into the core in form of sulphides has been suggested by Hutchison (1974) to explain the low concentration of Cr and Ti observed in many mantle xenoliths. However, undepleted nodules contain Ti in C 1 abundance relative to the other refractory oxyphile elements. Removal as oxides could be expected if the Earth's core contains a large amount of dissolved FeO as proposed by Ringwood (1977).

The moderately siderophile elements Ga, Cu, W, Ni, and Co are observed in C 1 abundance ratios in spite of their different metal/silicate partition coefficients. The high concentration of Ni in the Earth's mantle has been the concern of many scientists for a long time. Assuming a Ni concentration of 10% in the core, we should expect with a metal/silicate partition coefficient of 750 and under equilibrium conditions a Ni concentration in the mantle of 130 ppm. It has been argued that the partition coefficient is lowered due to high pressure or due to the presence of sulphur or oxygen. These possibilities cannot be ruled out; however, they seem unlikely. Because of the very similar (C 1 normalized) abundances of the elements Ga, Cu, W, Ni, and Co in the mantle we do not only require conditions to lower the partition coefficient for Ni, but conditions which make the partition coefficient of these five elements almost equal. We should also note that the moderately volatile elements K, Na, Rb, and F and the highly volatile element In are similar in abundance to the moderately siderophile elements.

The highly siderophile elements are again highly over abundant compared to the concentrations expected from their partition coefficients. Although less abundant than the moderately siderophile elements by more than an order of magnitude the highly siderophile elements Pd, Ir, Re, and Os appear again in almost chondritic abundances relative to each other. To summarize, we note the following trends in the elemental abundances in the mantle of the Earth (normalized to C 1 and Si):

1. Refractory oxyphile elements (Al, Ca, Ti, Sc, and most refractory trace elements) and Mg *enriched* $\times (1.3 \pm 0.15)$
2. V, Cr, Mn *depleted* $\times (0.25 \text{ to } 0.7)$
3. Fe and moderately siderophiles (Ga, Cu, W, Co, Ni) *depleted* $\times (0.1 \text{ to } 0.2)$
4. Moderately volatiles (Na, K, Rb, F, Zn) and the highly volatile element In *depleted* $\times (0.1 \text{ to } 0.2)$
5. Highly siderophiles (Ir, Os, Re, Au, etc.) *strongly depleted* $\times (0.002)$
6. Highly volatiles (Cd, Ag, I, Br, Cl, Te, Se, C) *strongly depleted* $\times (10^{-2} \text{ to } 10^{-4})$

The uniform enrichment of Mg and the refractory oxyphile elements relative to Si could also be interpreted in terms of a Si deficiency (Ringwood 1958; Wänke 1981). It was suggested that the missing Si went into the core in metallic form, which would indicate gross chemical disequilibrium between mantle and core. Alternatively, the high Mg/Si ratio found for the upper mantle might be compensated by a smaller ratio in the lower mantle (Liu 1979). The slightly higher abundances of Al, Ca, and refractory trace elements (average 1.3) as compared to Mg (1.18) might be real but could also reflect inhomogeneous distribution (less Al, Ca, etc. than Mg in the lower mantle). Lithium is variable in the nodules studied by Jagoutz et al. (1979), and its depletion follows La. Its small depletion in SC 1 (abundance 0.74) may be explained either by its incompatible character, by volatility or by a combination of both.

4 Accretion Sequence of the Earth

To explain the observed elemental abundance pattern of the Earth's mantle the following inhomogeneous two component accretion model has been proposed (Wänke 1981).

Accretion started with *highly reduced* material free of moderately volatile and volatile elements but containing all other elements in C 1 abundance ratios. Iron and all siderophile elements occurred as metals (W), Si partly as metal, Cr, Mn, and V as metals or sulphides. This is *component A*. Due to the high temperature reached during accretion (Safranov 1978; Kaula 1979) segregation of metal, i.e. core formation, will occur contemporaneously with accretion (Solomon et al. 1981).

After accretion of about two-thirds of the Earth and after core formation more and more *oxidized material* (Fe, Co, Ni, W as well as all other siderophile and oxyphile elements as oxides) was added, containing all elements including moderately volatile and at least some volatile elements (In, etc.) in C 1 abundances. This is *component B*.

As accretion proceeded, metal from component A, still present but in decreasing amounts, is thought to be responsible for complete extraction of highly siderophile elements (Ir, Au, etc.) into the core. In fact, an Fe-FeS alloy will be formed, as this metal will react with the material of component B. Ringwood (1983) has recently suggested that FeS, on reaction with FeO, will generate a low melting FeO-FeS eutectic. Noble metals would strongly partition into this FeO-FeS melt and become extracted from the mantle. Ringwood postulates that up to 5% of such an FeO-FeS melt would not significantly change the abundance of the moderately siderophile elements Ni, Co, Cu, W, and Ga.

Component B also contained volatile elements such as halogens, S, H_2O, and carbon. The latter occurred mainly in the form of organic compounds as found in C 1 chondrites which, during impact on the Earth, were transformed into H_2O and CO_2. The presence of H_2O as well as of Fe^{3+} were responsible that metallic Fe finally became unstable and that the highly siderophile elements contained in the last 0.2% of the mass added to the Earth remained in the mantle in chondritic abundance ratios. However, lack of chemical equilibrium between metallic iron and silicates must also be considered in this respect. We will return to this point in Smith et al., this volume.

The moderately volatile elements and volatile elements as well as moderately siderophile elements observed in the Earth's mantle are exclusively derived from component B. Thus, the similarity of the C 1 normalized abundances of moderately volatile (F, Na, K, Rb, etc.) and moderately siderophile elements (Ni, Co, Cu, Ga, W, etc.) in the Earth's mantle is readily explained. Assuming a C 1 composition for component B it is estimated that, in total, the Earth consists to about 85% of component A and 15% of component B.

Ringwood (1977, 1979) has previously proposed a model to build the inner planets from a reduced volatile-free and an oxidized volatile-containing component. He assumed homogeneous accretion in which both components are added simultaneously to the growing planet and are equilibrated with each other. In his homogeneous accretion model the high abundances of moderately siderophile elements in the Earth's mantle are explained by a decrease in the respective metal-silicate partition coefficients due to the presence of FeO in the metal phase. Consequently, Ringwood proposed oxygen to be the principal light element in the core. He argues and has recently presented experimental evidence (Jackson and Ringwood 1981; McCammon et al. 1983) that at temperatures above 2500 °C large amounts of FeO dissolve in metallic iron. The key problem in his model is that melting of the metal phase has to be delayed until the melting point has risen to at least 2500 °C which requires a pressure in the order of 50 GPa. However, due to the high accretion temperature (Kaula 1979) metal segregation will take place almost contemporaneously with accretion. Melting of the metal phase will be facilitated by an admixture of C, Si, P, and S in reduced form. That is why we prefer a mixture of these elements to have originally entered the core during accretion. After segregation of the major portion of metallic iron it may well be that considerable amounts of FeO from the lower mantle were transferred into the core because of disproportionation of FeO as proposed by Ringwood (1979). The resulting change of the FeO content of the Earth's mantle was briefly discussed by Jagoutz and Wänke (1982).

The new interpretation of the Hugoniot data for iron by Brown and McQueen (1982) may significantly reduce the density difference between liquid iron and that of the Earth's core. Consequently, the required amount of light elements admixed to FeNi is reduced. For example, 5 to 10% S would be sufficient compared to the previous estimate of 9 to 12% S (Ahrens 1979)

As outlined above, the observed chemical composition of the Earth's mantle − though under certain circumstances not really in contradiction with homogeneous accretion − seems to be explained in a more straightforward way by an accretion sequence in which the oxidized component is added in larger proportions only after the accretion of about two-thirds of the Earth's mass and after segregation of the metal phase. (i.e. after core formation). Ringwood (1983) has recently proposed a model in which the composition of the accreting matter also changes somewhat during accretion in the direction from component A to that of component B.

Anders (1965) applied a two component model originally formulated for meteorites (Anders 1964) to also explain the abundance patterns of planets but did not pursue this model in later publications (Ganapathy and Anders 1974).

The required change in oxygen fugacity of the accreting material fits into several models for the solar nebula. It is easy to visualize a scenario in which loss of H_2 from certain regions of the solar nebula is more rapid than that of H_2O which, for example, could be held back in form of ice grains.

The high depletion of all moderately volatile elements such as Na, K, F, Zn, etc. on Earth and other objects of the inner solar system is a more severe constraint. Temperatures in the order of 1 000 K are required to fractionate these elements from the more refractory ones. According to current models it is questionable that such temperatures had been reached in the region where the Earth formed (Cameron 1978). Small objects could have lost moderately volatile elements during or after accretion, but this seems impossible for large objects like the Earth.

We have previously tried to visualize a solar nebula in which the only fractionation of importance was that of water (Wänke and Rammensee 1981). Water would only be added in appreciable quantities to planetesimals in regions where it occurred in the form of ice or hydrates. Carbon seems to be the most abundant element present in solid form in interstellar matter, either in form of graphite (Mathis et al. 1977) and/or in form of "refractory" organic compounds (Greenberg 1982).

Hence, in planetesimals which accreted from interstellar material unaltered by a hot solar nebula, the abundance of carbon may be comparable to that of silicon. Depending on the accretion temperature, planetesimals would or would not contain H_2O. In the absence of water (inner part of the solar nebula) primary objects heated by the decay energy of ^{26}Al (Lee et al. 1976) and by chemical energy (Clayton 1980) become strongly reduced by carbon. Hence, not only FeNi etc. but also large amounts of Cr, Mn, V as well as a few percent of Si will be converted into metals or sulphides. However, as pointed out by Ringwood (1983), it may not be possible to produce a metal with 10% or more Si in this way.

Parallel to reduction, other highly volatiles will be lost from the primary objects, together with large quantities of CO and CO_2, while elements with condensation points >500 K would not be affected. However, collisions of objects and also explosions of objects due to the build-up of internal gas pressure will frequently lead to total disruption of primary objects. If these objects contain molten material, many droplets will be produced which loose their volatile and moderately volatile elements very efficiently (Wänke et al. 1981; Zook 1981). All that which is required then is a process which separates the devolatilized droplets from the fine smoke-like volatile-rich matter. Component A could be generated in this way.

Planetesimals containing large amounts of H_2O will only reach considerably lower temperatures since the heat sources will mainly be used up for the evaporation of H_2O and the reaction $C + 2H_2O \rightarrow CO_2 + 2H_2$. Highly oxidized planetesimals which retained most of their volatiles and which are compositionally similar to C 1 chondrites will result (component B). Thus we might have a scenario which fulfils the conditions of the accretion history of the Earth and which will result in an increase in the bulk oxygen content from planet Mercury to Mars.

We would like to underline that the two components as described here should only be considered as an approximation towards reality. The process responsible for the creation of component A (reduction accompanied by removal of volatile and moderately volatile elements) will not always lead to the visualized end products. However, it seems that the actual situation can be successfully described in the form of two components which physically should be addressed as end members of a fractionation sequence. We expect that many objects have not reached the reduction stage because of their high H_2O content, but were only heated to moderate temperatures. They will have lost only elements that are volatile under oxidizing conditions. As seen from Table 1 (left column) these elements include C, S, Se, Cl, Br, etc. but not In and Zn. The limited temperature will also prevent loss of elements like Na and K but may affect Cs, Pb, etc. Furthermore, we will have objects which are mixtures by themselves among the accreting matter.

Sun (1982) has tried to explain the abundance of elements in the Earth's mantle by dividing them into two major groups (siderophiles and lithophiles, i.e. core-mantle differentiation) and by fractionation of the two groups prior and during accretion by volatility-controlled processes. However, his appealing approach contains a number of inconsistencies. The siderophile elements Cu and Ga fall on the fractionation curve of lithophiles. New data by Newsom and Palme (1984) yield a Si-normalized mantle abundance of only 0.031 for the refractory siderophile element Mo. The fractionation curve for siderophile elements is based on P, its chemical homologues (As and Sb), and Ge. For the mantle abundance of these elements we refer to the discussion in Smith et al., this volume.

5 Consequences of an Inhomogeneous Accretion of the Earth

In the preceding chapter we have tried to summarize the evidence in favour of inhomogeneous accretion of the Earth as advocated previously by our group.

We will now investigate the consequences of inhomogeneous accretion of the Earth and will also try to find further evidence for this model. During the first phase of accretion component A is thought to be dominant, i.e. only small amounts of moderately volatile and volatile elements are added to the growing Earth. Water is not stable during this phase as it will react with metallic iron Fe^0 $+ H_2O \rightarrow FeO + H_2$. If metallic Si should be present we would have even lower oxygen fugacities. In any case, the relative abundance of FeO will certainly be considerably below the present value of the Earth's mantle. Carbon, if present, would also be in its elementary form.

Due to the energy of accretion, the Earth is heated to temperatures above the solidus of the silicates (Kaula 1979). As the temperature will be highest in regions close to the surface of the growing Earth, melting will begin from the outside. However, segregation of metal − starting already at low degrees of partial melting − will generate additional energy at greater depth. It is expected that from a time when the Earth has reached about 10% of its final mass, metal segregation and core formation will occur almost simultaneously with accretion.

After the Earth has reached about two-thirds of its present mass the mixing ratio component B/component A increases considerably. Together with the abundance of moderately volatile elements and metal oxides the abundance of H_2O and CO_2 also increases. The high temperatures reached during the accretion process (Kaula 1979) will result in an almost complete outgassing of the accreting matter, already severely devolatilized by the impact itself (Lange and Ahrens 1982, 1983), and only trace amounts of H_2O and CO_2 will remain inside the growing planet. However, it can be expected that carbon was present in reduced form in component A and, hence, the lower mantle may contain some carbon.

A thick atmosphere − H_2O being the main constituent − will develop and its thermal blanketing effect will lead to a magma ocean with a depth of the order of 10^2 km. Solubility of the atmospheric gases in this magma ocean governs the atmospheric pressure which, in turn, controls the depth of the magma ocean by its blanketing effect (Hayashi et al. 1979; Mizuno et al. 1980; Matsui and Abe 1984). Volatiles (Cl, Br, I, Tl, Pb, etc.) and incompatible elements will become enriched in the magma ocean relative to the solid mantle which will be very dry and devoid of volatiles. Afterwards the mantle will act as a sink rather than a source of volatiles (Arrhenius 1981). Until the Earth reaches about 50% of its present mass the mantle is low in FeO, which is later added by component B. A considerable fraction of FeO is possibly formed by the reaction of H_2O from component B with metal from component A, generating large amounts of H_2 which, on escape, might take heavier volatile species with it (see Srikantappa et al., this Vol.).

The energy of accretion retained in the whole Earth is about $1.8 \cdot 10^{31}$ J (Kaula 1979), the energy release during core formation is $1.5 \cdot 10^{31}$ J (Flasar and Birch 1973), while the energy produced during 4.55 Ga by the decay of K, Th, and U is $0.6 \cdot 10^{31}$ J. Hence, about 85% of the total energy of the Earth were released during accretion. As the estimates for the accretion time are in the 1% range of the age of the Earth (Wetherill 1978), the energy output during accretion was about 500 times the mean energy output by the decay of the radioactive elements. Today the turnover time of the Earth's mantle by convection is in the order of some 10^8 years. Therefore, we estimate the turnover time of the Earth's mantle during accretion to be less than 10^6 years, a value small relative to the accretion time of about $4 \cdot 10^7$ years (Wetherill 1978). Component B was apparently added to a vigorously convecting mantle, the outer layer of which was molten.

The highly convective mantle led to a nearly perfect homogenization of the silicate portion of component A with component B, except for those elements which, as a result of their incompatible character or their volatility, will concentrate in the magma ocean. The solid and the molten portion of the mantle convect independently because of the large difference in viscosity. As accretion progresses the solid/liquid boundary will continuously move upwards due to crystal settling at the bottom of the magma ocean. Hence all compatible elements of component B are readily mixed into the solid portion of the mantle. The enrichment of incompatible elements in the magma ocean must be compensated by only minor depletion of the solid mantle. Otherwise it is difficult to understand the existence of almost primitive mantle reservoirs as evidenced by primitive xenoliths. This may be accomplished by keeping the amount of melt small but also, and perhaps more likely, by incorporation of melt into the solid as trapped liquid. In this respect we disagree with the model proposed by Anderson (1982b), in which the composition of the solid portion is governed by a large degree of melting ($>15\%$).

When its final mass is reached we have an Earth which is already inhomogeneous for a number of elements. Volatile elements like the halogens, Pb, Tl, etc. will be strongly concentrated in the magma ocean. The reservoir of the magma ocean as a whole will also be enriched in incompatible elements. The enrichment may be somewhat higher for the moderately volatile elements which are supplied by component B towards the end of accretion. On crystallization the material of the original magma ocean was subjected to various additional fractionation processes. Some portions of it probably ended up in the crust, others were subducted and may in part be linked to ancient eclogites (Jagoutz et al. 1984).

From the moment at which plate tectonics and plate subduction start both refractory and volatile incompatible elements will be returned into the mantle via the oceanic crust but in part also from the continental crust in form of sediments that overlie the oceanic basalts (Armstrong 1968, 1981).

Subducting solid material will always loose most of its volatiles and also a considerable part of its incompatible elements. Thus, the mantle will always remain highly depleted in all volatile elements.

6 The Abundances of some Crucial Elements

According to the two component inhomogeneous accretion model described above, all moderately volatile and moderately siderophile elements should have equal abundances (concentrations normalized to C 1) in the Earth's mantle. However, the abundances of elements ranging between moderately and highly siderophile or having, under certain circumstances, high effective metal-silicate or sulphide-silicate partition coefficients may be influenced by equilibration with small amounts of metal, respectively sulphide. In the case of elements which are distributed inhomogeneously within the Earth's mantle our estimates might be considerably in error as we assumed the regions of the upper mantle sampled by the spinel-lherzolites to be representative for the whole mantle. Within this frame we will now discuss the abundances of some crucial elements.

Zinc. The estimated normalized abundance of 0.07 for Zn is considerably lower than that of other moderately volatile elements (like K, F, Cu). Zinc is a highly compatible element; it easily enters olivine and orthopyroxene. Hence, one can expect that Zn is evenly distributed over the whole mantle. In our approach to estimate the elemental abundances in the primitive mantle we have assumed that the concentrations observed in the upper mantle hold for the whole mantle. If the less compatible elements (Na, Cu, F, K, etc.) are slightly concentrated in the upper regions of the mantle, it is quite natural that we find lower abundances for compatible elements which distribute more easily over the whole mantle.

K, Rb, Cs. It seems that more than 50% of the total inventory of all highly incompatible elements (K, Rb, Cs, Ba) reside in the crust. Hence, the accuracy of the estimated abundances for the primitive mantle depends on the accuracy of the crustal estimates. Furthermore, the question of homogeneous or inhomogeneous distribution of these elements in the mantle is of consequence to their abundances in the crust-mantle system. From the amount of ^{40}Ar in the Earth's

Table 3. Concentrations of highly incompatible elements in spinel-lherzolite nodules[a]

	KH 1	PO 1	Ka 168	Fr 1	SC 1	Carb. chondr. type 1
K ppm	(6.5) 1.9[b]	(33)	10	8.1	138 127[b]	517
Rb ppm	0.018	0.13	0.075	0.048	0.276	2.06
Cs ppb	0.51	0.86	1.53	1.75	1.44	190
Sr ppm	5.62 5.0[b]	7.51	9.25	14.8	26.7 26.0[b]	8.6
Ba ppm	(2.38) 0.70[b]	(4.10)	2.05	1.91	2.60 2.40[b]	2.2
La ppm	0.051 0.042[b]	0.081	0.18	0.36	0.51 0.35[b]	0.245
U ppb	(8.1) 3.8[b]	(14.8)	9.1	22.2	59.7 47[b]	8.2
Cl ppm	0.5	1.4	1.93	1.4	0.95 0.50[b]	678
Br ppb	5.1	7.0	22.0	12.0	8.0 4.6[b]	2.53
I ppb	<2	–	<6	4.2	(40.5)	0.56
C ppm	73 56[b]	98 21[b]	–	103 37[b]	55 24[b]	35000
S ppm	20 14[b]	13 8[b]	–	<2 <2[b]	13 8[b]	58000
H_2O ppm	(660)	(740)	–	(1300)	(840)	–
K/Rb	361	254	133	165	500	251
K/U	550[b]	(2230)	1100	368	2700[b]	63000
Rb/Sr	0.0032	0.017	0.0081	0.0032	0.012	0.24
Rb/Cs	35	151	49	28	191	10.8
Ba/U	184[b]	277	225	86	51[b]	268
Cl/Br	98	200	88	117	119	268
Br/I	>2.6	–	>3.7	2.9	–	4.5

[a] Data from Jagoutz et al. (1979), Nonaka (1982), and this work; C 1 data: Palme et al. (1981). Values in parentheses for cases in which large surface contaminations are suspected.
[b] Denotes leached samples, see text.

atmosphere an absolute lower limit of an average concentration of K is found that would correspond to a K concentration of 117 ppm. Except for SC 1 (and SC 1*) all nodules recognized by Jagoutz et al. (1979) to be unaltered and representing primitive mantle material with respect to major and compatible trace elements contain surprisingly small amounts of K. Part of the K content of these nodules is, in fact, the result of contamination at the surface (see the leached sample KH 1* in Table 3), and the indigenous concentrations are even lower. Hence, our approach using the least depleted nodule to monitor todays' mantle abundances may lead to an overestimation of the primitive mantle abundances in the case of K but also in the case of other highly incompatible elements.

Ba. Contrary to the variation of La by about a factor of 8 between the least and most depleted spinel-lherzolites of Jagoutz et al. (1979), the concentration of

Ba, one of the most incompatible elements, are surprisingly constant (Table 3), even taking into account the lower concentration in the acid leached sample KH 1*. This might point towards metasomatic alterations and equilibration within large mantle volumes, caused by mobile fluid phases highly enriched in Ba and other incompatibles. The concentrations of Cs, Rb, K, and possibly U may be influenced in a similar way.

REE. The abundance of La (about 1.1) is significantly lower that that of the more compatible heavy REE (Yb = 1.48). This may point towards the existence of a reservoir in which La has smaller depletions than in the reservoir sampled by the most primitive spinel-lherzolites. As the majority of ultramafic xenoliths are considerably lower in La as FR 1 and SC 1, there must be huge reservoirs considerably depleted in the upper mantle, thus making the deficit in the LREE even larger. Anderson (1983) has tried to account for this apparent discrepancy by postulating the existence of a highly enriched reservoir (kimberlite composition), residing in the upper mantle and amounting in mass to one-fifth of the crust. The LILE (large ion lithophile elements) enriched liquids responsible for metasomatic alterations exhibited by many mantle xenoliths may in fact be related to this reservoir.

The REE patterns of the most primitive spinel-lherzolites studied by Jagoutz et al. (1979), i.e. SC 1, SC 1* and FR 1, show a flat maximum in the region between the elements Nd and Dy relative to the light REE and the heavy REE. For mantle + crust we find an overabundance of these elements. Again we have to remember that we have used samples from the upper mantle to represent the whole mantle. As discussed above me have to expect a migration of incompatible refractory elements from the lower into the upper mantle. Kimberlitic liquids may be responsible for this transport. In any case, addition of elements from the lower mantle will be highest for the LREE and only modest for elements heavier that Nd. However, the latter elements have been concentrated in the crust to a small degree only, thus their overabundance in the upper mantle may be explainable.

U. Taking the U concentrations measured in SC 1* and FR 1 at face value, we obtain abundances (above 1.8) higher than those of any other element. Assuming a C 1-normalized uranium abundance equal to that of other refractory elements and Mg ($\simeq$ 1.3), a mean U concentration of 21 ppb is obtained. This value comes close to the U content of FR 1 and would indicate that there are mantle regions that are almost devoid of K and considerably depleted in Ba and La, but that kept all their uranium. To avoid this unacceptable solution, we propose that the large concentration of U found in the least depleted reservoirs of the upper mantle are due to a transport of U from the lower into the upper mantle, thus overcompensating the extraction of U from the upper mantle into the crust. In this respect it is important to mention that the Pb isotopes in spinel-lherzolites (including one of the San Carlos area) plot in the MORB field, indicating high U concentrations in the upper mantle on a time-integrated basis (Zartman and Tera 1973).

U/Pb and Rb/Sr System. Compared to their variation in meteorites, the variations of Sr and Pb isotopes in the terrestrial mantle are very small. In meteorites the chemical abundances are mainly determined by volatility, hence the Rb/Sr and U/Pb ratios are anticorrelated. In consequence, the Sr and Pb isotopes are highly variable and are also anticorrelated at first order. In the terrestrial mantle anomalous radiogenic ^{206}Pb is observed. This anomalous Pb is an important constraint in all models for the evolution of the Earth. Volmer (1977) and Dupré and Allègre (1980) explained the variation in the Pb isotopes by continuous segregation of Pb with the sulphides to the core. Vidal and Dosso (1978) suggested that Rb might also be chalcophile and follow Pb into the core.

Segregation of a pure sulphide phase as observed in the case of Mars (Shergotty parent body; Burghele et al. 1983) seems highly unlikely for the Earth. As pointed out by Newsom and Palme (1984), it would strongly affect the abundance of other chalcophile elements like Mo in the terrestrial mantle, which is not observed. Hofmann and White (1982) suggested that some U from the upper continental crust may be transported back into the mantle by the river influx to the oceans, subsequent uptake by the oceanic crust and its subduction. In contrast, Ringwood (1982) suggested that U is retained in certain mantle minerals (e.g. garnet).

In our model of an inhomogeneous accretion the early Earth would start with a nearly homogeneous distribution of U over the whole mantle, only slightly affected by its enrichment in the material of the original magma ocean.

Lead is supplied by component B towards the end of accretion. As a volatile element it is concentrated in the magma ocean. Later Pb was gradually mixed into the mantle. Hence, the U/Pb ratio in the outer regions of the mantle is expected to have increased with time. In addition, U from the lower part of the mantle is transported upwards. Depending on the changes of the terrestrial convection system a continuous or episodic increase of the U/Pb-ratio could be explained by this scenario (Cumming and Richards 1975; Stacey and Kramers 1975). Transport of U from the deeper part to its source region could explain the anomalous ^{206}Pb/^{204}Pb ratio observed in MORB-basalts. For highly volatile elements it may be difficult to reach the lower mantle, hence the lead in the lower mantle possibly always has been, and still is, very radiogenic.

The Rb/Sr ratios of the spinel-lherzolites studied by Jagoutz et al. (1979) show only a relatively small range of variation (Table 3). The absolute values of these ratios are, however, between 3 to 10 times below the estimated value of the bulk Earth of about 0.029 (DePaolo and Wasserburg 1976; Hofmann and White 1983), indicating a strong Rb depletion even in the least depleted nodules. Part of the Rb measured in the nodules may be from surface contamination, thus making the actual depletion even larger. At least for the reservoir sampled by SC 1 most of the Rb depletion probably occurred at the time when the originally chondritic Sm/Nd ratio was disturbed, i.e. 800 Ma ago (Jagoutz et al. 1980).

Because of Rb concentration in the crust, the Rb abundance as well as the Rb/Sr ratio of upper and lower mantle are considerably below the values for the bulk Earth. In contrast, the upper mantle is not depleted in U and Th, having an overall higher and variable U/Pb ratio compared to the single stage Pb evolution of the Earth. Nonradiogenic Sr and highly radiogenic Pb observed in MORB and OIB (ocean island basalts) is in line with such a distribution of Rb and U relative to Sr, respectively Pb (Zindler et al. 1982). It is also possible that the concentrations of the most incompatible elements observed in MORB do not necessarily reflect their abundance in the MORB source regions, but may be influenced by interaction with fluid phases rich in highly incompatible elements.

FeO. Archaean basalts contain more FeO than modern basalts (Glikson 1979; Jagoutz and Wänke 1982). The FeO content of the mantle may have been lowered due to transfer of FeO from the lower mantle into the core. As discussed by Ringwood (1979) disproportionation of FeO may be the responsible process. Nickel and Co may be similarly affected.

Ge. The abundance of Ge (= 0.021), almost a factor of 5 below that of Ni, seems to argue strongly against an inhomogeneous accretion of the Earth. Germanium is only moderately volatile, but in the Earth's mantle it is less abundant than Zn or Cu.

We have explained the even higher depletion of Ir and other highly sidero-philes by the presence of small amounts of metal. Segregation of this metal lowered the abundance of Ir etc. drastically, but left the moderately siderophiles (Ni and Co) almost unaffected. Recent determinations of metal/silicate partition coefficients showed that the partition coefficient of Ge indeed falls between those of Ni and Ir. Schmitt and Wänke (1984) found that D(Ge) = 2500 for lg fO$_2$ = − 12.7 and 1300 °C. In addition to the partition coefficient it is a kinetic effect which allows the more volatile Ge to equilibrate faster and, hence, more complete with a metal phase than Ni or Ir. Rambaldi et al. (1978) and Rambaldi and Cendales (1979) observed the following metal/silicate concentration ratios in the equilibrated bronzite chondrite Pultusk: Ir = 48, Au = 128, Ni = 218, Ge = 1200. Hence, the concentration ratios do no follow the actual partition coefficients but merely reflect the volatility, i.e. the mobility of the elements.

Au. In connection with the kinetic effect discussed for Ge, it is worthwhile to note that the Au/Ir ratio in nodules analysed by Jagoutz et al. (1979) have a mean value of 0.13 ± 0.1 compared to the C1 Au/Ir ratio of 0.29. The nodules studied by Morgan et al. (1981) also show Au/Ir ratios considerably below the C1 ratio. Both Ir and Au are highly siderophile elements; however, Au is much more volatile than Ir, though not quite as volatile as Ge. We interpret the apparent depletion of Au relative to Ir as an indication that the more mobile Au found its way to equilibrate more effectively with metallic iron that Ir. In the case of Ge this effect should even be larger. In fact, Au is also more chalcophile than Ir. However, a trend towards depletion of chalcophile elements like that found for Mars (SPB, Burghele et al. 1983) is not observed for the Earth's mantle.

P. The abundances of P and its chemical homologues As and Sb are considerably below the abundances expected according to the inhomogeneous accretion model. Phosphorus has siderophile tendencies which are even larger if the metal is in liquid phase. As shown by Newsom and Drake (1983) the liquid-metal/liquid-silicate partition coefficient of P is 160 (for lunar conditions) which is in the range of that of Co. Phosphorus, contrary to Co, behaves as an incompatible element and will concentrate in partial melts. As pointed out by Newsom and Drake (1983), the actual extraction of an element by a metal phase is greatly enhanced in such cases, in which the element is concentrated in a partial melt or residual liquid of small degree. We must therefore expect a considerable depletion of P by even small amounts of metal.

7 Distribution of Volatiles in Earth's Mantle and Crust

Previous measurements of Cl and Br (Jagoutz et al. 1979) yielded very low concentrations of these elements in unaltered spinel-lherzolites (Cl = 0.4 – 2 ppm, Br = 5 – 13 ppb). Recently we found similarly low concentrations of iodine in these samples (I = 2 – 4 ppb) (Table 3). However, the Cl/Br and Br/I ratios match the corresponding C 1 values almost exactly. Hence, the assumption of C 1 ratios for Cl, Br, and I being valid for the whole Earth seems well justified as these ratios are also found in garnet-lherzolites and in crustal rock composites.

The water and carbon contents in spinel-lherzolites are in the range of 0.066 – 0.13% and 21 – 103 ppm, respectively. Extremely low concentrations of S were found in these samples (8 – 20 ppm). We have listed all data on water in parentheses in Table 3 as considerable takeup of water by the olivine-rich nodules is to be expected at the surface. Hence, the data probably do not reflect the true abundance of water in the upper mantle.

Partial loss of sulphur, halogens and other volatiles during eruption of the ultramafic nodules as observed for subaerial basalts (Unni and Schilling 1978; Moore and Fabbi 1971; Sakai et al. 1982) cannot be excluded. In particular the S and Se abundances in Table 2, based on the observed concentrations in SC 1, may have been lowered considerably by such a process.

In order to investigate such a partial loss in the case of the halogens, we compare, in Table 4, the concentrations of incompatible volatile and some non-volatile incompatible elements in the spinel-lherzolite PA 15 A with the concentrations of these elements in the basanite Pa 53 B. It was shown by Zindler and Jagoutz (unpubl. data) that the spinel-lherzolite is in isotopic equilibrium, for Nd and Sr, with the basanite PA 53 B. As can be seen from Table 4 the basalt/lherzolite ratio for Cl is 234, i.e. between that of Sr and La. For the elements F, Sr, La, Ba the enrichment factors found for basanite PA 53 B almost match respective values obtained by Schilling et al. (1980) who compared basalts from different sections of the mid-atlantic ridge. In the case of Cl Schilling et al. (1980) found an enrichment factor equal to that of the most incompatible elements Ba, Rb, Cs. We suspect that the Cl concentration measured in PA 15 A may still be

Table 4. Comparison of various incompatible elements in spinel-lherzolite PA 15 A and its host basanite PA 53 B from San Carlos (Both samples were kindly supplied by M. Prinz, see also Frey and Prinz 1978)

	F	Sr	La	K	Ba	Cl
Basanite PA 53 B	ppm 652	856	40.4	20150	380	234
Spinel-lherzolite PA 15 A	ppm 5.5	6.27	0.11	28.8	0.54	1.0
Ratio Pa 53 B/PA 15 A	119	137	355	699	703	234

too high because of incomplete removal of the surface contamination. The untreated sample contained 2.6 ppm Cl, but leaching with hot water lowered the Cl concentration to 1.0 ppm. The conclusion that most of the total inventory of halogens reside in the Earth's crust seems well justified. Certainly one cannot expect that an element like Cl is concentrated in the crust to a smaller degree as, for example, Ba or K.

The abundances of a number of volatile species in the Earth are considerably below those which would be introduced by the homogeneous or inhomogeneous two component accretion models, assuming C1 abundances for these volatile elements in the component containing oxidized volatiles.

Loss of an early atmosphere has been discussed by several authors (Walker 1982; Sekiya et al. 1980) who suggested depletion factors in the order of 10^5. It might well be that component B was depleted for the most volatile elements relative to C1 prior to accretion. Neglecting such a depletion, we can estimate upper limits for the abundance of volatile elements before their hypothetical loss. Taking the SC 1 concentration at face value, we find more than 90% of the Earth's total inventory of halogens to reside in the crust. In the case of K and Ba the crustal contribution is in the order of 50%. For the equally incompatible but volatile elements Cl, Br, and I the crustal portion must be at least equal or more likely higher. A C 1 abundance in component B would yield 102 ppm Cl for the whole Earth compared to 8 ppm Cl as derived from Table 2. Hence, for the halogens a depletion factor due to escape after accretion in the order of 10 is obtained. A H_2O concentration of 7.2% in C 1 chondrites (for the 15% of component B) yields an abundance of 1.1% H_2O for the Earth. This amount exceeds the crustal H_2O abundance by about a factor of 30. However, H_2O could have been used up by the reaction $Fe + H_2O \rightarrow FeO + H_2$ (Ringwood 1979), raising the FeO content from 3.5% (15% component B) to 7.8%, a value not unreasonable for the early Earth (Jagoutz and Wänke 1982). The large amounts of H_2 will be lost in a hydrodynamic escape process, which also removed heavier volatile species very efficiently (Hayashi et al. 1979; Walker 1982). The loss of volatiles must have been mass independent as Cl, Br, and I are present on the Earth in C 1 abundance ratios (Table 2).

Arrhenius (1981) assumed that the Earth's mantle was completely devoid of volatiles after accretion and acted as a sink rather than a source of volatiles. According to his model the oceans and the Earth's atmosphere observed today originated from outgassing of the solid matter during accretion. The volatiles were reintroduced into the mantle after core formation by the subduction of oceanic plates.

Acknowledgements. We would like to express our thanks to I. Kiefer and B. Spettel for their devoted help in this investigation. All neutron activation analyses were carried out in the TRIGA-research reactor of the Institut für Anorganische Chemie und Kernchemie der Universität Mainz. We wish to thank the staff of the TRIGA-reactor. We are indebted to A. E. Ringwood, S.-S. Sun, and G. Wetherill for their constructive reviews which improved the paper considerably. This work was carried out within the Forschergruppe Mainz supported by the Deutsche Forschungsgemeinschaft.

References

Ahrens TJ (1979) Equation of state of iron sulfides and constraints on the sulfur content of the Earth. J Geophys Res 84:985 – 998

Anders E (1964) Origin, age and composition of meteorites. Space Sci Rev 3:583 – 714

Anders E (1965) Chemical fractionations in meteorites. NASA Contract Rep CR-299

Anderson DL (1982a) Isotopic evolution of the mantle: A model. Earth Planet Res Lett 57:13 – 24

Anderson DL (1982b) Chemical composition and evolution of the mantle. In: Akimoto S, Manghnani MH (eds) High-pressure research in geophysics, advances in Earth and planetary sciences, Vol 12, Japan, pp 301 – 318

Anderson DL (1983) Chemistry of the primitive mantle. Lunar und Planetary Science-XIV, pp 5 – 6, Lunar and Planetary Institute, Houston

Armstrong RL (1968) A model for the evolution of strontium and lead isotopes in a dynamic Earth. Rev Geophys 6:175 – 199

Armstrong RL (1981) Radiogenic isotopes: the case for crustal recycling on a near-steady-state no-continental-growth Earth. Philos Trans R Soc Lond A 301:443 – 472

Arrhenius G (1981) In: Stiller H, Sagdeev RZ (eds) Advances in space research (planetary interiors) 1:37 – 48

Brown JM, McQueen RG (1982) The equation of state for iron and the Earth's core. In: Akimoto S, Manghnani MH (eds) High-pressure research in geophysics, advances in Earth and planetary sciences, Vol 12, Japan, pp 611 – 623

Burghele A, Dreibus G, Palme H, Rammensee W, Spettel B, Weckwerth G, Wänke H (1983) Chemistry of shergottites and the shergotty parent body (SPB): Further evidence for the two component model of planet formation. Lunar and Planetary Science-XIV, pp 80 – 81, Lunar and Planetary Institute, Houston

Cameron AGW (1978) Physics of the primitive solar accretion disk. Moon Planets 18:5 – 40

Clayton DD (1980) Chemical energy of cold-cloud aggregates: The origin of meteoritic chondrules. Astrophys J 239:L37 – L41

Cumming GL, Richards JR (1975) Ore lead isotope ratios in a continuously changing Earth. Earth Planet Sci Lett 28:155 – 171

DePaolo DJ, Wasserburg GJ (1976) Inferences about magma sources and mantle structure from variations of $^{143}Nd/^{144}Nd$. Geophys Res Lett 3:743 – 746

Dreibus G, Wänke H (1979) On the chemical composition of the Moon and the eucrite parent body and a comparison with the composition of the Earth. Lunar and Planetary Science-X, pp 315 – 317, Lunar and Planetary Institute, Houston

Dupré B, Allègre CJ (1980) Pb-Sr-Nd isotopic correlation and the chemistry of the North Atlantic mantle. Nature 286:17 – 22

Flasar FM, Birch F (1973) Energetics of Core formation: A correction. J Geophys Res 78:6101 – 6103

Frey F, Prinz M (1978) Ultramafic inclusions form San Carlos, Arizona: Petrologic and geochemical data bearing on their petrogenesis. Earth Planet Sci Lett 38:129 – 179

Ganapathy R, Anders E (1974) Bulk compositions of the Moon and Earth, estimated from meteorites. Proc 5th Lunar Sci Conf, Geochim Cosmochim Acta, Suppl 5:1181 – 1206

Glikson AY (1979) Siderophile and lithophile trace-element evolution of the Archaean mantle. BMR J Aust Geol Geophys 4:253 – 279

Greenberg JM (1982) What are comets made of? A model based on interstellar dust. In: Wilkening LL (ed) Comets, University of Arizona Press, Tucson, Arizona, pp 131 – 163

Hayashi C, Nakazawa K, Mizundo H (1979) Earth's melting due to the blanketing effect of the primordial dense atmosphere. Earth Planet Sci Lett 43:22 – 28

Hofmann AW, White WM (1982) Mantle plumes from ancient oceanic crust. Earth Planet Sci Lett 57:421 – 436

Hofmann AW, White WM (1983) Ba, Rb and Cs in the Earth's mantle. Z Naturforsch 38a:256 – 266

Hunt JM (1972) Distribution of carbon in crust of Earth. Bull Am Assoc Petrol Geol 56:2273 – 2277

Hutchison R (1974) The formation of the Earth. Nature 250:556 – 558

Irving AJ (1978) A review of experimental studies of crystal/liquid trace element partitioning. Geochim Cosmochim Acta 42:743 – 770

Jackson I, Ringwood AE (1981) High-pressure polymorphism of the iron oxides. Geophys J R Astr Soc 64:767 – 783

Jagoutz E, Palme H, Baddenhausen H, Blum K, Cendales M, Dreibus G, Spettel B, Lorentz V, Wänke H (1979) The abundances of major, minor and trace elements in the Earth's mantle as derived from primitive ultramafic nodules. Proc 10th Lunar Planet Sci Conf, Geochim Cosmochim Acta, Suppl 11:2031 – 2050

Jagoutz E, Carlson RW, Lugmair GW (1980) Equilibrated Nd-unequilibrated Sr isotopes in mantle xenoliths. Nature 286:708 – 710

Jagoutz E, Wänke H (1982) Has the Earth's core grown over geologic times? Lunar and Planetary Science-XIII, pp 358 – 359, Lunar and Planetary Science Institute, Houston

Jagoutz E, Dawson JB, Hoernes S, Spettel B, Wänke H (1984) Anorthositic oceanic crust in the Archaean Earth. Lunar and Planetary Science-XV, pp 395 – 396, Lunar and Planetary Institute, Houston

Kallemeyn GW, Wasson JT (1981) The compositional classification of chondrites – I. The carbonaceous chondrite groups. Geochim Cosmochim Acta 45:1217 – 1230

Kaula WM (1979) Thermal evolution of Earth and Moon growing by planetesimal impacts. J Geophys Res 84:999 – 1008

Lange ML, Ahrens T (1982) The evolution of an impact-generated atmosphere. Icarus 51:96 – 120

Lange ML, Ahrens T (1983) Shock-induced CO_2-production from carbonates and a proto-CO_2-atmosphere on the Earth. Lunar and Planetary Science XIV, pp 419 – 420, Lunar and Planetary Institute, Houston

Lee T, Papanastassiou DA, Wasserburg GJ (1976) Demonstration of ^{26}Mg excess in Allende and evidence for ^{26}Al. Geophys Res Lett 3:41 – 44

Liu L-G (1979) On the 650-km discontinuity. Earth Planet Sci Lett 42:202 – 208

Mathis JS, Rumpl W, Nordsieck KN (1977) The size distribution of interstellar grains. Astrophys J 217:425 – 433

Matsui T, Abe Y (1984) The formation of an impact-generated H_2O atmosphere and its implication for the early thermal history of the Earth. Lunar and Planetary Science-XV, pp 517 – 518, Lunar and Planetary Institute, Houston

McCammon CA, Ringwood AR, Jackson I (1983) Thermodynamics of the system Fe-FeO-MgO at high pressure and temperature and a model for formation of the Earth's core. Geophys J R Astr Soc 72:577 – 595

Mizuno H, Nakazawa K, Hayashi C (1980) Dissolution of the primordial rare gases into the molten Earth's material. Earth Planet Sci Lett 50:202 – 210

Moore JG, Fabbi BP (1971) An estimate of the juvenile sulfur content of basalt. Contrib Mineral Petrol 33:118 – 127

Morgan JW, Wandless GA, Petrie RK, Irving AJ (1981) Composition of the Earth's upper mantle – 1. Siderophile trace elements in ultramafic nodules. Tectonics 75:47 – 67

Newsom HE, Drake MJ (1983) Experimental investigation of the partitioning of phosphorus between metal and silicate phases: Implication for the Earth, Moon, and eucrite parent body. Geochim Cosmochim Acta 47:93 – 100

Newsom HE, Palme H (1984) The depletion of siderophile elements in the Earth's mantle: New evidence from molybdenum and tungsten. Lunar and Planetary Science-XV, pp 607 – 608, Lunar and Planetary Institute, Houston

Nonaka J (1982) Über die Häufigkeit von bisher wenig untersuchten Elementen im Erdmantel. Thesis, Universtität Mainz

Palme H, Suess HE, Zeh HD (1981) Abundances of the elements in the solar system. In: Schaifers K, Voigt HH (eds) Landoldt-Börnstein Vol 2, (Astronomy and astrophysics), Springer, Berlin Heidelberg New York pp 257 – 273

Rambaldi ER, Cendales M, Thacker R (1978) Trace element distribution between magnetic and non-magnetic portions of ordinary chondrites. Earth Planet Sci Lett 40:175 – 186

Rambaldi ER, Cendales M (1979) Moderately volatile siderophiles in ordinary chondrites. Earth Planet Sci Lett 44:397 – 408

Rammensee W, Wänke H (1977) On the partition coefficient of tungsten between metal and silicate and its bearing on the origin of the moon. Proc Lunar Sci Conf 8th, Geochim Cosmochim Acta, Suppl 8:399 – 409

Ringwood AE (1958) The constitution of the mantle – III; Consequences of the olivine-spinel transition. Geochim Cosmochim Acta 15:195 – 212

Ringwood AE (1966) Mineralogy of the Mantle. In: Hurley P (ed) Advances in Earth science, MIT Press, Boston, pp 357 – 398

Ringwood AE (1970) Origin of the Moon: The precipitation hypothesis. Earth Planet Sci Lett 8:131–140

Ringwood AE (1971) Core-mantle equilibrium: Comment on a paper by R Brett. Geochim Cosmochim Acta 35:223–230

Ringwood AE (1977) Composition of the core and implications for origin of the Earth. Geochem J 11:111–135

Ringwood AE, Kesson SE (1977) Basaltic magmatism and the bulk composition of the Moon II. Siderophile and volatile elements in Moon, Earth and chondrites. Implications for lunar origin. Moon 16:425–464

Ringwood AE (1979) Origin of the Earth and Moon. Springer, Berlin Heidelberg New York

Ringwood AE (1982) Phase transformations and differentiation in subducted lithosphere: Implications for mantle dynamics, basalt petrogenesis, and crustal evolution. J. Geol. 90:611–643

Ringwood AE (1983) Geochemical relationships between the Earth's core and mantle. Lunar and Planetary Science-XIV, pp 646–647, Lunar and Planetary Institute, Houston

Safranov VS (1978) The heating of the Earth during its formation. Icarus 33:3–12

Sakai H, Casadevall TJ, Moore JG (1982) Chemistry and isotope ratios of sulfur in basalts and volcanic gases at Kilauea Volcano, Hawaii. Geochim Cosmochim Acta 46:729–738

Schilling J-G, Bergeron MB, Evans R (1980) Volatiles in the mantle beneath the North Atlantic. Philos Trans R Soc Lond A 297:147–178

Schmitt W, Wänke H (1984) Experimental determination of metal/silicate-partition coefficients of P, Ga, Ge, and W as function of oxygen fugacity. Lunar and Planetary Science-XV, pp 724–725, Lunar and Planetary Institute, Houston

Sekiya M, Nakazawa K, Hayashi C (1980) Dissipation of the rare gases contained in the primordial Earth's atmosphere. Earth Planet Sci Lett 50:197–201

Solomon SC, Ahrens TJ, Cassen PM, Hsui AT, Minear JW, Reynolds RT, Sleep NH, Strangway DW, Turcotte DL (1981) Chapter 9: Thermal histories of the terrestial planets. In: Basaltic volcanism on the terrestrial planets, pp 1129–1234, Pergamon, New York

Stacey JS, Kramers JD (1975) Approximation of terrestrial lead isotope evolution by a two-stage model. Earth Planet Sci Lett 26:207–221

Spettel B, Jagoutz E (1981) Granatlherzolite und Mantelzusammensetzung. Fortschr Miner 59:259–260

Sun S-S (1982) Chemical composition and the origin on the Earth's primitive mantle. Geochim Cosmochim Acta 46:179–192

Sun S-S (1984) Geochemical characteristics of Archaen ultramafic and mafic volcanic rocks: Implications for mantle composition and evolution. This Vol. 25–46

Unni CK, Schilling J-G (1978) Cl and Br degassing by volcanism along the Reykjanes Ridge and Iceland. Nature 272:5648–5652

Vidal P, Dosso L (1978) Core formation: catastrophic or continuous? Sr and Pb isotope geochemistry constraints. Geophys Res Lett 5:169–172

Volmer R (1977) Terrestrial lead isotopic evolution and formation time of the Earth's core. Nature 270:144–147

Walker JCG (1982) The earliest atmosphere of the Earth. Precambrian Res 17:147–171

Wänke H, Baddenhausen H, Dreibus G, Jagoutz E, Kruse H, Palme H, Spettel B, Teschke F (1973) Multielement analyses of Apollo 15, 16, and 17 samples and the bulk composition of the Moon. Proc 4th Lunar Sci Conf, Geochim Cosmochim Acta, Suppl 4:1461–1481

Wänke H (1981) Constitution of terrestrial planets. Philos Trans R Soc Lond A 303:287–302

Wänke H, Dreibus G, Jagoutz E, Palme H, Rammensee W (1981) Chemistry of the Earth and the significance of planets and meteorite parent bodies. Lunar and Planetary Science-XII, pp 1139–1140, Lunar and Planetary Institute, Houston

Wänke H, Rammensee W (1981) Primary and secondary objects: A new concept of the early days of the solar nebula. Meteoritics 16:397–398

Weckwerth G (1983) Anwendung der instrumentellen β-Strektrometrie im Bereich der Kosmochemie, insbesondere zur Messung von Phosphorgehalten. Thesis, Universität Mainz

Wedepohl KH (1975) The contribution of chemical data to assumptions about the origin of magmas from the mantle. Fortschr Miner 52:141–172

Wedepohl KH (1981) Der primäre Erdmantel (Mp) und die durch die Krustenbildung verarmte Mantelzusammensetzung (Md). Fortschr Miner 59:203–205

Wetherill GW (1978) Accumulation of the terrestrial planets. In: Gehrels T (ed) Protostars and planets, University Arizona Press, Tucson, Arizona, p 565

Williams DL, Von Herzen RP (1974) Heat loss from the Earth: New estimate. Geology (Boulder) 2:327 – 328

Zartman RE, Tera F (1973) Lead concentration and isotopic composition in five peridotite inclusions of probable mantle origin. Earth Planet Sci Lett 20:54 – 66

Zindler A, Jagoutz E, Goldstein S (1982) Nd, Sr and Pb isotopic systematics in a three-component mantle: A new perspective. Nature 298:519 – 523

Zook HA (1981) On a new model for the generation of chondrites. Lunar and Planetary Science-XII, pp 1242 – 1244, Lunar and Planetary Institute, Houston

Geochemical Characteristics of Archaean Ultramafic and Mafic Volcanic Rocks: Implications for Mantle Composition and Evolution

S.-S. Sun[1]

Contents

1	Introduction	26
2	Chemical Composition of the Earth's Primitive Mantle	27
2.1	Effect of the Mantle-Core Differentiation Process	30
2.2	Effect of Meteorite Bombardment after Mantle-Core Differentiation	32
2.3	Possible Secular Variation of Total FeO Abundance in the "Primitive Mantle"	33
3	Chemical and Isotopic Heterogeneities in the Archaean Mantle	34
3.1	Isotopic Heterogeneity	34
3.2	REE Heterogeneity	35
3.3	Major Element Heterogeneity	36
4	Geochemical Characteristics of Archaean Mafic and Ultramafic Volcanic Rocks and Their Tectonic Implications	36
5	Effect of Archaean Volcanic Activity on the Chemistry of the Continental Lithosphere and Archaean Upper Mantle	40
6	Effect of the Archaean Tectonism and Magmatism on Mantle Evolution	41
	References	42

Abstract

A constant TiO_2/P_2O_5 ratio of 10, estimated for the primitive mantle from 3.8 Ga to the present, suggests that the mantle-core fractionation process, at least for the upper mantle, was completed before 3.8 Ga ago. Moreover, a significant degree of mantle-crust and/or intra-mantle differentiation in the pre-Archaean (>3.8 Ga) is indicated by Nd and Sr isotopic data.

Chondritic ratios for refractory lithophile elements (e.g. Al, Ti, Zr, Y, REE) are commonly observed among non-Barberton type (Al-undepleted) Archaean komatiites and high magnesian basalts with chondritic REE patterns. Barberton type (Al-depleted) peridotitic komatiites and differentiates (in contrast to contemporary tholeiitic rocks and non-Barberton type komatiites) commonly found in 3.5 – 3.8 Ga terrains have Al/Ti values only half the chondritic ratio and are depleted in heavy REE and Sc. This difference might be due to different magma generation processes (preferred) or chemical and mineralogical stratification in the early Archaean mantle. Available Hf isotope data do not support the view that such heterogeneity in the mantle is related to core-mantle differentiation occurring ~ 4.5 Ga ago.

Comparatively small-scale heterogeneities in major and trace element abundances and isotopic ratios inferred for the sources of Archaean mafic and ultra-

---

1 Division of Petrology and Geochemistry, Bureau of Mineral Resources, Geology and Geophysics, PO Box 378, Canberra ACT 2601, Australia

Archaean Geochemistry (ed. by A. Kröner et al.)
© Springer-Verlag Berlin Heidelberg 1984

mafic volcanic rocks are similar in degree to those estimated from Recent mantle-derived volcanic rocks. Many of the Archaean mafic and ultramafic volcanic rocks exhibit geochemical characteristics similar to typical modern MORB (light REE depleted) showing coherency between major and trace element abundances. However, some high magnesian basalts are characterized by light REE enrichment and a decoupling of major and some commonly incompatible elements such as Ti, Zr, Nb, REE, and P, suggestive of a mantle enrichment process related to eclogite melting. The occurrence of eclogite may be related to either subduction in the Archaean or sinking of underplated mafic to ultramafic igneous rocks from beneath the continental crust.

Underplating by residual peridotite beneath the Archaean continental crust and subsequent mantle metasomatism could significantly modify the trace element and isotopic compositions of the sub-continental upper mantle. Reactivation of this material could provide a major source for Proterozoic to Recent continental basalts.

Because chemical and isotopic heterogeneities are such ancient features within the mantle, a deeper understanding of the present-day mantle requires a more detailed knowledge of pre-Archaean and Archaean mantle processes.

1 Introduction

Chemical and isotopic studies of Archaean mafic and ultramafic volcanic rocks are relevant to a number of important questions including:

1. the composition of the bulk Earth, core and the primitive mantle;
2. the chemical effect of core-mantle differentiation;
3. the early history of the actively evolving mantle-crust system;
4. the chemical effect of meteorite bombardment events in the early history of the Earth;
5. chemical and isotopic heterogeneities in the Archaean mantle;
6. boundary conditions for mantle evolution models.

They also offer an insight into magma generation processes and possible tectonic environments in which the Archaean volcanic rocks were emplaced.

The early history of the Earth's mantle was complicated by several processes such as heterogeneous accretion of the Earth from meteoritic material, mantle-core differentiation, extensive partial melting, intra-mantle mineralogical and chemical differentiation, formation of the protocrust, recycling of crustal material back into the mantle and continuous influx of meteoritic material before and after core formation. These processes are likely to have produced considerable long term and short term mineralogical and chemical heterogeneities in the pre-Archaean mantle. To counter-balance these processes, vigorous mantle convection at that time offers a potential homogenization mechanism.

Conceptionally it would be very useful if a unique primitive mantle composition after core formation could be estimated. Such a composition could then offer important boundary conditions for a discussion of cosmochemistry of the Earth and to evaluate the geochemical effect of core-mantle differentiation.

Modification of such primitive mantle composition could involve processes such as extraction and addition of melts, recycling of crustal material, mantle mixing and mantle metasomatism under different tectonic environments. With reasonable assumptions on mantle mineralogy, melting conditions and mineral-melt distribution coefficients during partial melting, estimates for the geochemical consequences of these processes could be made.

Physical conditions in the Archaean mantle could have been considerably different from the present in many aspects, and magma generation processes could also have been more complicated (e.g. Nisbet and Walker 1982). There might have been more than one tectonic environment where ultramafic and mafic magmas were generated. Geochemical characteristics of these magmas could offer useful information of the enrichment-depletion processes in the mantle, residual mineralogy and possible tectonic environments.

Chemical and isotopic heterogeneities observed in the Archaean mantle could be the results of both long term and short term processes. The long term processes include those mentioned earlier for the pre-Archaean period and their continuation into Archaean time, whereas the short term processes could include intra-mantle (such as in the Low Velocity Zone) differentiation, mantle crust interaction and dynamic melting (Langmuir et al. 1977) which took place shortly before or during magma generation. As a result of Archaean tectonic activities and related magmatism, the chemistry of the lithosphere underneath the continental crust could be extensively modified by underplating of mantle material and interaction of the Archaean magmas with the lithosphere (e.g. Weaver and Tarney 1981). Further melting of these parts of the upper mantle at a later time could produce mafic and ultramafic magmas bearing geochemical and isotopic finger prints of such processes.

Physical and chemical processes that operated in the Archaean mantle also had a decisive effect on the subsequent chemical and isotopic evolution of the convective mantle regions which later served as magma sources for younger volcanics.

In this paper, data and ideas relevant to the above mentioned concepts will be critically evaluated. New approaches that might lead to a better understanding of mantle chemistry and evolution will be proposed. Such understanding will also help to improve our knowledge concerning the origin and evolution of the continental crust. Throughout this paper, emphasis will be made on necessity of high quality data on well selected samples. Only when this condition is fulfilled will significant progress become possible.

2 Chemical Composition of the Earth's Primitive Mantle

On the basis of the chemistry of Archaean and modern mafic and ultramafic volcanic rocks (e.g. Nesbitt and Sun 1976; Sun and Nesbitt 1977; Sun et al. 1979), modern ultramafic mantle nodules (e.g. Jagoutz et al. 1979), alpine peridotites and ultramafic tectonites of ophiolite complexes Sun (1982) has recently made an estimate for the composition of the Earth's primitive mantle (= unfractionated mantle material after core formation). It was suggested that chemical composi-

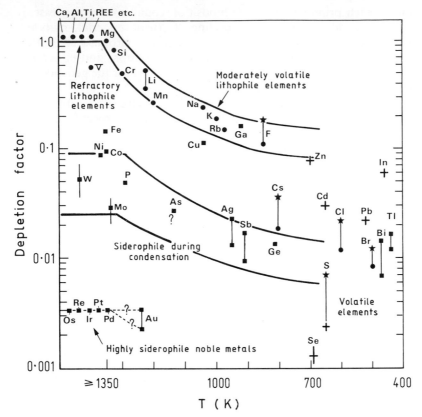

Fig. 1. Elemental depletion factors versus solar condensation temperature for the Earth's primitive mantle. Depletion factor = abundance of element in the primitive mantle/Cl chondrite relative to Mg. ■ = siderophile, ● = lithophile, + = chalcophile during solar condensation. *Stars* used for some elements represent possible upper limits. Modified after Fig. 3 of Sun (1982). New V value of 57 ppm for Cl chondrite (Anders and Ebihara 1982) is used to replace 42 ppm used in Sun (1982). New W and Mo values are from Newsom and Palme (1984)

tion of the inferred primitive mantle has essentially not changed since 3.8. Ga ago to the present, although there is a suggestion that the Archaean magma sources in the mantle might have had a somewhat higher Σ FeO content than the modern upper mantle (e.g. Glikson 1971, 1979; Sun and Nesbitt 1977; Jagoutz and Wänke 1982; *see also Wänke et al., this Vol., eds.*).

As shown in Fig. 1 on a carbonaceous chondrite normalized plot with condensation temperature $>700°K$, the element abundances for the primitive mantle appear to show two roughly parallel trends which are defined by the lithophile and siderophile elements, respectively. The lithophile element trend might represent a volatility controlled pattern for the Earth before core formation, whereas the siderophile trend represents the effect of core formation. Available Ti/Ga and Ti/P data would suggest that these two trends probably have not changed since the early Archaean. A constant Ti/P ratio observed among mafic and ultra-

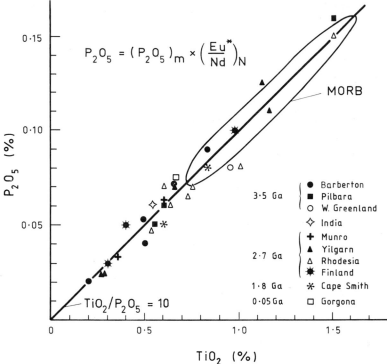

Fig. 2. TiO_2 versus P_2O_5 plot to estimate the TiO_2/P_2O_5 ratio for the primitive mantle. Measured $(P_2O_5)_m$ value for each sample was corrected by a chondrite-normalized Eu^*/Nd ratio of that sample to account for deviation of its REE pattern and TiO_2/P_2O_5 ratio from the primitive mantle with chondritic REE abundances (see Fig. 2 of Sun 1982). Data source: Nesbitt and Sun (1976), Nesbitt et al. (1979), Sun and Nesbitt (1978, and unpubl. data), Sun et al. (1979), Jahn et al. (1980, 1982), Schwarz and Fujiwara (1977), Gansser et al. (1979) and other literature data

mafic rocks since 3.8 Ga ago to the present (Fig. 2) would also argue for completion of core formation before 3.8 Ga ago.

Data presented in Fig. 1 are, however, open to alternative interpretations. It is possible that the lithophile and siderophile element trends are results of a fortuitous combination of processes occurring before and after core formation. For example, the lithophile element trend from Mg through Si, Cr, Li to Mn might have resulted from core formation but not from the volatility effect, i.e. these elements might have had C1 chondrite relative abundances before core formation. Large amounts of V, Cr, and Mn could have been taken into the core along with FeO (e.g. Dreibus and Wänke 1979), whereas Si was taken into the lower mantle (e.g. Liu 1982). If this interpretation is correct, then Li depletion in the primitive mantle (Fig. 1) has to be explained by the same mechanism. Li appears to behave as a chalcophile element in enstatite chondrite under very reducing conditions (e.g. Shima and Honda 1967). Concentration of sulphide into the Earth's core could explain this Li depletion. A possible scenario for earth-forming processes involves three stages of inhomogeneous accretion of the Earth and core formation (e.g. Morgan et al. 1981; Wänke 1981; Newsom and Palme 1984; *see also Wänke et al., this Vol., eds.*):

1. 85 – 90% of the Earth accreted from reduced chondritic components. Wänke (1981) suggested that these components were refractory and did not contain volatile elements. Simultaneous iron core formation depleted basically all siderophile elements but with some retention of FeO in the silicate mantle.
2. Addition of 10 – 15% more oxidized chondritic components took place after core formation was completed. These components account for abundances of siderophile elements (e.g. Ni, Co, W, Mo, Cu, Ga, P, Ag, Sb, Ge) and volatile elements in the mantle. Extraction of a very small amount ($< 1\%$) of metal into the core from the partially molten mantle could severely deplete highly siderophile elements (e.g. Ir, Pd, Au, Pt), whereas moderately siderophile elements such as W, Mo, P, Ag, Ge would be depleted to different extents. Ni, Co, Cu, and Ga remained in the silicate mantle.
3. A final accretion of ~1% of chondritic material brought in highly siderophile elements. Retention of a small amount of metals in the mantle during the second stage process could also contribute partly to siderophile element abundances in the mantle.

The systematic patterns observed in Fig. 1 could serve as firm boundary conditions for any model concerning the cosmochemical origin and geochemical evolution of the Earth. For example, on the basis of the lithophile element trend in Fig. 1 one could confidently conclude that K and Rb are not abundant and S is not the major light element in the core (e.g. Ringwood 1977). The data could also be used for a comparison of the chemical composition of the Earth and other planetary bodies in the solar system.

2.1 Effect of the Mantle-Core Differentiation Process

During the very early history of the Earth, and probably within the first 100 Ma, the bulk of the Earth's core was formed through segregation of iron from a mixture of silicates, oxides and iron (e.g. Ringwood 1977). Through this process a large quantity of siderophile elements and practically all noble metals were concentrated into the core. Dreibus and Wänke (1979) suggested that some Mn, Cr, and V (lithophile elements) were also incorporated into the core along with some FeO. Considerable variation in ratios between lithophile and siderophile elements such as Ti/P, Al/Ga, Si/Ge, and U/Pb could be generated through heterogeneous accretion and core formation, if subsequent mixing through mantle convection was not effective. Depending on the density contrast and the thermal conditions in the mantle during core formation, Liu (1979, 1982) proposed that, as a result of density contrasts among mantle minerals under high pressure and temperature conditions, mineralogical and chemical stratification in the mantle could be generated during core formation. The lower mantle will be enriched in minerals with perovskite structure and be richer in Si and Al than the upper mantle. This process could explain the depletion of Si (~15%) relative to the refractory elements such as Mg in the estimated "primitive mantle" (Fig. 1), although the relatively volatile character of Si might also be responsible for this depletion. On the other hand, Nisbett and Walker (1982) suggest that a melt layer of ultramafic composition could have existed within the Archaean upper mantle and acted as the source for peridotitic komatiites. Solidification of this melt layer might result in chemical and mineralogical stratification in the upper mantle (e.g. Anderson 1980). To counter-balance these processes, vigorous mantle convection associated with core formation and during the early history of the Earth offers a homogenization mechanism.

The net result of these competing processes could be evaluated quantitatively by a detailed study of spatial and secular variation of elemental ratios such as Ti/Zr, Ti/P, Al/Ga, Si/Ge, and U/Pb in mafic and ultramafic volcanic rocks erupted since the early Archaean to the present. Available data on Ti/Zr, Ti/P, Al/Ga, and U/Pb (based on Pb isotopic data) for inferred primitive mantle fail to indicate any obvious large scale secular variation (e.g. Sun 1982). Unless the lower mantle has never been directly involved in volcanism on the Earth's surface, the apparent constant ratios observed would favour the idea of a grossly homogeneous (upper?) mantle.

It is important to note, however, that early Archaean komatiites of Barberton type commonly have Al_2O_3/TiO_2 ratios only half that of chondrites (Fig. 3), and on a chondritic normalized plot they show heavy REE and Sc depletion relative to the middle REE and Ti (e.g. Sun and Nesbitt 1978a; Jahn et al. 1982). Nevertheless, basaltic and komatiitic samples with close to chondritic Al_2O_3/TiO_2 ratios are common in the early Archaean (e.g. Smith and Erlank 1982). In contrast, komatiites with low Al_2O_3/TiO_2 ratios are rarely found in the late

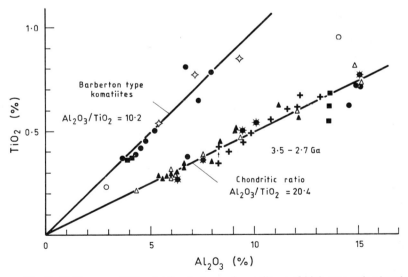

Fig. 3. Al_2O_3 versus TiO_2 plot for Archaean komatiites and high magnesian basalts. Barberton-type komatiites have Al_2O_3/TiO_2 ratios close to 10, whereas other samples have values close to chondrite (~ 20). Data source and symbols are the same as in Fig. 2

Archaean greenstone belts (e.g. Sun and Nesbitt 1978a; Schulz 1982). They generally have ratios close to chondrites. Such contrast could be caused by long term mineralogical and chemical stratification of the mantle (as discussed earlier) or magma generation processes involving garnet separation (Green 1975; Sun and Nesbitt 1978a; Ohtani 1984). On the basis of high pressure experimental work, Ohtani (1984) suggested that komatiites derived from a partially molten mantle diapir originating at great depth (200 – 400 km) with density controlled garnet settling at ≥ 200 km will be Al-poor (Barberton type); whereas Al-undepleted komatiites were probably derived from a diapir originating at shallower depth (≤ 200 km) without garnet separation. To distinguish between the two alternatives, it is necessary to apply isotope studies. Available Nd isotope data for Archaean mafic and ultramafic volcanics do not offer a solution to this problem, as no distinction has been reconized between Barberton and non-Barberton type komatiites. The Lu-Hf decay scheme could be useful to test the possibility of long term fractionation of heavy REE (Lu) from middle REE (may be represented by Hf) associated with Barberton type komatiites. Inferences could be derived from ^{176}Hf/^{177}Hf data for early Archaean felsic volcanic rocks reported by Patchett et al. (1981) and Patchett (1982). These data give no indication of long term and large scale fractionation of Lu/Hf from a chondritic ratio in their source regions which were ultimately derived from the mantle.

2.2 Effect of Meteorite Bombardment after Mantle-Core Differentiation

There is general consensus that, during core formation, practically all noble metals in the mantle at that time were incorporated into the core, and their presence in the mantle is due to later falling of meteorites. About 1% (as compared to 0.4% for the present day crust) of meteorite input after the time of core formation ($\sim$ 4.4 Ga?) till about 3.8. Ga (major lunar cratering period) could account for the noble metal abundances now observed in mantle samples (e.g. Chou 1978). Heterogeneous distribution of these meteorite showers could cause considerable mantle chemical heterogeneity in volatile, chalcophile and siderophile elements, especially in noble metals. It might also affect the Pb isotope evolution in different parts of the Earth, especially in the continental crust and upper mantle.

Available Pb isotope data in the literature (e.g., Richards et al. 1981; Tilton 1983; Chauvel et al. 1983), although subject to model-dependent interpretation, suggest that only a limited degree of long term integrated U/Pb ratio heterogeneity (< 8%?) has existed in the Archaean mantle. The limited amount of noble metal data reported in the literature for 2.7 Ga old greenstone samples and modern ultramafic nodules as summarized in Sun (1982) also fail to indicate a large scale of heterogeneity in the Earth's mantle.

A detailed study of Ti/Re ratios in mantle-derived ultramafic and mafic melts of different ages could help greatly to better evaluate the question concerning heterogeneous distribution of noble metals in the Earth's mantle. During large degrees of mantle melting, Re apparently behaves like an incompatible lithophile element similar to Ti, and therefore, the measured Ti/Re ratios could offer a good indication of noble metal abundances in the magma sources.

2.3 Possible Secular Variation of Total FeO Abundance in the "Primitive Mantle"

In Fig. 2 it was shown that the TiO_2/P_2O_5 ratio (10 ± 1) in the primitive mantle apparently has not changed since the early Archaean. This constancy implies that core formation had the same effect on mantle sources of these analyzed samples, consequently no secular variation in the total FeO content in the primitive mantle of different ages is expected. In contrast to this conclusion, Σ FeO estimates for modern upper mantle based on fertile ultramafic xenoliths, alpine peridotites and ultramafic tectonites associated with ophiolite complexes consistently suggest a value of $8.3 \pm 0.2\%$ (e.g. Ringwood 1975; Jagoutz et al. 1979; Sun 1982; Ernst and Piccardo 1979), whereas model-dependent estimates for the Archaean upper mantle based on komatiite chemistry generally give higher Σ FeO values of 9.0 to 9.3% (e.g. Bickle et al. 1977; Sun and Nesbitt 1977). A plot of MgO vs FeO for komatiites and high magnesian basalts from the early Archaean to the present (Fig. 4) also suggests that the Archaean magma sources might have had a considerably higher Σ FeO content than modern mantle domains which produce mid-ocean ridge basalts (MORB) with both light REE enriched and depleted patterns (*see also Wänke et al., this Vol., eds.*).

The paradox between the constant Ti/P ratio and a seemingly secular variable Σ FeO abundance in the primitive mantle could be due to a deficiency in our models of magma generation processes and/or due to a mechanism which could decouple FeO from Ti, P, (W and Mo?) and other lithophile elements such as REE in the mantle. A mechanism such as transfer of FeO from the lower mantle to the core (e.g. Jagoutz and Wänke 1982) might not affect the Ti/P ratio. However, if this mechanism was responsible, available geochemical data discussed earlier would require that mantle-wide convection took place during the Archaean and that growth of Earth's iron core by this mechanism had no effect on chondritic relative abundances of the highly siderophile elements observed in the primitive mantle (unlikely?). Other possible processes, which may account for

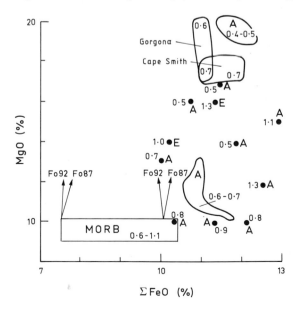

Fig. 4. Total FeO versus MgO plot for komatiites and high magnesian basalts of various ages and modern mid-oceanic ridge basalts (MORB). Komatiites include samples from Archaean (*A*) greenstone belts, the Proterozoic (1.8 Ga.) Cape Smith belt, Canada, and early Tertiary Gorgona Island. High magnesian basalts include samples from Archaean greenstone belts and early Tertiary (*E*) picrites from the Baffin Bay area. TiO_2 abundances in these samples are also shown as references for degree of partial melting. Effects of addition of olivine Fo 92 and Fo 87 to MORB are indicated by *arrows*. Data source: Glikson (1979), Arndt and Nesbitt (1982), Echeverria (1980), Clarke (1970) plus references for Fig. 2

differences of Σ FeO in Archaean and modern mantle include heterogeneous distribution of subducted oceanic crust (FeO rich) and residual lithosphere (FeO poor) back into (Archaean and modern) mantle, and variations in Mg/Fe distribution coefficients between melt and residual minerals under different P-T melting conditions. Further research is needed to resolve this paradox.

3 Chemical and Isotopic Heterogeneities in the Archaean Mantle

On the basis of $^{87}Sr/^{86}Sr$ initial ratios observed in Archaean basalts and anorthosites, Jahn and Nyquist (1976) suggested that a long term regional heterogeneity of the Rb/Sr ratio has existed in the Archaean mantle. Similarly, Sun and Nesbitt (1977) used REE data for Archaean mafic and ultramafic volcanic rocks to suggest that large scale trace element heterogeneities similar to those observed in the modern mantle existed in the Archaean. On the basis of major element data, Glikson (1971, 1979) suggested that the Archaean mantle was richer in FeO than the modern upper mantle. Chemical and isotopic heterogeneities in the mantle could be produced by long term and short term processes such as mantle-core-crust differentiation, mantle-crust interaction, subduction of oceanic crust and intra-mantle differentiation under different tectonic environments.

It is important to point out that some observed variations might also be produced by post-magmatic alteration. For example, negative Eu anomalies are commonly found in Archaean ultramafic and mafic volcanic rocks. Sun and Nesbitt (1978a) suggested that they are most likely due to Eu^{2+} (behaves like mobile element Sr) removal under reducing crustal environment. A possible alternative explanation is that the negative Eu anomalies reflect the real mantle source character and are related to the formation of an anorthositic Earth's protocrust similar to the situation on the Moon. This explanation is not favoured. In addition to lack of consistent Eu anomalies in Archaean mafic to ultramafic volcanics pointed out by Sun and Nesbitt (1978a), Taylor and McLennan (1981) also showed that REE data for the earliest Archaean sediments are not consistent with an anorthositic protocrust.

3.1 Isotopic Heterogeneity

Accurate Sr and Nd isotope data now available in the literature (e.g. Jahn and Nyquist 1976; Fletcher and Rosman 1982; McCulloch and Compston 1981; Chauvel et al. 1984) firmly indicate that long term chemical heterogeneities existed in the Archaean mantle. $^{87}Sr/^{86}Sr$ initial ratios for 2.7 Ga old greenstone rocks, anorthosites and clinopyroxene separates of worldwide distribution range from 0.7005 to 0.7015. Although the cause of this variation is not well defined, corresponding variations in $^{143}Nd/^{144}Nd$ initial ratios have been observed. For example, greenstone samples from Vermillion (Minnesota) and Kambalda (W. Australia) have low initial $^{87}Sr/^{86}Sr$ ratios of 0.7005 to 0.7007. They have ε_{Nd} of about $+4$ (McCulloch and Compston 1981; Jahn and Nyquist 1976; Ashwal et al. 1981; Chauvel et al. 1984). In contrast, samples from Zimbabwe and Abitibi (Canada) have higher initial $^{87}Sr/^{86}Sr$ ratios of 0.7010 to 0.7012 (Jahn and Nyquist 1976; Zindler et al. 1978; Machado et al. 1983; Chauvel et al. 1983), and lower ε_{Nd} values (0 to 2.5). If these heterogeneities were allowed to exist since 4.5 Ga ago and developed to the present, Jahn and Nyquist (1976) and McCulloch and Compston (1981) showed that they could explain the variations observed among modern oceanic basalts from mid-ocean ridges and oceanic islands.

Recently, Cattell et al. (to be published) and Chauvel et al. (1984) reported large variations in $^{143}Nd/^{144}Nd$ initial ratios for different late Archaean mafic to ultramafic metavolcanics from Newton Township, Canada ($\varepsilon_{Nd} = 4.2$ to $+1.6$) and Kambalda, W. Australia ($\varepsilon_{Nd} = -2.4$ to $+4.4$). These data strongly suggest mantle source isotopic heterogeneities and/or contamination by sialic crust within a single greenstone belt.

Assuming that the primitive Earth's mantle had exactly the same average Sm/Nd ratio as average chondrites (Jacobsen and Wasserburg 1980), the $^{143}Nd/^{144}Nd$ data ($\varepsilon_{Nd} +1$ to $+2$) for metavolcanics from the 3.5 Ga old Pilbara (Western Australia) and Barberton (South Africa) and the 3.8 Ga old Isua (West Greenland, see McCulloch and Compston 1981) greenstone belts would suggest that their mantle sources had a long term light REE depleted history which could have resulted from melt extraction, intra-mantle differentiation or formation of the protocrust long before 3.5 Ga ago.

Lead isotopic heterogeneities in the Archaean mantle have also been established through several recent detailed studies of greenstone metavolcanics (e.g. Tilton 1983; Chauvel et al. 1983, 1984). A variation of integrated μ_1 values of about 8% ($\mu_1 = 7.8$ to 8.4) from 4.6 Ga to the late Archaean (~ 2.7 Ga) is indicated by available data. However, no correlation between μ_1 and ε_{Nd} values has yet been established. It is important to point out that, as in the case of Sr and Nd isotope studies, crustal contamination may also be responsible for some of the Pb isotope variation especially those observed in layered intrusive mafic-ultramafic complexes.

3.2 REE Heterogeneity

It is now well established that the range of variation of La/Sm ratios existing in the mantle sources of Archaean volcanics is similar to that estimated for the modern mantle (e.g. Sun and Nesbitt 1977; Jahn et al. 1980). Similar to the modern basalts, such variation could be due to either short term or long term fractionation processes operating in the Archaean mantle. For example, REE patterns of mafic and ultramafic volcanic rocks from Kambalda, Western Australia and Newton Township, Canada show a large range of variation from severely light REE depleted to light REE enriched (Arth et al. 1977; Sun and Nesbitt 1978a; McCulloch and Compston 1981). Since these volcanic rocks were produced by large degrees of partial melting, their REE patterns would reflect the mantle source character (e.g. Sun and Nesbitt 1977). Detailed Nd isotope studies on these two areas (e.g. Cattell et al. to be published; Chauvel et al. 1984) indicated quite different initial $^{143}Nd/^{144}Nd$ ratios for different rock types. Consequently, such variation could be due to long term differences of Sm/Nd ratios in their source regions. However, Chauvel et al. (1984) pointed out that such variation could also be generated by involvement of ancient crustal material through lithosphere subduction or crustal contamination.

Variations in abundance of other incompatible elements such as Ti, Zr, Y, Nb, Rb, Sr, Th, U, P, and Ba are expected to be associated with the REE heterogeneity. A detailed and systematic study of Archaean mafic and ultramafic volcanics such as in the case of Munro Township komatiites and tholeiites (Arndt and Nesbitt 1982) is required to accurately assess the covariant abundance relationship among these elements including REE.

3.3 Major Element Heterogeneity

It is difficult to accurately evaluate the extent of major element heterogeneity in the Archaean mantle using chemical composition of Archaean mafic and ultramafic volcanics, because the melt composition is controlled by melting processes and the residual mineralogy as well as source composition. For example, variation in TiO_2 abundances in komatiites of similar MgO content (Fig. 1, in Sun 1982) could be due to Ti abundance variation in the source regions (favoured by the author) but could also be due to different degrees of partial melting under different P, T conditions. Nevertheless, on the basis of present knowledge about Archaean mantle conditions, such as active thermal and magmatic activities, it is logical to expect some degree of heterogeneity in the mantle at that time.

As discussed earlier (Sec. 2.1) the cause of large variations in the Al abundance of Archaean komatiites with similar MgO and Ti contents has been a major issue in the study of komatiites (e.g. Sun and Nesbitt 1978a; Jahn et al. 1982). This has possible implications for the long term chemical and mineralogical stratification in the early Archaean mantle.

On a smaller scale, the inferred major and trace element heterogeneity in the Archaean mantle appears to have been documented in some well studied areas such as Pilbara, Western Australia (Glikson 1979; Glikson and Hickman 1981), and Barberton, South Africa (e.g. Viljoen and Viljoen 1969; Jahn et al. 1982). The well pronounced nature of such variation is likely to be related to short term mantle processes in a vigorously dynamic Archaean mantle. These could involve processes such as dynamic melting (Langmuir et al. 1977), mixing of partial melting products (e.g. eclogite) and residues (e.g. refractory harzburgite) back into the mantle melting domains, and second stage melting of refractory mantle material.

4 Geochemical Characteristics of Archaean Mafic and Ultramafic Volcanic Rocks and Their Tectonic Implications

There is a general consensus that basalts occurring in different modern tectonic environments commonly have distinctive geochemical characteristics (Figs. 5 and 6),

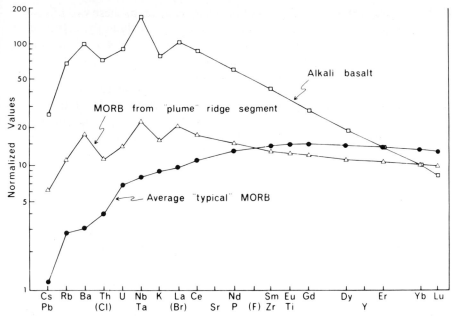

Fig. 5. Normalized abundance patterns for moderately to highly incompatible elements in basalts from mid-oceanic ridges and alkali basalts from ocean islands (after Fig. 18a of Sun 1980). For refractory lithophile elements such as Ti, Zr, REE, U, Th, Nb, Sr, and Ba ordinary chondrite values are used for normalization. For moderately volatile elements such as K, Rb, and P values estimated for the primitive mantle (Sun 1982) are used

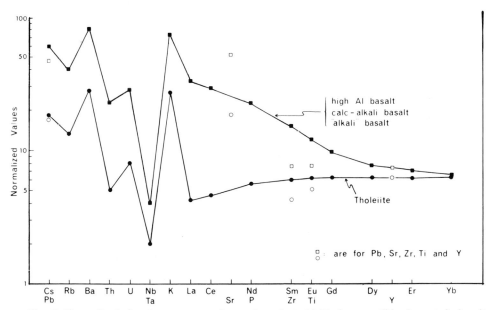

Fig. 6. Normalized abundance patterns for moderately to highly incompatible elements in basalts from island arcs (after Fig. 18b of Sun 1980)

although disagreement exists with regard to the interpretation of these variations. Factors most likely to be responsible for such differences include the effect of oceanic lithosphere subduction under island arcs, different mechanisms for chemical enrichment or depletion taking place under different tectonic environments, and different partial melting processes (e.g. Ringwood 1975; Sun 1980). In island arc environments dehydration and partial melting of the subducted oceanic crust are considered responsible for the relative overabundance of Rb, Sr, Ba, K, Cs, and Pb and depletion of Nb, Ta, and Ti in the basalts (Fig. 6). Hydrous melting conditions under the island arc has also been suggested to be responsible for depletion of Ni in arc basalts (e.g. Hart and Davis 1978).

Archaean komatiites and high magnesian basalts are believed to be generated by large degrees of partial melting of a pyrolite-type source with a residue consisting of olivine with or without some orthopyroxene (e.g. Green 1981). Under these conditions, practically all the incompatible elements (e.g. REE, Ti, Zr, Y, Nb, Ta, P) and major elements such as Al and Ca will be concentrated in the melt. Consequently, the incompatible element ratios of the melt reflect the source character.

There are two basic patterns observed among Archaean mafic volcanic rocks (Fig. 7). Samples of the first type show coherence of major and trace elements. On a chondritic normalized plot they have flat or light REE depleted, heavy REE flat patterns and TiO_2/P_2O_5 ratios close to 10. Chondritic ratios are commonly observed among refractory lithophile elements such as Ti/Zr, Ti/Y, Ti/heavy REE, Al/Ti, and Ca/Ti. This regularity is basically the same as that observed in modern 'typical' MORB (Fig. 5). Elements which are mobile during alteration and metamorphism (e.g. Rb, U, Ba, K) are not considered here, although available data on least altered samples do not indicate much deviation from Fig. 5.

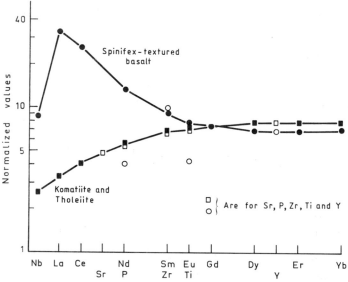

Fig. 7. Two distinct normalized abundance patterns for "immobile" moderately to highly incompatible elements in Archaean komatiites and high magnesian basalts. See text for data source

Samples with the above characteristics are very common in most Archaean green-stone belts (e.g. Nesbitt and Sun 1976; Nesbitt et al. 1979; Jahn et al. 1980; Arndt and Nesbitt 1982).

Samples of the second type show light REE enrichment and decoupling of major and incompatible trace elements. This kind of sample is best illustrated by the high magnesian spinifex-textured basalts (STB) from the ~ 2.7 Ga old Negri volcanics in Pilbara, Western Australia (Sun and Nesbitt 1978a; Nesbitt and Sun 1980). They have a flat heavy REE pattern and lower chondrite normalized Ti and Zr values than the middle REE. There is a distinct Nb depletion relative to light REE. This type of sample is silica-rich and generally has a Al_2O_3/TiO_2 ratio considerably higher and Ti/Zr ratio considerably lower than the chondrite ratios of ~21 and 110, respectively. These geochemical characteristics are similar to modern boninites (e.g. Sun and Nesbitt 1978b). Enrichment of Rb, Ba, and K are commonly observed but not well defined due to secondary mobility problems. The occurrences in several localities in Yilgarn and Pilbara, Western Australia (Sun and Nesbitt 1978b) and in Barberton, South Africa (Jahn et al. 1982), would suggest that this type of volcanic rock is common in the Archaean.

Other examples of major/trace elemental decoupling are tholeiites from Theo's flow in Munro Township, Canada (Arndt and Nesbitt 1982). One picrite sample from this flow shows continuous enrichment from heavy REE to light REE. Abundances of Nb, Ti, Zr, and P are coherently similar to the light REE depleted samples occurring in the same sequence. However, their abundances are only ≤1/2 of those expected from their REE concentration (Arth et al. 1977; Arndt and Nesbitt 1982). These samples have Al_2O_3/TiO_2 ratios considerably lower than chondrites. In this regard (and on account of heavy REE depletion) they are similar to the Barberton type komatiites.

A third type of pattern, which is rarely observed among Archaean samples, was proposed by Jahn et al. (1982). It is similar to the second type discussed above except that it has a chondrite normalized Gd/Yb ratio of less than one and a very high Al_2O_3/TiO_2 ratio (≥40?). It also shares geochemical characteristics of modern boninites.

A literature survey indicates that the degree of elemental decoupling in Archaean basalts depends on the extent of light REE enrichment. The extreme cases are found in some light REE highly enriched basalts from Finland (Jahn et al. 1980), Minnesota (Schulz et al. 1979) and northern China (Wang 1982).

The main geochemical features of these light REE enriched Archaean high magnesian basalts are similar to those of modern island arc basalts (Fig. 6). However, a distinct difference seems to be in the Sr abundance. Unlike island arc basalts, 'fresh' Archaean basalts of this type do not show an obvious Sr overabundance. Source enrichment processes of these Archaean high magnesian basalts are very likely to be related to melting of eclogite. Addition of the melting product to a pyrolite-type source could generate the observed elemental pattern. In this connection it is noted that eclogite melting processes have been considered to be important in the generation of Archaean calc-alkaline basalts with severe depletion in heavy REE, Ta, and Nb (e.g. Condie 1976).

The tectonic implications of eclogite melting in the Archaean mantle should be evaluated within the context of other geological constraints. On the basis of

REE data, Arth and Hanson (1975) suggested that Archaean tonalites and trond-jemites with severe heavy REE depleted patterns were generated by partial melting of eclogite. Geophysical and geological models for Archaean terrains generally indicate that there may not be enough mafic and ultramafic material in the lower crust to account for the observed quantity of felsic igneous rocks (e.g. Drummond et al. 1981; Schulz 1980). A mechanism capable of removing the residue after eclogite partial melting is therefore required. A lithosphere subduction type process for generation of these felsic rocks is favoured by some authors (e.g. Tarney and Windley 1981). However, thermal condition in the Archaean are believed to have been quite different from those of today, since radiogenic heat production in the Archaean was two to four times that of the present (Basaltic Volcanism Study Project 1981). Mantle convection and volcanic activity must also have been more intense at that time. Green (1975, 1981) suggested that ocean crust subduction probably did not take place in the Archaean for lack of the gabbro to eclogite transformation, believed to provide the gravity pull required for modern subduction. In contrast to this opinion, Sleep and Windley (1982) suggested that the Archaean oceanic crust was considerably thicker than its modern counterpart and lithosphere subduction was operative in the Archaean. An independent mechanism for eclogite formation could be related to extensive underplating of komatiitic magmas in the Archaean due to their density contrast with the continental crust (e.g. Fyfe 1978). Conversion of this underplated material to eclogite could cause delamination and sinking back into the mantle. Melts produced from this material will have chemical characteristics of eclogite melting. This process could also modify parts of the neighbouring mantle domains which might serve as magma sources at a later time.

Although available geochemical data seem to suggest that basalts of Archaean greenstone belts were not formed in island arc environments, it remains an unsolved problem whether or not lithosphere subduction existed in the Archaean. Tarney et al. (e.g. 1976, 1981) have proposed a subduction model to account for the evolution of Archaean greenstone-gneiss terrains. In this model the greenstone belts were formed in back arc environments, whereas gneiss terrains represent the root zones of island arcs or a cordillera. To fully evaluate this model, more integrated geological studies would be necessary in addition to detailed geochemical work on mafic and ultramafic rocks in the gneiss terrains.

5 Effect of Archaean Volcanic Activity on the Chemistry of the Continental Lithosphere and the Archaean Upper Mantle

If crustal underplating by ultramafic to mafic magmas and mantle residual peridotite was indeed an important mechanism for the growth of the Archaean crust and lithosphere (e.g. Brooks et al. 1976; Jordan 1981; Fyfe 1978), we would expect to observe a variety of mafic and ultramafic rock types in the subcrustal lithosphere and in the lower crust. Furthermore, the chemistry of the continental lithosphere could be strongly modified by continuous magmatic processes and mantle metasomatism (e.g. Bailey 1982). This concept could be evaluated by detailed chemical and isotopic studies of ultramafic and eclogitic mantle nodules brought up to the surface in kimberlite pipes (e.g. Kramers 1977; Allègre et al. 1982) and post-Archaean continental basalts generated by reactivation of the continental lithosphere (e.g. Brooks et al. 1976; Menzies et al. 1982).

Archaean magma generation processes such as eclogite melting could also modify the chemistry of magma source domains in the convective upper mantle. Some geological observations seem to support such a mechanism. For example, Weaver and Tarney (1981) showed that late Archaean and early Proterozoic mafic dykes in the continental interior and related to crustal tension are commonly enriched in light REE, Rb, Ba, Th, K, and Rb but do not show equivalent enrichment in other incompatible high field strength ions such as Nb, Ta, and P. In this respect they resemble subduction zone arc basalts. However, in similarity with Archaean mafic volcanics such as spinifex-textured basalts, they do not have an obvious Sr overabundance. These chemical characteristics suggest that they were derived from mantle sources which have previously been affected by Archaean mantle enrichment processes related to eclogite melting. It is necessary to point out, however, that the possibility of a metasomatic enrichment process operating under the subcontinental lithosphere (without subduction) and different from that operating under the oceanic lithosphere has not yet been critically evaluated.

Reactivation of parts of the sub-continental lithosphere formed in the Archaean could be responsible for magmatic activity at later times. Sheraton and Black (1981) suggested that 2.4 and 1.2 Ga old Proterozoic tholeiite dykes in east Antarctica are likely to have been generated in this way. Similarly, many other Proterozoic tholeiites such as those from Keweenawan ($\sim$ 1.1 Ga) also show Nb depletion relative to other incompatible elements (Basaltic Volcanism Study Project 1981). Pb, Sr, and Nd isotope studies on these basalts (Leeman 1977; Dosso and Rama Murthy 1982) suggest that they were generated from reactivated sub-continental mantle formed in the Archaean. Recent isotope studies (e.g. Doe et al. 1982; Menzies et al. 1982) also suggest that reactivation of ancient sub-continental lithosphere, triggered by lithosphere subduction or continental rifting, could be a major mechanism for generation of much of the modern continental basalts.

6 Effect of Archaean Tectonism and Magmatism on Mantle Evolution

On the basis of the discussions presented above, it is expected that Archaean and pre-Archaean magmatic and mantle-crust interaction processes have extensively modified the chemical composition of the primitive mantle and have caused long-lasting effects on the chemical and isotopic composition of the Earth's upper mantle.

If an extensive mafic to ultramafic protocrust existed in the early Archaean and if the generation of abundant Archaean tonalite and trondjemite required partial melting of a mafic source (e.g. Glikson 1970; Arth and Hanson 1975) vast volumes of mafic and ultramafic rocks, far exceeding the volume of greenstones in Archaean terrains, must have returned to the mantle through a yet undefined process. This recycled material could have modified the chemical composition of the convecting mantle through partial melting and mechanical mixing, resulting in chemical and mineralogical heterogeneity in the upper mantle. On the basis of chemical and mineralogical variations observed in alpine peridotites, ultramafic tectonites of ophiolite complexes and peridotites occurring on the ocean floor, Frey (1984) concluded that upper mantle lherzolites have heavy REE contents ranging from 1.5 to 3.0 (commonly 2.0) times ordinary chondrites. However, the extent and scale of such heterogeneity have not been defined. Ringwood (1982) suggested that much of the returned mafic material might eventually sink into the lower mantle and create chemical and mineralogical heterogeneities between the upper and lower mantle. Recycling of continental material back into the mantle has also been proposed by Armstrong (1981). Seawater-rock interaction in the Archaean is expected to have been more extensive than at present (e.g. Fyfe 1978). Lithosphere subduction, if it existed in the Archaean, would have had a profound impact on the chemical and isotopic evolution of the Earth's mantle. Armstrong (1981) suggested that a steady state relationship between continental crust and upper mantle has been established since late Archaean times. Whether or not some parts of the primitive mantle have survived to the present day is yet to be demonstrated. One possible candidate is the source for Koolau Series tholeiites of Oahu Island, Hawaii. In addition to primitive $^3He/^4He$ ratios found in the Hawaiian basalts, these tholeiites also have Pb, Sr, Nd, and Hf isotopic compositions close to those expected for 'primitive' mantle (e.g. Stille et al. 1983; Sun 1984).

Several mantle evolution models have been proposed recently to account for the observed Pb, Sr, and Nd isotopic variations in modern volcanic rocks (e.g. O'Nions et al. 1979; Armstrong 1981; Allègre et al. 1980; De Paolo 1980; Zindler et al. 1982; Hofmann and White 1982). Whether or not these models are relevant to the geological history and could properly describe the evolution of the Earth's mantle require further evaluation. In addition to other geological constraints, integrated isotopic and chemical studies of volcanic rocks from Archaean terrains are indispensible and might prove to be critical for a refinement of present mantle evolution models.

Acknowledgements. I thank G. N. Hanson and R. W. Nesbitt who introduced me to Archaean studies 10 years ago when the IGCP Project No. 92 was first initiated by A. Y. Glikson. This paper benefits from discussions with A. Y. Glikson, L.-G. Liu, R. S. Taylor, D. Whitford and constructive reviews by B.-M. Jahn, A. Kröner and an anonymous reviewer. This paper is published with the permission of the Director, Bureau of Mineral Resources, Geology and Geophysics, Canberra.

References

Allègre CJ, Brevart O, Dupre B, Minster J-F (1980) Isotopic and chemical effects produced in a continuously differentiating convecting Earth mantle. Philos Trans R Soc Lond A 297: 447 – 477
Allègre CJ, Shimizu N, Rousseau DC (1982) History of the continental lithosphere recorded by ultramafic xenoliths. Nature 296: 732 – 735
Armstrong RL (1981) Radiogenic isotopes: the case for crustal recycling on a near-steady-state no-continental-growth Earth. Philos Trans R Soc Lond A 301: 443 – 472

Anders E, Ebihara M (1982) Solar-system abundances of the elements. Geochim Cosmochim Acta 46: 2363 – 2380

Anderson DL (1980) Early evolution of the mantle. Episodes 3: 3 – 7

Arndt NT, Nesbitt RW (1982) Geochemistry of Munro Township basalts. In: Arndt NT, Nisbet EG (eds) Komatiites. Allen and Unwin, London, pp 309 – 329

Arth JG, Hanson GN (1975) Geochemistry and origin of the early Precambrian crust of northeast Minnesota. Geochim Cosmochim Acta 39: 325 – 362

Arth JG, Arndt NT, Naldrett AJ (1977) Genesis of Archaean komatiites from Munro Township, Ontario: trace element evidence. Geology (Boulder) 5: 590 – 594

Ashwal LD, Wooden JL, Morrison DA, Shih C-Y, Wiesmann H (1981) The Bad Vermilion Lake anorthosite complex, Ontario: Sr and Nd isotopic evidence for depleted Archaean mantle. Abst with program, Geol Soc Am 13: 399

Bailey DK (1982) Mantle metasomatism – continuing chemical change within the Earth. Nature 296: 525 – 530

Basaltic Volcanism Study Project (1981) Basaltic volcanism on the terrestrial planets. Pergamon, New York

Basu AR, Tatsumoto M (1980) Nd-isotopes in selected mantle-derived rocks and minerals and their implications for mantle-evolution. Contrib Mineral Petrol 75: 43 – 54

Bickle MJ, Ford CE, Nisbet EG (1977) The petrogenesis of peridotitic komatiites: Evidence from high-pressure melting experiments. Earth Planet Sci Lett 37: 97 – 106

Brooks C, James DE, Hart SR (1976) Ancient lithosphere: Its role in young continental volcanism. Science 193: 1086 – 1094

Cattell A, Krogh TE, Arndt NT (to be published) Conflicting Sm-Nd whole rock and U-Pb zircon ages for Archaean lavas from Newton Township, Abitibi Belt, Ontario. Earth Planet Sci Lett

Chauvel C, Dupré B, Jenner GA (1984) Kambalda greenstone belt: 2700 or 3200 Ma old? Earth Planet Sci Lett

Chauvel C, Dupré B, Todt W, Arndt NT, Hofmann AW (1983) Pb and Nd isotopic correlation in Archaean and Proterozoic greenstone belts. Eos Trans Am Geophys Union 64: 330

Chou CL (1978) Fractionation of siderophile elements in the Earth's upper mantle. Proc Lunar Sci Conf 9: 219 – 230

Clarke DB (1970) Tertiary basalts of Baffin Bay: possible primary magma from the mantle. Contrib Mineral Petrol 25: 203 – 224

Condie KC (1976) Trace-element geochemistry of Archaean greenstone belts. Earth-Sci Rev 12: 393 – 417

De Paolo DJ (1980) Crustal growth and mantle-evolution: inferences from models of element transport and Nd and Sr isotopes. Geochim Cosmochim Acta 44: 1185 – 1196

Doe BR, Leeman WP, Christiansen RL, Hedge CE (1982) Lead and strontium isotopes and related trace elements as genetic tracers in the upper Cenozoic rhyolite-basalt association of the Yellowstone Plateau volcanic field. J Geophys Res 87: 4785 – 4806

Dosso L, Rama Murthy V (1982) Precambrian continental mantle evolution: Keweenawan volcanism of the north shore of Lake Superior. (Abstract) 5th Internat Conf Geochron Cosmochron Isotope Geol 82

Dreibus G, Wänke H (1979) On the chemical composition of the moon and the eucrite parent body and a comparison with the composition of the Earth, the case of Mn, Cr and V. (Abstract) Lunar Sci Conf 10: 315 – 317

Drummond BJ, Smith RE, Horwitz RC (1981) Crustal structure in the Pilbara and northern Yilgarn blocks from deep seismic sounding. In: Glover JE, Groves DI (eds) Archaean geology. Spec Publ Geol Soc Aust 7: 33 – 42

Echeverria LM (1980) Tertiary or Mesozoic komatiites from Gorgona Island, Columbia: Field relations and geochemistry. Contrib Mineral Petrol 73: 253 – 266

Ernst WG, Piccardo GB (1979) Petrogenesis of some Ligurian peridotites – I. Mineral and bulk rock chemistry. Geochim Cosmochim Acta 43: 219 – 237

Fletcher IR, Rosman KJR (1982) Precise determination of initial εNd from Sm-Nd isochron data. Geochim Cosmochim Acta 46: 1983 – 1987

Frey FA (1984) Rare earth element abundances in upper mantle rocks. In: Henderson P (ed) Rare element geochemistry Elsevier, Amsterdam, pp 153 – 203

Fyfe WS (1978) The evolution of the Earth's crust: Modern plate tectonics to ancient hot spot tectonics? Chem Geol 23: 89 – 114

Gansser A, Dietrich VJ, Cameron WE (1979) Paleogene komatiites from Gorgona Island. Nature 278:545–546

Glikson AY (1970) Geosynclinal evolution and geochemical affinities of early Precambrian systems. Tectonophys 9:397–433

Glikson AY (1971) Primitive Archaean element distribution patterns: Chemical evidence and geotectonic significance. Earth Planet Sci Lett 12:309–320

Glikson, AY (1979). Siderophile and lithophile trace element evolution of the Archaean mantle. BMR J Aust Geol Geophys 4:253–279

Glikson AY, Hickman AH (1981) Geochemical stratigraphy and petrogenesis of Archaean basic-ultrabasic volcanic units, eastern Pilbara block, Western Australia. Spec Publ Geol Soc Aust 7:287–300

Green DH (1975) Genesis of Archaean peridotite magmas and constraints on Archaean geothermal gradients and tectonics. Geology (Boulder) 3:15–18

Green DH (1981) Petrogenesis of Archaean ultramafic magmas and implications for Archaean tectonics. In: Kröner A (ed) Precambrian plate tectonics. Elsevier, Amsterdam, pp 469–489

Hart SR, Davis KE (1978) Nickel partitioning between olivine and silicate melt. Earth Planet Sci Lett 40:203–219

Hofmann AW, White WM (1982) Mantle plumes from ancient oceanic crust. Earth Planet Sci Lett 57:421–436

Jacobsen SB, Wasserburg GJ (1980) Sm-Nd isotopic evolution of chondrites. Earth Planet Sci Lett 50:139–155

Jagoutz E, Palme H, Baddenhausen H, Blum K, Cendales M, Dreibus G, Spettel B, Lorenz V, Wänke H (1979) The abundances of major, minor and trace elements in the Earth's mantle as derived from primitive ultramafic nodules. Proc Lunar Sci Conf 10:2031–2050

Jagoutz E, Wänke H (1982) Has the Earth's core grown over geologic time? Lunar and Planetary Science XIII, pp 358–359, Lunar and Planetary Science Institute, Houston

Jahn BM, Auvray B, Blais S, Capdevila R, Cornichet J, Vidal F, Hameurt J (1980) Trace element geochemistry and petrogenesis of Finnish greenstone belts. J Petrol 21:201–244

Jahn BM, Gruau G, Glikson AY (1982) Komatiites of the Onverwacht Group, S. Africa: REE geochemistry, Sm/Nd age and mantle evolution. Contrib Mineral Petrol 80:24–40

Jahn BM, Nyquist LE (1976) Crustal evolution in the early earth-moon system: Constraints from Rb-Sr studies. In: Windley BF (ed) Early history of the Earth. Wiley, New York, pp 55–76

Jordan TH (1981) Continents as a chemical boundary layer. Philos Trans R Soc Lond A 301:359–373

Kramers JD (1977) Lead and strontium isotopes in Cretaceous kimberlites and mantle derived xenoliths from Southern Africa. Earth Planet Sci Lett 34:419–431

Kramers JD, Smith CB, Lock NP, Harmon RS, Boyd FR (1981) Can kimberlites be generated from an ordinary mantle? Nature 291:53–56

Langmuir C, Bender JF, Bence A, Hanson GN, Taylor SR (1977) Petrogenesis of basalt from the FAMOUS area: Mid-Atlantic ridge. Earth Planet Sci Lett 36:133–156

Leeman WP (1977) Pb and Sr isotopic study of Keweenawan lavas and inferred 4 b.y. old lithosphere beneath part of Minnesota. Abs with program, Geol Soc Am 9:1068

Liu L-G (1979) On the 650 km seismic discontinuity. Earth Planet Sci Lett 42:202–208

Liu L-G (1982) Chemical inhomogeneity of the mantle: geochemical consideration. Geophys Res Lett 9:124–126

Machado N, Brooks C, Hart SR (1983) Isotopic characteristics of the Archaean mantle: Nd and Sr from the Abitibi belt Eos Trans Am Geophys Union 64:330

McCulloch MT (1982) Identification of Earth's earliest differentiates. (Abstract) 5th Internat Conf Geochron Cosmochron Isotope Geol, pp 244–245

McCulloch MT, Compston W (1981) Sm-Nd age of Kambalda and Kanowna greenstones and heterogeneity in the Archaean mantle. Nature 294:322–327

Menzies MA, Leeman WP, Hawkesworth CJ (1982) Grossly heterogeneous sub-continental mantle in the source regions of Cenozoic volcanic rocks from the western USA: Sr and Nd isotopic evidence. (Abstract) 5th Internat Conf Geochron Cosmochron Isotope Geol, pp 250–251

Morgan JW, Wandless GA, Petrie RK, Irving AJ (1981) Composition of the Earth's upper mantle. – I. Siderophile trace elements in ultramafic nodules. Tectonophys 75:47–67

Nesbitt RW, Sun S-S (1976) Geochemistry of Archaean spinifex textured peridotites and magnesian and low magnesian tholeiites. Earth Planet Sci Lett 31:433–453

Nesbitt RW, Sun S-S (1980) Geochemical features of some Archaean and post-Archaean high-magnesian-low-alkali liquides. Philos Trans R Soc Lond A 297:365 – 381

Nesbitt RW, Sun S-S, Purvis AC (1979) Komatiites: Geochemistry and genesis. Can Mineral 17:165 – 186

Newsom HE, Palme H (1984) The depletion of siderophile elements in the Earth's mantle: new evidence from molybdenum and tungsten. Earth Planet Sci Lett

Nisbet EG, Walker D (1982) Komatiites, and the structure of the Archaean mantle. Earth Planet Sci Lett 60:105 – 116

Ohtani E (1984) Generation of komatiite magma and gravitational differentiation in the deep upper mantle. Earth Planet Sci Lett 67:261 – 272

O'Nions RK, Evensen NM, Hamilton PJ (1979) Geochemical modelling of mantle differentiation and crustal growth. J Geophys Res 84:6091 – 6101

Patchett PJ (1982) Hf isotopes and mantle evolution. (Abstract) 5th Internat Conf Geochron Cosmochron Isotope Geol, pp 305 – 306

Patchett PJ, Kouvo O, Hedge CE, Tatsumoto M (1981) Evolution of continental crust and mantle heterogeneity: evidence from Hf isotopes. Contrib Mineral Petrol 78:279 – 297

Richards JR, Fletcher IR, Blockley JG (1981) Pilbara galenas: precise isotopic assay of the oldest Australian leads; model ages and growth-curve implications. Mineral Deposita 16:7 – 30

Ringwood AE (1975) Composition and Petrology of the Earth's mantle. McGraw-Hill

Ringwood AE (1977) Composition of the core and implications for the origin of the earth. Geochem J 11:111 – 135

Ringwood AE (1982) Phase transformation and differentiation in subducted lithosphere: Implications for mantle dynamics, basalt petrogenesis and crustal evolution. J Geol 90:611 – 644

Schulz KJ (1980) The magmatic evolution of the Vermillion greenstone belt, NE Minnesota. Precammrian Res 11:215 – 245

Schulz KJ (1982) Magnesian basalts from the Archaean terranes of Minnesota. In: Arndt NT, Nisbet EG (eds) Komatiites. Allen and Unwin, London, pp 171 – 186

Schulz KJ, Smith IEM, Blanchard D (1979) The nature of Archaean alkalic rocks from the southern portion of the Superior privince. Eos Trans Am Geophys Union 60:410

Schwarz EJ, Fujiwara Y (1977) Komtiitic basalts from the Proterozoic Cape Smith Range in northern Quebec, Canada. Geol Assoc Can Spec Pap 16:193 – 201

Sheraton JW, Black LP (1981) Geochemistry and geochronology of Proterozoic tholeiite dykes of east Antarctica: Evidence for mantle metasomatism. Contrib Mineral Petrol 78:305 – 317

Shima M, Honda M (1967) Distributions of alkali, alkaline earth und rare earth elements in component minerals of chondrites. Geochim Cosmochim Acta 31:1995 – 2006

Sleep NH, Windley BF (1982) Archaean plate tectonics: constraints and inferences. J Geol 90:363 – 379

Smith HS, Erlank AJ (1982) Geochemistry and petrogenesis of komatiites from the Barberton greenstone belt, South Africa. In: Arndt NT, Nisbet EG (eds) Komatiites. Allen and Unwin, London, pp 347 – 397

Stille P, Unruh DM, Tatsumoto MC (1983) Pb, Sr, Nd and Hf isotopic evidence of multiple sources for Oahu, Hawaii basalts. Nature 304:25 – 29

Sun S-S (1980) Lead isotopic study of young volcanic rocks from mid-ocean ridges, ocean islands and island arcs. Philos Trans R Soc Lond A 297:409 – 446

Sun S-S (1982) Chemical composition and origin of the earth's primitive mantle. Geochim Cosmochim Acta 46:179 – 192

Sun S-S (1984) Some geochemical constraints on mantle evolution models. In: Proceedings 27th Internat Geol Congress, Moscow, VNU BV, Netherlands

Sun S-S, Nesbitt RW (1977) Chemical heterogeneity of the Archaean mantle, composition of the earth and mantle evolution. Earth Planet Sci Lett 35:429 – 448

Sun S-S, Nesbitt RW (1978a) Petrogenesis of Archaean ultrabasic and basic volcanics: Evidence from rare earth elements. Contrib Mineral Petrol 65:301 – 325

Sun S-S, Nesbitt RW (1978b) Geochemical regularities and genetic significance of ophiolitic basalts. Geology (Boulder) 6:689 – 693

Sun S-S, Nesbitt RW, Sharaskin A Y (1979) Geochemical characteristics of mid-ocean ridge basalts. Earth Planet Sci Lett 44:119 – 138

Tarney J, Dalziel IWD, de Wit MJ (1976) Marginal basin 'Rocas Verdes' complex from S. Chile: a model for Archaean greenstone belt formation. In: Windley B F (ed) The early history of the Earth. Wiley, New York, pp 131 – 146

Tarney J, Windley BF (1981) Marginal basins through geological time. Philos Trans R Soc Lond A 301:217–232

Taylor SR, McLennan SM (1981) The composition and evolution of the continental crust: rare earth element evidence from sedimentary rocks. Philos Trans R Soc Lond A 301:381–399

Tilton GR (1983) Evolution of depleted mantle: The lead perspective. Geochim Cosmochim Acta 47:1191–1197

Viljoen MJ, Viljoen RP (1969) Evidence for the existence of a mobile extrusive peridotitic magma from the Komati Formation of the Onverwacht Group. Geol Soc S Afr Spec Publ 2:87–112

Wang K (1982). REE content and its tectonic implications of the ancient metamorphic rocks from Taipingzhai, Qiansi County, Hebei province. Sci Geol Sin:144–151

Wänke H (1981) Constitution of terrestrial planets. Philos Trans R Soc Lond A 303:287–302

Wänke H, Dreibus G, Jagoutz E (1984) Mantle chemistry and accretion history of the Earth. This Vol., 1–24

Watt JP, Ahrens TJ (1982) The role of iron partitioning in mantle composition, evolution und scale of convection. J Geophys Res 87:5631–5644

Weaver BL, Tarney J (1981) The Scourie dyke suite: Petrogenesis and geochemical nature of the Proterozoic sub-continental mantle. Contrib Mineral Petrol 78:175–188

Zindler A, Brooks C, Arndt NT, Hart S (1978) Nd and Sr isotope data from komatiitic and tholeiitic rocks of Munro Township, Ontario. US Geol Surv Open File Rep 78-701:469–471

Zindler A, Jagoutz E, Goldstein S (1982) Nd, Sr and Pb isotopic systematics in a three-component mantle: a new perspective. Nature 298:519–523

Archaean Sedimentary Rocks and Their Relation to the Composition of the Archaean Continental Crust

S. M. McLennan and S. R. Taylor[1]

Contents

1	Sedimentary Rocks and Crustal Composition	48
1.1	Origin and Significance of Eu-Anomalies in Sedimentary Rocks	49
2	Trace Elements in Archaean Sedimentary Rocks	50
2.1	Rare Earth Elements	51
2.2	La-Th Systematics	53
2.3	Th-U	53
2.4	Ferromagnesian Trace Elements	53
2.5	Boron	54
3	Provenance of Archaean Sedimentary Rocks	54
3.1	Tectonic Models	54
3.2	Petrological Evidence	56
3.3	Comparison of Archaean and Phanerozoic Greywackes	58
3.4	Archaean Sedimentary Rocks as Crustal Samples	61
4	Mixing Models for Archaean Sedimentary Rocks	61
4.1	Archaean Bimodal Suite	61
4.2	Rare Earth Elements	62
4.3	Mixing Models	64
5	Composition and Nature of the Archaean Crust	65
	References	68

Abstract

The composition of post-Archaean terrigenous clastic sedimentary rocks is very uniform for several insoluble trace elements (e.g. REE, Th, Sc) and is thought to reflect the composition of the exposed upper continental crust. In contrast, Archaean sedimentary rocks display highly variable trace element abundances, but on average have lower levels of incompatible elements (Th, LREE), lower La/Yb ratios and lack the Eu depletion characteristic of post-Archaean sedimentary rocks. Early Archaean terrigenous sedimentary rocks also have high abundances of Cr and Ni. An examination of tectonic, sedimentological and trace element evidence indicates that all of the abundant Archaean igneous lithologies play a role in the provenance of Archaean sedimentary rocks. It is concluded that, although caution is warranted, information about the nature of the Archaean upper crust can be obtained from the sedimentary data. Systematic variations of trace element patterns (e.g. REE, La/Sc, Co/Th) in Archaean sedimentary rocks provide persuasive evidence for a dominant origin from the common Archaean bi-

---

1 Research School of Earth Sciences, Australian National University, G.P.O. Box 4, Canberra, ACT 2600, Australia

Archaean Geochemistry (ed. by A. Kröner et al.)
© Springer-Verlag Berlin Heidelberg 1984

modal mafic-felsic igneous suite. On average, a 1 : 1 mix is indicated for the Archaean upper crust. Such a model results in a considerably less differentiated composition compared to the present day upper crust. The general lack of Eu depletion in Archaean sedimentary rocks and evidence for garnet fractionation during the formation of the felsic end member indicates that formation of both basic and felsic components of the Archaean crust occurred at mantle depths (>40 km). In contrast, the formation of the latest Archaean and post-Archaean upper crusts took place by intra-crustal melting, forming K-granites and granodiorites with Eu depletion.

1 Sedimentary Rocks and Crustal Composition

The problem of establishing the composition of the present upper continental crust is greatly simplified by the fact that sedimentary processes provide a natural sampling mechanism of the exposed crust. Elements which are relatively insoluble in natural waters and have very short residence times are transferred virtually quantitatively into clastic sedimentary sequences. Those insoluble elements which are strongly fractionated among clastic sediments, such as Zr (due to heavy mineral concentrations in sandstones), are of limited value in estimating crustal composition. Elements which are less fractionated during sedimentary processes, including the rare earth elements (REE), Th and Sc, are very useful for such estimates. Thus, the remarkable uniformity of REE patterns in post-Archaean terrigenous sedimentary rocks (e.g. Nance and Taylor 1976) contrasts strongly with the considerable variability observed in igneous source rocks and attests to the efficiency of mixing during transportation and sedimentation.

It is beyond the scope of this paper to discuss the effects of weathering, sedimentation and metamorphism on the chemical composition of sedimentary rocks; a complete discussion is given in Taylor and McLennan (1984). Processes of diagenesis and most types of metamorphism generally do not affect REE patterns (e.g. Chaudhuri and Cullers 1979; Muecke et al. 1979). Some fractionation of the REE does occur during sedimentation, primarily due to heavy mineral separation, but these effects are minor for shales, and the fine-grained portion of sedimentary rocks is found consistently to reflect their provenance (e.g. Cullers et al. 1979). The processes of weathering also fractionate the REE, but this fractionation occurs within the weathering profile and is mineralogically controlled; the ultimate transport of the REE is primarily a mechanical process (e.g. Nesbitt 1979).

The REE patterns for terrigenous clastic sedimentary rocks are similar in Australia (PAAS), Europe (ES) and North America (NASC), and this worldwide similarity is reinforced by the observation that comparable patterns are found in glacially-derived loess (Taylor et al. 1983a) of Pleistocene age from North America, China, Europe and New Zealand (Fig. 1). Accordingly, it may be safely concluded that the average REE pattern in sedimentary rocks is typical of the exposed upper continental crust. In this laboratory, we have adopted the values for post-Archaean average Australian shale (PAAS) which are reduced by

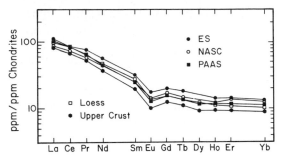

Fig. 1. Chondrite-normalized REE diagram of various estimates of average shale: *ES* European Shale Composite (Haskin and Haskin 1966); *NASC* North American Shale Composite (Haskin and Haskin 1966); *PAAS* Post-Archaean Average Australian Shale (Nance and Taylor 1976). Also shown is an average of ten Pleistocene loess samples from USA, China and New Zealand (Taylor et al. 1983a). It is concluded that such patterns are typical of post-Archaean clastic sedimentary rocks and are taken to be parallel to the average REE pattern of the upper continental crust. The upper crust values adopted in this paper are derived by lowering PAAS values by 20% (Taylor and McLennan 1981). The Eu-depletion seen in the upper crust indicates that intracrustal partial melting was the dominant process in its formation

20% (Fig. 1) to account for sedimentary rocks with lower REE abundances, such as sandstones and carbonates (see Taylor and McLennan 1981).

Exceptions to the general REE patterns shown in Fig. 1 do exist in the post-Archaean clastic sedimentary record. Presently available studies indicate that such exceptions are restricted to first-cycle volcanogenic sedimentary rocks deposited at island arcs (Condie and Snansieng 1971; Nance and Taylor 1977; Bhatia 1981). In these cases, the REE patterns simply reflect the immediate source, typically calc-alkaline andesite, as would be expected. We shall return to this point below.

The regularities in upper crustal composition seen in sedimentary REE patterns extend back to the early Proterozoic but are not typically observed in Archaean sedimentary rocks. Archaean sedimentary rocks are characterized by lower REE abundances, lower La/Yb and most importantly no Eu-anomaly (see below). If Archaean sedimentary REE are representative of the exposed crust, then such differences have many implications for the composition and nature of the Archaean crust and place severe constraints on the mechanisms and timing of crustal growth and differentiation (McLennan et al. 1980; Taylor and McLennan 1981; McLennan and Taylor 1982). In this paper, we address the problem of relating the chemical composition of Archaean sedimentary rocks to the composition and nature of the Archaean crust and remark on some of the implications for crustal evolution.

1.1 Origin and Significance of Eu-Anomalies in Sedimentary Rocks

Understanding the origin of the Eu depletion relative to the other chondrite-normalized REE in post-Archaean terrigenous clastic sedimentary rocks (Fig. 1) is crucial to any interpretations of crustal composition and evolution. The most significant observation in this regard is that all major post-Archaean sediment types (sandstones, mudstones, carbonates) are characterized by Eu depletion of approximately comparable magnitude (McLennan 1981). The only important sedimentary rock types which do not have Eu depletion are some of the volcanogenic sediments deposited in island-arc environments (Nance and Taylor 1977; Bhatia 1981). No common sedimentary rock is *characterized* by Eu enrichment. Similarly, river and seawater also show Eu depletion (e. g. Martin et al. 1976; Elderfield and Greaves 1982). This is very strong evidence that the upper continental crust must be similarly depleted in Eu. Any non-crustal hydrothermal additions to the hydrosphere or sedimentary record would be expected to be enriched in Eu (Graf 1977; Courtois and Treuil 1977; Kerrich and Fryer 1979).

Common igneous rocks derived from the mantle (e.g. MORB, island arc basalts and andesites) are not characterized by Eu-anomalies (Bence et al. 1980; Basaltic Volcanism 1981), and Eu-anomalies which have been noted usually can be readily attributed to late stage addition or removal of feldspar or hydrothermal alteration (Sun and Nesbitt 1978). Accordingly, if the continental crust is derived from the mantle, then it follows that the bulk continental crust is not anomalous with respect to Eu. Thus, any suggestion that Eu-anomalies seen in post-Archaean clastic sedimentary rocks are due to oxidation-reduction processes during weathering or the breakdown of feldspar are unfounded since there is no significant upper crustal reservoir with the complementary Eu enrichment. The most reasonable explanation is that the Eu depletion in sedimentary rocks, and hence that of the upper continental crust, is due to chemical fractionation within the continental crust, related to partial melting and production of granitic rocks. The residual material in the lower continental crust would thus contain the complementary Eu enrichment, the magnitude of which would depend on the relative proportions of the upper and lower crust (Taylor and McLennan 1981, Taylor et al. 1983b). Eu released during weathering will be oxidized to Eu^{3+} and hence behave like the other trivalent REE. The presence of an Eu anomaly is thus the signature of earlier events in a more reducing igneous environment.

2 Trace Elements in Archaean Sedimentary Rocks

The trace-element geochemistry of Archaean sedimentary rocks is less well understood than post-Archaean sedimentary rocks, but available data indicate several important differences, particularly in REE characteristics. Such differences may have implications for the composition of the Earth's early crust. For the past several years this laboratory has been accumulating geochemical data on Archaean sedimentary rocks, primarily shales, from Australia, Greenland and South Africa in order to evaluate these differences (Nance and Taylor 1977; Bavinton and Taylor 1980; McLennan et al. 1983a, b, 1984).

In Fig. 2 the chemical composition of the average post-Archaean shale (on a carbonate-free basis) is compared to the average Archaean shale. In terms of major elements, Archaean shales are enriched in Na, Ca, Mg, Fe, and depleted in Si, K. This is consistent with the findings of McLennan (1982) who examined the

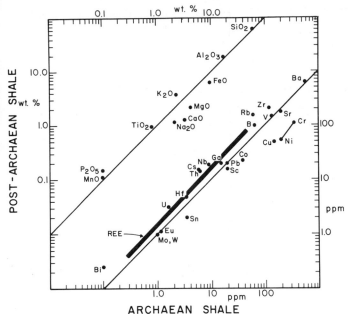

Fig. 2. Comparison diagram of average post-Archaean shale and average Archaean shale. Both compositions are on a carbonate-free basis. *Solid block field* encompasses REE. *Diagonal lines* represent equal composition. Note that the Archaean composition is depleted in most incompatible elements and enriched in most ferromagnesian elements, particularly Cr and Ni

bulk compositions of all clastic sedimentary rocks through time. In terms of trace elements Archaean shales are generally enriched in the ferromagnesian trace elements, particularly Cr and Ni. The REE and other large ion lithophile (LIL) elements are depleted in Archaean shales.

2.1 Rare Earth Elements

Two features stand out in the REE patterns of Archaean sedimentary rocks. Firstly, REE patterns are highly variable from nearly flat with REE depletion ($La_N/Yb_N < 1$) to very steep with $La_N/Yb_N > 25$. Secondly, although local exceptions exist, Archaean shales do not possess Eu anomalies. In Fig. 3 some typical REE patterns of Archaean sedimentary rocks are shown. By far the most common types of patterns are those shown in Fig. 3a, with intermediate La_N/Yb_N and no Eu-anomalies. Such patterns are found in the Yilgarn and Pilbara Blocks of Western Australia (Nance and Taylor 1977; Bavinton and Taylor 1980; McLennan 1981; McLennan et al. 1983a), the Fig Tree and Moodies Groups of South Africa (Wildeman and Condie 1973; McLennan et al. 1983b) and in West Greenland (McLennan et al. 1984). Of the common terrestrial igneous rocks, such patterns are most similar to calc-alkaline andesites, but Archaean andesites with such REE patterns are relatively rare (e.g. Jahn and Sun 1979). The data are equally consistent with sedimentary rocks being derived from a mixture of the well documented Archaean bimodal igneous suite of mafic volcanics and tonalites, trondhjemites, dacites (Barker and Peterman 1974; Barker and Arth 1976; Barker et al. 1981).

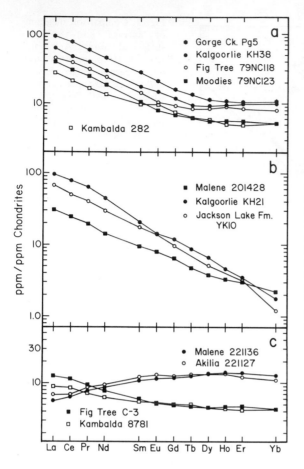

Fig. 3a–c. Chondrite-normalized REE diagram showing some typical Archaean sedimentary rocks; **a** most typical Archaean sedimentary REE pattern showing intermediate La/Yb ratios and no Eu-depletion. Such patterns would be consistent with a source of calc-alkaline andesites or a mixture of mafic and felsic igneous rocks; **b** patterns with very high La/Yb ratios and no Eu-anomalies. Such patterns are considerably less common than those shown in **a** and are probably derived almost exclusively from Na-rich granitic rocks (tonalites, trondhjemites) or felsic volcanics; **c** flat REE patterns with slight LREE enrichment or depletion. Such patterns are uncommon and are probably derived almost exclusively from mafic volcanics. Data sources include Wildeman and Condie (1973), Nance and Taylor (1977), Bavinton and Taylor (1980), Jenner et al. (1981), McLennan et al. (1983a, b, 1984)

Less common are REE patterns with very high or very low La/Yb. Steep patterns such as those shown in Fig 3b are found in the Yilgarn and Pilbara Blocks (Nance and Taylor 1977; McLennan et al. 1983a), West Greenland (McLennan et al. 1984) and the Slave Province of Canada (Jenner et al. 1981). Such patterns are similar to felsic volcanics (mainly dacites) and Na-rich granitic-rocks (tonalites, trondhjemites) commonly found in Archaean terrains (e.g. O'Nions and Pankhurst 1978; Jahn and Sun 1979). Patterns such as those shown in Fig. 3c, with low La/Yb ratios, are least abundant and are very similar to typical Archaean mafic volcanics (e.g. Sun and Nesbitt 1978).

Rarely, Archaean sedimentary rocks display positive or negative Eu-anomalies. Positive anomalies are generally restricted to plagioclase-rich samples (Nance and Taylor 1977) or can be ascribed to hydrothermal activity (Bavinton and Taylor 1980). Boak et al. (1982) recently described a suite of metamorphosed Archaean sedimentary rocks from West Greenland with significant negative Eu-anomalies (see also McLennan et al. 1984). It was suggested that these were volcanogenic sedimentary rocks derived from felsic volcanics with comparable REE patterns (Boak et al. 1982).

2.2 La-Th Systematics

The geochemical coherence between Th and the light rare earth elements (LREE) in sedimentary rocks was documented by McLennan et al. (1980). This correlation, considered to be due to the incompatible nature of these elements and to their similar behaviour during sedimentary processes, was used successfully to estimate Archaean and post-Archaean upper crustal abundances for Th and several other elements. The most recent estimate of the average La/Th ratio in Archaean sedimentary rocks is 3.5 ± 0.3 (at 95% confidence) which is indistinguishable from an earlier quoted value of 3.6 ± 0.4 (McLennan et al. 1980). This is higher than post-Archaean sedimentary La/Th ratios (2.7 ± 0.2) and consistent with a more mafic Archaean upper crust (McLennan et al. 1980).

2.3 Th-U

The abundances of both Th and U in Archaean shales (Th = 6.3 ± 1.4 ppm; U = 1.6 ± 0.4 ppm) are much lower than post-Archaean shales (Th = 14.6 ± 1.2; U = 3.1 ± 0.4; McLennan and Taylor 1980). Th/U ratios in Archaean shales are quite variable but average 3.9 ± 0.5, which is lower than in post-Archaean shales (Th/U = $4.5 - 5.5$; McLennan and Taylor 1980) and is indistinguishable from typical igneous ratios. The Archaean values for Th, U and Th/U quoted here are all higher than those found in the preliminary study of McLennan and Taylor (1980) which did not include data from the Pilbara or South Africa.

2.4 Ferromagnesian Trace Elements

Danchin (1967) first noted the extreme levels or Cr and Ni in Fig Tree shales (average Cr = 860 ppm; Ni = 500 ppm). Our data confirm these high levels and also find consistently high Cr and Ni values in shales of the Moodies Group, South Africa, and Gorge Creek Group, Pilbara Block. On the other hand, many other sequences, such as at Kalgoorlie, Yilgarn Block (Taylor 1977) and in the Canadian Shield (Cameron and Garrels 1980) are only slightly enriched in Cr and Ni. Thus, it appears that the high levels or Cr and Ni may be restricted to early Archaean sequences. The level of enrichment for Cr and Ni seen in these Archaean sedimentary sequences does not extend to the other ferromagnesian trace elements V, Co or Sc (see Fig. 2), resulting in anomalously high Cr/V and Ni/Co ratios. Taylor (1977) suggested that high Cr and Ni in Archaean sedimentary rocks would be evidence for a bimodal origin, rather than an island-arc origin, since young island-arc volcanics generally have undergone olivine and spinel fractionation, drastically reducing Cr and Ni abundances. Some caution is warranted, however, since in some areas it can be shown that the Cr and Ni abundances in Archaean shales are too high for any realistic source composition (either basic or ultramafic) and that some enrichment process during weathering or sedimentation has probably occurred (e.g. McLennan et al. 1983a).

2.5 Boron

The abundance of B in Archaean shales is highly variable, ranging from <5 ppm to in excess of 300 ppm. In some cases high B content can be related to the presence of metamorphic tourmaline (e.g. West Greenland; McLennan et al. 1984), but in most instances this is not the case. Archaean sedimentary sequences with average B in excess of 75 ppm include the Gorge Creek Group (McLennan et al. 1983a), Fig Tree Group (McLennan et al. 1983b) and the Kalgoorlie sequence (McLennan 1981).

The use of B as a palaeosalinity indicator is not infallible (e.g. Shaw and Bugry 1966) and must be approached with considerable caution, especially for Archaean sedimentary rocks. For example, in modern marine environments B uptake in sedimentary rocks is greatly influenced by the presence of illite, which may absorb twice as much B as montmorillonite or chlorite (Degens 1965). Thus, in the Archaean where illite is rare in shales (see McLennan et al. 1983a for exception), low B contents may not reflect a non-marine environment. Keeping these caveats in mind it has commonly been observed that a B content in argillaceous sedimentary rocks in excess of 80 – 100 ppm is associated with marine environments. Igneous rocks seldom have more than 15 – 20 ppm B. Whether the high B seen in many Archaean shales indicates a marine environment, similar to that found today, for the deposition of some Archaean sedimentary rocks must be considered uncertain. However, many of the processes of B enrichment in sedimentary rocks, possibly including salinity effects, must have occurred.

3 Provenance of Archaean Sedimentary Rocks

The average Archaean sedimentary REE pattern mimics that of modern calc-alkaline volcanic rocks, both in slope and in the absence of a europium anomaly. Similarly, many Archaean sedimentary REE patterns are similar to Phanerozoic first-cycle volcanogenic sediments deposited in fore-arc basins. However, in the previous sections it has been suggested that the REE data could also be explained through a mixture of a bimodal mafic-felsic source. An important question thus arises as to the precise nature of the provenance of these and other Archaean clastic sedimentary rocks and how it relates to the composition of the upper Archaean crust. In this section we address this question by examining Archaean tectonic models and well documented Archaean sedimentary rock sequences in which provenance can be established with some certainty.

3.1 Tectonic Models

Some understanding of the origin of Archaean sedimentary rocks can be gained through examining the tectonic models of Archaean greenstone belt development. These have been reviewed by Condie (1981a). Some workers have attempted to give generalized models to explain the common features of all greenstone belts (e.g. Fyfe 1974; Hargraves 1976, 1981; Tarney et al. 1976; Condie 1980; Goodwin 1981; Tarney and Windley 1981). A prime example of this approach is the 'back-arc basin' model of Tarney et al. (1976). Perhaps the most successful models are those which approach each greenstone belt separately using detailed field and sedimentological data (e.g. Walker and Pettijohn 1971; Eriksson 1980a, 1982a, b; Dimroth et al. 1982, 1983a, b). If there is any insight at all to be gained from modern sediments, it is that no single tectonic environment will explain the nature of all Archaean sedimentary rocks.

Models of intracratonic rifting or warping with greenstone belts being deposited between sialic continental blocks have been proposed for the Yellowknife Supergroup (Henderson 1981) and greenstone belts of the Yilgarn Block (Gee et al. 1981). In contrast, Dimroth and co-workers (Dimroth et al. 1982, 1983a, b) have suggested an evolving fore-arc basin setting for parts of the Abitibi Belt. Eriksson (1977, 1978, 1979, 1980a, b, 1981, 1982a, b) has made detailed examinations of the Fig Tree, Moodies and Gorge Creek Groups of South Africa and the Pilbara Block. These sedimentary rocks appear to be synorogenic (Eriksson pers. comm.) and were deposited at an evolving continental margin. In the case of the Fig Tree and Moodies Groups there is no evidence for sediment contribution from an island arc (Eriksson 1980a). In the Archaean of West Greenland the development of tectonic models is hampered by high metamorphic grade and limited exposure. At Isua there is some evidence that most of the sedimentary rocks were deposited at a site remote from sialic crust (Nutman et al. 1984). On the other hand, at least some of the Malene supracrustals lie on Amîtsoq Gneiss basement and were in part derived from it (Nutman and Bridgwater 1983).

Recently, it has been suggested that there may be some fundamental differences in the tectonic settings of early (> 3.3 Ga) and late (2.8 – 2.6 Ga) Archaean greenstone belts. Lowe (1982) has noted that sedimentary rocks deposited in the lower, volcanic sequences of early Archaean greenstone belts (e.g. Warrawoona Group, Pilbara; Onverwacht Group, Barberton Mountain Land) lack a terrigenous component and suggested there is no evidence for a sialic basement. Others (e.g. Kröner 1982, 1984) maintain the existence of a sialic basement for these greenstone belts. For the most part the interpretation of geochronological data from early Archaean greenstone belt volcanics and associated granite-granite gneiss bodies is equivocal (e.g. Kröner 1982; Bickle et al.1983), but recent dating in the Barberton Mountain Land suggests that the Ancient Gneiss Complex may be younger than the Onverwacht Group (Carlson et al. 1983). In any case, it is clear that the provenance of the upper, sedimentary sequences of early Archaean greenstone belts (e.g. Gorge Creek Group, Pilbara; Fig Tree and Moodies Groups, Barberton Mountain Land) were strongly influenced by granite-granite gneiss debris (Eriksson 1980b, 1981, 1982a, b).

In contrast, Lowe (1982) has suggested that late Archaean greenstone belts were underlain by, or in close proximity to, sialic basement (see also Baragar and McGlynn 1976). Sediments deposited in the volcanic sequences of late Archaean greenstone belts were derived dominantly from felsic volcanics with a minor terrigenous component (Lowe 1982), whereas sedimentary rocks found in the thick sedimentary sequences commonly contain a significant granite-granite gneiss component (see Lowe 1980 for review).

3.2 Petrological Evidence

Traditionally, the method of studying the provenance of clastic sedimentary rocks has been to examine the petrography of sandstones (Pettijohn 1975; Pettijohn et al. 1973). Rock and mineral fragments provide the least equivocal evidence of source rock lithology. In Table 1 some typical petrographic analyses of Archaean sandstones are listed. The rock fragment data must be considered of limited value since in most cases they comprise less than 15% of the rock. It is unlikely that much of the matrix is primary but rather results from diagenesis, mainly involving the alteration of lithic fragments (e.g. Whetten and Hawkins 1970; Pettijohn et al. 1973).

Table 1. Petrography of some Archaean sandstones

	1[a]	2	3	4	5	6	7	8	9
Quartz	8.5	13.8	20.4	24.9	24.6	26	29.5	45.2	63.5
Plagioclase	24.7	18.9	7.2	12.6	4.1	3	16.2	3.7	7.7
K-feldspar	0.6	–	0.3	–	–	10	0.3	24.4	2.1
Rock fragments					(1.4)[c]			(2.9)[c]	
Felsic volcanic (including porphyry)	40.2	6.2	–	7.7	28.0	0.9	14.2	–	–
Intermediate-mafic volcanic	2.0	tr.	–	3.5	–	0.6	tr.	–	–
Granite-granite gneiss	1.4	tr.	–	1.3	–	0.3	tr.	–	–
Sedimentary and metamorphic (including chert)	–	tr.	–	6.0	–	13	–	11.0	–
Matrix (including recrystallized matrix)	12.2	59.1	71.2	37.0	34.6	36	37.2	12.8	22.1
Other	10.4	2.1	1.0	7.0	7.2	10	1.3	–	4.7
Framework components:									
Q	11.0	35.5	73.1	44.5	42.3	48.3	49.0	51.8	86.6
F	32.7	48.6	26.9	22.5	7.1	24.3	27.4	32.2	13.4
R	56.3	15.9	–	33.0	50.6	27.5	23.6	15.9	–
Provenance[b]	FI	FM(G)	SG	FMG	F(±M)	GFM	G(SFM)	G(FM)	G or S

[a] 1) Average of 10 greywackes from Vermilion District, Minnesota (Ojakangas 1972). 2) Average of 14 greywackes from Gamitagama Lake, Superior Province (Ayres 1983). 3) Average of 23 greywackes from Wyoming (Condie 1967a). Recalculated to exclude undifferentiated quartz and feldspar. 4) Average of 9 greywackes from Burwash Fm., Slave Province (Henderson 1975). 5) Average of 5 lithic arenites from Jackson Lake Fm., Slave Province (Henderson 1975). 6) Average of 21 greywackes from Fig Tree Group, South Africa (Condie et al. 1970). 7) Average of 6 greywackes from Minnitaki Basin, Superior Province (Walker and Pettijohn 1971). 8) Average of 15 sandstones from Moodies Group, South Africa (Eriksson 1980b). Sedimentary rock fragments are all chert; other rock fragments are undifferentiated. 9) Average of 7 greywackes from North Spirit Lake, Canada (Donaldson and Jackson 1965).

[b] Provenance Key: *F* felsic volcanics; *M* mafic volcanics; *I* intermediate volcanics; *G* granite-granite gneiss; *S* recycled sedimentary; listed in decreasing importance with minor components in parentheses.

[c] Undifferentiated rock fragments

Most Archaean sedimentary sequences contain abundant sedimentary rock fragments and less commonly metamorphic fragments. A conventional interpretation of these data would be that some basement recycling had taken place. Again, caution is warranted since in some cases (e.g. Henderson 1975) there is good evidence that the sedimentary rock fragments are of intraformational origin. An interesting feature is the predominance of felsic volcanic fragments over intermediate (andesitic) fragments (McLennan 1984).

Feldspar is not normally an unequivocal indicator of provenance (Pettijohn et al. 1973). Eriksson (1980b) suggested that the relatively high abundances of K-feldspar (along with other petrographic evidence) indicated a significant granitic source for the Fig Tree and Moodies Groups (Table 1, columns 6 and 8). Late Archaean sandstones contain abundant plagioclase feldspar, in excess of K-feldspar, and this is commonly cited as evidence for a volcanic source (Folk 1968). Although this interpretation may be valid for young sedimentary rocks, some care must be taken in interpreting Archaean sedimentary rocks. Unlike post-Archaean terrains, Na-rich granitic rocks (tonalites, trondhjemites) are very abundant in the Archaean and form part of the Archaean bimodal igneous suite (Barker and Arth 1976). The production of K-rich granitic rocks generally occurred in the later stages of greenstone belt development and commonly post-dates much of the sedimentary sequences (e.g. Anhaeusser and Robb 1981; Condie 1981a).

In terms of provenance the most interesting feature of Archaean sandstones is the high abundance of quartz. Excluding samples from the Vermilion District, quartz comprises between about 35 − 85% of the framework grains of the Archaean sandstones listed in Table 1. In their pioneering study of the North Spirit Lake area Donaldson and Jackson (1965) pointed out that high quartz content in Archaean clastic sedimentary rocks was inconsistent with a simple volcanic source and that a recycled quartz-rich sedimentary provenance or quartz-rich granitic provenance was called for. It is now almost universally accepted that volcanic rocks are not a significant source of quartz in extensive sandstone bodies (e.g. Donaldson and Jackson 1965; Folk 1968; Walker and Pettijohn 1971; Pettijohn et al. 1973; Henderson 1975; Pettijohn 1975; Eriksson 1980a). Note that the Jackson Lake Formation (Table 1, column 5) with abundant sand-sized quartz is not regionally extensive (Henderson 1975). Recently, Ayres (1983) has questioned this interpretation for greywackes from the Gamitagama Lake greenstone belt (Table 1, column 2) and has argued for a dominant felsic volcanic provenance in spite of the moderate quartz content. Thus, additional work is required to constrain more rigorously the amount of sand-sized quartz grains that felsic volcanics can supply to extensive Archaean sandstone bodies.

An overall view of the petrographic data would appear to indicate that Archaean clastic sedimentary rocks are derived from a multitude of igneous lithologies and in some cases may have undergone some degree of sedimentary recycling. In some cases, such as the Vermilion District (Table 1, column 1), Gamitagama Lake greenstone belt (column 2) and Jackson Lake Formation (column 5), a major felsic volcanic component is indicated (Ojakangas 1972; Ayres 1983; Henderson 1975). In others, such as the South Pass greenstone belt (column 3), Minnitaki Basin (column 7), Moodies Group (column 8) and North Spirit Lake (column 9), volcanogenic detritus is scarce or absent, and the provenance is dom-

inated by sialic or recycled terrigenous debris (Condie 1967a; Walker and Pettijohn 1971; Eriksson 1980b; Donaldson and Jackson 1965). The remaining sequences, including the Burwash Formation (column 4) and Fig Tree Group (column 6), are derived from a mixed plutonic-volcanic provenance (Henderson
1975; Condie et al. 1970).

McLennan (1984) reviewed the major element geochemistry of Archaean
greywackes. He noted that Archaean quartz-intermediate greywackes had a more
mafic composition than Phanerozoic quartz-intermediate greywackes. Archaean
varieties typically had about 5% less SiO_2 and had FeO + MgO contents greater
than 8% and commonly greater than 10% which contrasts with the less than 7%
for Phanerozoic varieties. It was suggested that these differences were due to a
significant component of mafic volcanics in Archaean greywackes which has de-
graded into the chloritic matrix (thus explaining the scarcity of mafic volcanics
rock fragments).

3.3 Comparison of Archaean and Phanerozoic Greywackes

A general similarity exists between the major element chemical composition of
Archaean greywackes and some young greywackes and has led to suggestions
that they are indistinguishable (Condie 1967a, b, 1981b; Pettijohn 1972). This
was cited as evidence that there is no substantial difference in the composition of
the post-Archaean and Archaean upper crusts. Considerably more data are now
available on young greywackes and their modern equivalents (Crook 1974;
Schwab 1975; Maynard et al. 1982; Bhatia 1981), and the general picture emerg-
ing is that young greywackes and modern deep sea sands have extremely variable
compositions which are related to the tectonic environment (island arc, continental arc, passive margin, etc.). Accordingly, relating the composition of young
greywackes *alone* to that of the "average" exposed crust is not simple, and comparisons of Archaean greywackes with young greywackes are similarly complicated.

In Table 2 petrographic data for some post-Archaean greywackes are listed.
Trace element data are available on a number of these sequences, so some comparisons with Archaean greywackes can be made. On the basis of framework
mineralogy and major element chemistry, Crook (1974) has divided greywackes
into quartz-poor ($\leq 15\%$ Q, $K_2O/Na_2O \ll 1$), quartz-intermediate ($15 - 65\%$ Q,
$K_2O/Na_2O < 1$) and quartz-rich ($> 65\%$ Q, $K_2O/Na_2O > 1$) varieties, for which
there may be tectonic implications (e.g. Crook 1974; Schwab 1975; Bhatia 1981).
Comparison of Tables 1 and 2 show that Archaean greywackes appear to cover
the range from quartz-poor to quartz-rich greywackes, with quartz-intermediate
varieties being most common (see also McLennan 1984). The most striking
difference in the petrographic character of Archaean and Phanerozoic greywackes is that volcanic rock fragments are dominated by felsic varieties in the
former and andesitic varieties in the latter. Some of the REE data for Phanerozoic greywackes are displayed in Fig. 4. Where data are available associated argillaceous rocks show similar REE patterns to greywackes, although absolute abundances may vary slightly (Nathan 1976; Bhatia 1981). The important observation

Table 2. Petrography of some Phanerozoic greywackes

	1[a]	2	3	4	5	6
Quartz	0.4	9.0	29.2	31.9	45.8	80.2
Plagioclase	12.7	10.4	2.9	8.4	4.6	1.0
K-feldspar	–	7.1	2.1	2.9	–	9.4
Rock fragments						
Felsic volcanic (including porphyry)	–	1.8	–	–	–	–
Intermediate-mafic volcanic	63.6	23.3	–	–	2.0	–
Granitic	–	2.7	–	–	–	1.0
Sedimentary and metamorphic (including chert)	–	20.4	30.6	8.9	4.9	1.0
Matrix	20.3	21.6	35.2	38.7	36.7	–
Other	3.0	3.8	–	7.4	6.0	7.3
Framework components:						
Q	0.5	12.0	45.1	61.2	79.9	86.6
F	16.6	23.4	7.7	21.7	8.0	11.2
R	82.9	64.5	47.2	17.1	12.0	2.2

[a] 1) Average of 31 volcanogenic, quartz-poor greywackes, Baldwin Fm. (Devonian), Australia (Chappell 1968). 2) Average of 7 quartz-poor greywackes, Gazelle Fm (Silurian) USA (Condie and Snansieng 1971). 3) Average of 5 quartz-intermediate greywackes, Martinsburg Fm (Ordovician), USA (McBride 1962). 4) Average of 5 quartz-intermediate greywackes, Duzel Fm. (Ordovician), USA (Condie and Snansieng 1971). 5) Average of 12 quartz-rich greywackes, Greenland Gp. (Ordovician), New Zealand (Laird 1972). 6) Average of 92 modern deep sea sands, North Atlantic (Hubert and Neal 1967)

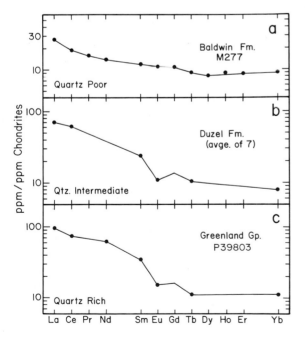

Fig. 4. Chondrite-normalized REE diagram for some typical Phanerozoic greywackes. Note that only quartz-poor greywackes show REE characteristics similar to Archaean sedimentary rocks. Petrographic evidence shows that these rocks were derived almost exclusively from calc-alkaline andesites, in contrast to Archaean sedimentary rocks. Other greywackes show REE patterns more closely comparable to other post-Archaean sedimentary rocks (see Fig. 1). Data sources include Condie and Snansieng (1971), Nathan (1976), Nance and Taylor (1977)

Table 3. Provenance characteristics of selected Archaean sedimentary sequences

Sequence	Approx. age (Ga)	Suggested provenance	Reference
1) Isua supracrustals-Akilia association (West Greenland)	3.8	Not thoroughly studied. Dominantly felsic volcanic with variable admixture of mafic rocks.	Nutman et al. 1984; McLennan et al. 1984
2) Fig Tree Group (South Africa)	3.4	Varies with geography and stratigraphic height. Dominant lithologies include felsic and mafic (to ultramafic) volcanics, granitic rocks, chert and metasedimentary rocks. The proportion of granitic detritus increases significantly towards the top.	Condie et al. 1970; Eriksson 1980b; McLennan et al. 1983b
3) Moodies Group (South Africa)	3.4	Similar to the upper part of the Fig Tree Group but with a greater proportion of granitic material.	Eriksson 1980b; McLennan et al. 1983b
4) Gorge Creek Group (Western Australia)	3.4	Not documented in detail. Sandstone petrography indicates complex source of granitic rocks, felsic-mafic volcanics, chert, iron formation and recycled quartz arenite.	Eriksson 1981, 1982a, b; McLennan et al. 1983a
5) Malene Supracrustals (West Greenland)	>3.05	Not well documented. Probably complex in detail. Field evidence suggests at least partly laid down on Amîtsoq gneiss basement.	Nutman and Bridgwater 1983; Dymek et al. 1983; McLennan et al. 1984
6) Kambalda (Yilgarn Block) (Western Australia)	2.8 – 2.7	Mixture of mafic volcanics and Na-rich granitic rocks. Role of felsic volcanics not fully assessed.	Bavinton 1979, 1981; Bavinton and Taylor 1980
7) Kalgoorlie (Yilgarn Block) (Western Australia)	2.8 – 2.7	Dominantly from variable mixtures of felsic-mafic volcanics. Associated Na-rich granitic rocks could also be an important component.	Glikson 1971; Nance and Taylor 1977
8) Yellowknife Supergroup (Canada)	2.7 – 2.6	Widespread turbidites (Burwash-Walsh Formations) derived from felsic volcanics, granite-granite gneiss and mafic volcanics. Restricted fluvial facies (Jackson Lake Formation) derived from variable mixtures of felsic and mafic volcanics.	Henderson 1975; Jenner et al. 1981
9) South Pass Greenstone Belt (USA)	2.7 – 2.6	Recycled quartz-rich metasedimentary rocks and granitic rocks.	Condie 1967a

here is that Archaean-like REE patterns are restricted to quartz-poor volcanogenic greywacke terrains in the Phanerozoic. As pointed out above, the dominant volcanic lithology in such greywackes is andesitic.

3.4 Archaean Sedimentary Rocks as Crustal Samples

The available data indicate that many Archaean clastic sedimentary sequences were deposited in intracratonic basins or at the edges of sialic blocks. In addition, Archaean clastic sedimentary rocks were derived from a number of different sources including felsic to mafic volcanic rocks, granitic rocks and lesser recycled sedimentary and metamorphic rocks. First cycle volcanogenic sedimentary rocks are not uncommon but, in contrast to Phanerozoic volcanogenic sediments, were derived predominantly from highly differentiated felsic volcanic sources. Typically, such felsic volcanics are geochemically indistinguishable from the tonalite-trondhjemite plutonic suite and may be genetically related to it (see following sections). Certainly, the trace element composition of Archaean sedimentary rocks derived from felsic volcanics is indistinguishable from those derived from plutonic rocks (cf. Table 1 and Fig. 3). In Table 3 we list the provenance characteristics of the Archaean sedimentary sequences which have been examined in some detail for their trace element characteristics. An important observation is the diverse and lithologically complex nature of the provenance among the various sequences.

From this analysis we conclude that Archaean sedimentary rocks were derived from a wide and varied provenance. Included as a major component are geochemically highly differentiated igneous rocks comprising felsic volcanics or plutonics. Thus, we consider such sedimentary rocks can be used as a sample of the Archaean exposed crust. The highly variable nature of REE patterns in Archaean sedimentary rocks contrasts with the uniformity seen in most post-Archaean sedimentary REE patterns and, accordingly, considerably more uncertainty is associated with formulating average compositions.

4 Mixing Models for Archaean Sedimentary Rocks

4.1 Archaean Bimodal Suite

Early Archaean igneous terrains are comprised of a bimodal suite of mafic volcanics and felsic volcanics (generally dacitic) or trondhjemite-tonalite with a distinct scarcity of intermediate rock types (Barker and Peterman 1974; Barker and Arth 1976; Barker et al. 1981). For the purposes of geochemical modelling, the trondhjemites-tonalites are indistinguishable from the felsic volcanics and are generally considered to be their plutonic equivalent (Barker and Peterman 1974). In some Archaean terrains a full range of igneous lithologies, including andesites, are found (Baragar and Goodwin 1969; Jahn and Sun 1979). In some cases Archaean andesites show geochemical characteristics indistinguishable from Cenozoic calc-alkaline andesites (Taylor and Hallberg 1977), but the volumetric abundance of such rocks in Archaean greenstone belts is not great.

4.2 Rare Earth Elements

The rare earth element data for Archaean sedimentary rocks were summarized above. Such data would be consistent with sediment derivation from calc-alkaline andesitic volcanoes (e.g. Jakes and Taylor 1974), but this is inconsistent with the petrographic nature of Archaean sediments and the general paucity of andesitic rocks in most Archaean terrains. The data can be explained equally well through a mixture of the Archaean bimodal igneous suite (see above). In Fig. 5 two typical REE patterns for Archaean Na-rich felsic igneous rocks (trondhjemite-tonalite, dacite) and mafic volcanics are shown. A great deal of variability in REE characteristics are shown in both lithologies. For example, Archaean felsic volcanics locally may have well developed negative Eu-anomalies where plagioclase has been a fractionating phase (e.g. Condie 1976), or mafic volcanics may show some LREE enrichment or depletion (e.g. Sun and Nesbitt 1978), but the patterns shown here are typical. It can be seen that the entire range of sedi-

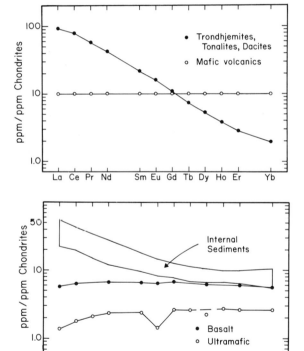

Fig. 5. Chondrite-normalized REE diagram showing typical patterns for the Archaean bimodal suite. Considerable variations are seen in both compositions; for example, Archaean mafic volcanics may show LREE enrichment or depletion. However, patterns similar to those displayed are most typical. The entire range of Archaean sedimentary REE patterns shown in Fig. 3 can be derived from variable mixtures of these two end member compositions

Fig. 6. Chondrite-normalized REE diagram for some Archaean supracrustal rocks from Kambalda, Yilgarn Block, Western Australia. The 'Internal Sediments' are found as thin, laterally restricted lenses entirely within ultramafic volcanic rocks. REE patterns are characterized by considerable LREE enrichment and no Eu-anomalies. Also shown are REE patterns of the enclosing ultramafic volcanics and underlying basaltic rocks. It can be seen that the sedimentary rocks cannot have been derived from the local environment and that a distal, LREE enriched component (probably Na-rich granitic rocks) is called for. Data sources include Sun and Nesbitt (1978), Bavinton (1979), Bavinton and Taylor (1980)

mentary REE patterns displayed in Fig. 3 can be explained by variable mixtures of these end member compositions.

A unique example where mixing can be shown to have occurred from geochemical evidence is at Kambalda, Yilgarn Block, Western Australia. Here, a sequence of metamorphosed sedimentary rocks termed the 'Internal Sediments' are found entirely within a succession of ultramafic volcanic rocks, at flow boundaries (Bavinton 1979, 1981; Bavinton and Taylor 1980), which in turn lies on a sequence of basaltic rocks (Footwall Basalts). The sedimentary rocks clearly occur in localized lenses a few hundred metres in extent. The REE data, however, are obviously inconsistent with an ultramafic or mafic volcanic provenance (Fig. 6), and there are no other lithologies within the volcanic sequence which could act as a source to give the LREE-enriched patterns (younger 'hanging wall basalts' which overlie the sediments do have LREE enriched patterns). Accordingly, felsic rocks not from the immediate vicinity must have also contributed detritus to these sedimentary rocks, a conclusion also supported by major element data (Bavinton 1979, 1981; Bavinton and Taylor 1980). The most likely candidates are Na-rich granitic rocks (Bavinton 1979, 1981) which are found

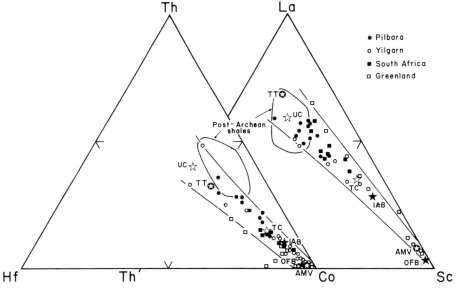

Fig. 7. Ternary diagrams of Th-Hf-Co and La-Th-Sc for Archaean shales and metasedimentary rocks (from Greenland). Abbreviations as follows: *UC* present-day upper continental crust (Taylor and McLennan 1981); *TC* present-day total continental crust predicted by the andesite model (Taylor and McLennan 1981); *IAB* typical island-arc basalt; *OFB* typical ocean floor basalt; *TT* typical Archaean felsic igneous rock (tonalite, trondhjemite, dacite); *AMV* typical Archaean mafic volcanic. Also shown is the field of post-Archaean shales from Australia (Nance and Taylor 1976). Post-Archaean shales show relatively low dispersion with respect to these parameters and plot around or near the present-day upper crust composition. In contrast, Archaean sedimentary rocks form linear trends, indicative of two component mixing. Archaean mafic volcanics and felsic igneous rocks are viable end members. Data sources for Archaean sedimentary rocks include Nance and Taylor (1977), Bavinton and Taylor (1980), McLennan (1981), McLennan et al. (1983a, b, 1984)

around Kambalda and are of approximately equal ages to the volcanics (see McCulloch and Compston 1981).

4.3 Mixing Models

Other trace element data strongly support a two component mixing model. In Fig. 7 the Archaean shale data are plotted on two ternary diagrams of Th-Hf-Co and La-Th-Sc. Post-Archaean shales from Australia (Nance and Taylor 1976) tend to plot in fairly restricted fields around or near the present-day upper continental crust (UC; Taylor and McLennan 1981). The slight displacement of the post-Archaean shales from the upper crust, away from the Hf apex, probably results from the concentration of zircon in sandstones. On the other hand, the Archaean data plot on linear arrays and the fact that such linear trends occur on more than one ternary diagram is taken as good evidence for two component mixing. The data are consistent with end members being Archaean felsic igneous rocks (felsic volcanics, trondhjemites-tonalite; TT) and Archaean mafic volcanics (AMV). The present-day upper continental crust (UC) could also be an end

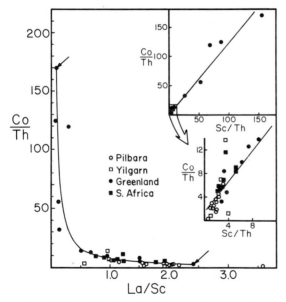

Fig. 8. Plot of Co/Th vs La/Sc for Archaean sedimentary rocks. Superimposed is a mixing line using the extreme sedimentary compositions as end members as indicated by the *arrows* (sample with highest La/Sc ratio could not be used in generating mixing line because Co is below detection limits, and only a maximum Co/Th ratio can be calculated). The data correspond well to the mixing line. *Inset* shows plot of Co/Th vs Sc/Th for these data. If two component mixing is involved, the data should fall on a straight line on such a plot (Langmuir et al. 1978). The correlation coefficient for the regression line plotted is statistically significant, and it is concluded that the source of these Archaean sedimentary rocks is best explained by mixing of two components. The most likely candidates are mafic volcanics (high Co/Th, low La/Sc) and felsic igneous rocks (low Co/Th, high La/Sc) of the Archaean bimodal suite

member on such a diagram, but this possibility is excluded by the general lack of negative Eu-anomalies in Archaean sedimentary rocks.

Langmuir et al. (1978) have developed a general mixing equation for geochemical data. When dealing with igneous rocks all interelement relations must conform to the mixing model in an internally consistent manner, and the 'fit' to the mixing equations should be close. However, application of such mixing models to determine the source of sedimentary rocks require some allowances which do not necessarily apply to igneous systems. For example, secondary processes such as weathering, sedimentary transport and diagenesis can affect the distribution of many elements. Also, end member compositions cannot be specified, but only characterized, since a great deal of compositional variation would be expected in source rocks, related to the local igneous history (for example, between plutons). In Fig. 8 the Archaean shale data are plotted on a ratio-ratio plot of Co/Th vs La/Sc. Superimposed is a mixing line which uses two of the extreme sedimentary compositions as end members. Considering the above caveats, the agreement to the mixing line is excellent. It can be shown that such agreement may be fortuitous, and a good test is to plot one of the ratios against the ratio of the denominators; a distinct correlation should result if mixing is involved (Langmuir et al. 1978). In Fig. 8 such a plot is also given (Co/Th vs Sc/Th), and the correlation between these ratios is significant and adds considerable support to the mixing model. There is also a good correlation of these data with the REE patterns. Thus, samples with high Co/Th and low La/Sc ratios also have flat REE patterns, indicative of mafic volcanics. Samples with low Co/Th and high La/Sc ratios have the steeper REE patterns characteristic of felsic volcanics and Na-rich granitic rocks.

5 Composition and Nature of the Archaean Crust

Trace element relationships in Archaean sedimentary rocks provide persuasive evidence for an origin through two component mixing of the Archaean bimodal igneous suite. Thus the similarity in major element composition of Archaean greywackes and Phanerozoic greywackes noted by several authors (e.g. Condie 1967b; Pettijohn 1972) is fortuitous and results from the inability of major elements to distinguish between mixtures of mafic and felsic igneous rocks and intermediate compositions such as andesites. Only when the full spectrum of Archaean sedimentary compositions are considered is the evidence for mixing apparent.

The overall conclusion, based on evidence from petrography, tectonic settings and geochemistry, is that some trace element relationships in Archaean sedimentary rocks can be used to provide information about the nature of the exposed Archaean crust, although the variability in composition calls for caution. The best estimate for the average REE abundances in Archaean shales, based on all of the available data from this laboratory, is listed in Table 4. The average REE pattern has $\sum REE = 105$ ppm, $La_N/Yb_N = 6.8$ and $Eu/Eu^* = 0.99$. This pattern is somewhat more enriched than the previously reported Archaean aver-

Table 4. Rare earth elements in Archaean shales (in ppm)

	Archaean shale	AAS	PAAS
La	20	12.6	38
Ce	42	26.8	80
Pr	4.9	3.13	8.9
Nd	20	13.0	32
Sm	4.0	2.78	5.6
Eu	1.2	0.92	1.1
Gd	3.4	2.85	4.7
Tb	0.57	0.48	0.77
Dy	3.4	2.93	4.4
Ho	0.74	0.63	1.0
Er	2.1	1.81	2.9
Tm	0.30	0.26	0.40
Yb	2.0	1.79	2.8
Lu	0.31	0.27	0.43
$\sum$REE	104.9	70.3	183.0
Eu/Eu*[a]	0.99	1.00	0.66
La_N/Yb_N	6.8	4.8	9.2

[a] Eu* is the interpolated value for no chondrite-normalized anomaly

age (AAS; see Table 4) which was determined on the limited data base from the Yilgarn Block, Western Australia. The new average differs from the post-Archaean shales (PAAS; Table 4) in having lower $\sum$REE, La/Yb and no Eu-anomaly.

McLennan et al. (1980) documented the coherence of Th with the LREE and showed how these systematics could be used to make estimates for other trace element abundances in the upper crust. Our best estimate of average La/Th in Archaean shales is 3.5 ± 0.3 (see above). If we assumed that the average REE pattern of Archaean shales is equivalent to the Archaean exposed crust, then we can derive an upper crustal Th value of 5.7 ppm. These are probably upper limits, since Archaean sedimentary lithologies with lower REE abundances such as quartzites and carbonates have not been considered. From the Th concentration we can also derive values for U = 1.5 ppm (Th/U = 3.8), K = 1.5% (K/U $= 10^4$) and Rb = 50 ppm (K/Rb ~300). These values for incompatible elements are about half the concentrations in the present-day upper continental crust (Table 5) and indicate the exposed Archaean crust was considerably less differentiated. In terms of the bimodal model such a composition is about equivalent to a 1:1 mixture of Archaean mafic volcanics and felsic igneous rocks.

Taylor and McLennan (1984) have attempted to model the bulk composition of the total Archaean crust using the bimodal model. A reasonable model is a 2:1 mix of mafic and felsic end members (Table 5). This would indicate that the Archaean upper crust was more silica-rich, due to a near-surface concentration of felsic rocks. The composition obtained from this model is very similar to the

Table 5. Abundances of some incompatible elements in the Archaean and present-day continental crust

	Archaean		Present-day[a]		Primitive[a] mantle
	UC	TC[a]	UC[b]	TC	
Rb (ppm)	50	28	112	32	0.55
Th (ppm)	5.7	2.9	10.7	3.5	0.064
U (ppm)	1.5	0.75	2.8	0.91	0.18
K (%)	1.5	0.75	2.80	0.91	0.0180
La (ppm)	20	15	30	19	0.55
Yb (ppm)	2.0	2.2	2.2	2.2	0.37
La_N/Yb_N	6.8	4.6	9.2	4.9	–
Eu/Eu*	1.0	1.0	0.65	1.0	1.0

[a] From Taylor and McLennan (1984).
[b] *UC* upper crust; *TC* total crust.

average composition of Archaean volcanic rocks from the Superior Province (Goodwin 1977). Values for heat-producing elements (K = 0.75%, Th = 2.9 ppm, U = 0.75 ppm) provide a crustal contribution to heat flow of about 14 m Wm^{-2} for a 30 km crust and about 19 m Wm^{-2} for a 40 km crust. These values agree very well with the measured values for Archaean heat flow provinces of 14 ± 2 m Wm^{-2} (Morgan 1984), which are derived from only a limited data base. From Table 5 we can see that the bulk composition of the Archaean crust is somewhat less depleted in incompatible elements than the present-day total crust.

The sedimentary data clearly show that the mechanisms for differentiating the post-Archaean and Archaean crust were fundamentally different. The post-Archaean upper crust is characterized by a negative Eu-anomaly with fairly flat HREE. This indicates that the major agent of crustal differentiation was partial melting, probably of an upper amphibolite grade intermediate composition at crustal depths (<40 km) where plagioclase is a dominant residual phase. Such processes can be traced back to the late Archaean (earlier to about 3 Ga in South Africa), when K-rich granitic rocks of intracrustal origin became widespread (Taylor and McLennan 1981, 1984). In contrast, the earlier Archaean upper crust did not have a Eu-anomaly on average, and the felsic end member of the bimodal suite was characterized by steep REE patterns with HREE-depletion indicative of garnet fractionation. The most important mechanism for forming the felsic part of the crust during most of the Archaean probably was partial melting of amphibolite or eclogite grade mafic rocks at depths below plagioclase stability (presently 40 km) and possibly at depths greater than 60 km (e.g. O'Nions and Pankhurst 1978). Thus, in contrast to the differentiation of the late Archaean and post-Archaean crust which is dominantly an intracrustal process, the differentiation of the early Archaean crust took place dominantly in the upper mantle.

The present-day growth of the continental crust probably results from the accretion of island-arc material onto the edge of continents (Taylor 1967), and accordingly andesite has been used to characterize total continental crustal compositions (Taylor and McLennan 1981). Elsewhere (e.g. Taylor and McLennan 1981, 1984; McLennan and Taylor 1982) we have argued that most of the growth of continents occurred during a relatively short interval during the late Archaean. The mechanism of this growth is not certain, but there is no a priori reason to assume that it was related to island-arc volcanism. The composition proposed in the modelling presented above (see also Taylor and McLennan 1984) is somewhat more mafic than average andesites and, accordingly, we would propose that the overall composition of the present-day continental crust is best modelled by 75% Archaean crust and 25% andesite in order to account for the large amount of crustal growth during the late Archaean. This model is discussed fully elsewhere (Taylor and McLennan 1984).

Acknowledgements. Former colleagues at RSES P. Jakeš, W. Nance, and O. Bavinton have added greatly to our understanding of Archaean sedimentary rocks. We are also grateful to K. Eriksson, J. Henderson, A. Kröner, V. McGregor, and S. Moorbath for providing many of the samples we have analyzed and for endeavouring to keep our interpretations consistent with the geological relations.

References

Anhaeusser CR, Robb LJ (1981) Magmatic cycles and the evolution of the Archean granitic crust in the eastern Transvaal and Swaziland. Spec Publ Geol Aust 7:457 – 467

Ayres LD (1983) Biomodal volcanism in Archean greenstone belts exemplified by greywacke composition, Lake Superior Park, Ontario. Can J Earth Sci 20:1168 – 1194

Baragar WRA, Goodwin AM (1969) Andesites and Archean volcanism of the Canadian Shield. In: McBirney AR (ed) Proceedings of the Andesite Conference. Oregon Dept Geol and Mineral Indust Bull 65:121 – 142

Baragar WRA, McGlynn JC (1976) Early Archean basement in the Canadian Shield: a review of the evidence. Geol Surv Can Pap 76 – 14

Barker F, Arth JG (1976) Generation of trondhjemitic-tonalitic liquids and Archean bimodal trondhjemite-basalt suites. Geology (Boulder) 4:594 – 600

Barker F, Peterman ZE (1974) Bimodal tholeiitic-dacitic magmatism and the early Precambrian crust. Precambrian Res 1:1 – 12

Barker F, Arth JG, Hudson T (1981) Tonalites in crustal evolution. Philos Trans R Soc Lond A 301:293 – 303

Basaltic Volcanism Study Project (1981) Basaltic volcanism on the terrestrial planets. Pergamon, New York

Bavinton AO (1979) Interflow sedimentary rocks from the Kambalda ultramafic sequence: their geochemistry, metamorphism and genesis. PhD thesis, The Australian National University, Canberra

Bavinton OA (1981) The nature of sulphidic metasediments at Kambalda and their broad relationships with associated ultramafic rocks and nickel ores. Econ Geol 76:1606 – 1628

Bavinton AO, Taylor SR (1980) Rare earth element abundances in Archean metasediments from Kambalda, Western Australia. Geochim Cosmochim Acta 44:639 – 648

Bence AE, Grove TL, Papike JJ (1980) Basalts as probes of planetary interiors: constraints on the chemistry and mineralogy of their source regions. Precambrian Res 10:249 – 279

Bhatia MR (1981) Petrology, geochemistry and tectonic setting of some flysch deposits. PhD thesis, The Australian National University, Canberra

Bickle MJ, Bettenay LF, Barley ME, Chapman HJ, Groves DI, Campbell IH, Laeter JR de (1983) A 3500 Ma plutonic and volcanic calc-alkaline province in the Archaean east Pilbara block. Contrib Mineral Petrol 84:25 – 35

Boak JL, Dymek RF, and Gromet LP (1982) Early crustal evolution: constraints from variable REE patterns in metasedimentary rocks from the 3800 Ma Isua supracrustal belt, West Greenland. Lunar and Planetary Science XIII, Lunar and Planetary Institute, Houston, pp 51 – 52

Cameron EM, Garrels RB (1980) Geochemical comparisons of some Precambrian shales from the Canadian Shield. Chem Geol 28:181 – 197

Carlson RW, Hunter DR, Barker F (1983) Sm-Nd age and isotopic systematics of the bimodal suite, ancient gneiss complex, Swaziland. Nature 305:701 – 704

Chappell BW (1968) Volcanogenic greywackes from the Upper Devonian Baldwin Formation, Tamworth-Burraba District, New South Wales. J Geol Soc Aust 15:87 – 102

Chaudhuri S, Cullers RL (1979) The distribution of rare-earth elements in deeply buried Gulf Coast sediments. Chem Geol 24:327 – 338

Condie KC (1967a) Geochemistry of early Precambrian greywackes from Wyoming. Geochim Cosmochim Acta 31:2135 – 2149

Condie KC (1967b) Composition of the ancient North American crust. Science 155:1013 – 1015

Condie KC (1976) Trace element geochemistry of Archean greenstone belts. Earth-Sci Rev 12:393 – 417

Condie KC (1980) Origin and early development of the earth's crust. Precambrian Res 11:183 – 197

Condie KC (1981a) Geochemical and isotopic constraints on the origin and source of Archaean granites. Spec Publ Geol Soc Aust 7:469 – 479

Condie KC (1981b) Archean greenstone belts. Elsevier, New York

Condie KC, Snansieng S (1971) Petrology and geochemistry of the Duzel (Ordovician) and Gazelle (Silurian) Formations, northern California. J Sediment Petrol 41:741 – 751

Condie KC, Macke JE, Reimer TO (1970) Petrology and geochemistry of Earth Precambrian graywackes from the Fig Tree Group, South Africa. Geol Soc Am Bull 81:2759 – 2776

Courtois C, Treuil M (1977) Distribution des terres rares et de quelques éléments en trace dans des sédiments récents des fosses de la Mer Rouge. Chem Geol 20:57 – 72

Crook KAW (1974) Lithogenesis and geotectonics: the significance of compositional variations in flysh arenites (greywackes). In: Dott RH, Shaver RH (eds) Modern and ancient geosynclinal sedimentation. Soc Econ Paleontol Mineral Spec Pub 19:304 – 310

Cullers R, Chaudhuri S, Kilbane N, Koch R (1979) Rare-earths in size fractions and sedimentary rocks of Pennsylvanian-Permian age from the mid-continent of the U.S.A. Geochim Cosmochim Acta 43:1285 – 1301

Danchin RV (1967) Chromium and nickel in the Fig Tree Shale from South Africa. Science 158:261 – 262

Degens ET (1965) Geochemistry of sediments. Prentice-Hall

Dimroth E, Imreh L, Rocheleau M, Goulet N (1982) Evolution of the south-central part of the Archean Abitibi belt, Quebec. Part 1: Stratigraphy and paleogeographic model. Can J Earth Sci 19:1729 – 1758

Dimroth E, Imreh L, Goulet N, Rocheleau M (1983a) Evolution of the south-central segment of the Archean Abitibi belt, Quebec. Part II: Tectonic evolution and geomechanical model. Can J Earth Scie 20:1355 – 1373

Dimroth E, Imreh L, Goulet N, Rocheleau M (1983b) Evolution of the south-central segment of the Archean Abitibi belt, Quebec. Part III: Plutonic and metamorphic evolution and geotectonic model. Can J Earth Sci 20:1374 – 1388

Donaldson JA, Jackson GD (1965) Archean sedimentary rocks of North Spirit Lake area, northwestern Ontario. Can J Earth Sci 2:622 – 647

Dymek RF, Weed R, Gromet LP (1983) The Malene metasedimentary rocks on Rypeø, and their relationship to Amîtsoq gneisses. Rapp Grønlands Geol Unders 112:53 – 69

Elderfield H, Greaves MJ (1982) The rare earth elements in seawater. Nature 296:214 – 219

Eriksson KA (1977) Tidal deposits from the Archaean Moodies Group, Barberton Mountain Land, South Africa. Sediment Geol 18:255 – 281

Eriksson KA (1978) Alluvial and destructive beach facies from the Archaean Moodies Group, Barberton Mountain Land, South Africa and Swaziland. In: Miall AD (ed) Fluvial Sedimentology. Can Soc Petrol Geol Mem 5:287 – 311

Eriksson KA (1979) Marginal marine depositional processes from the Archaean Moodies Group, Barberton Mountain Land, South Africa: evidence and significance. Precambrian Res 8:153 – 182

Eriksson KA (1980a) Hydrodynamic and paleogeographic interpretation of turbidite deposits from the Archean Fig Tree Group of the Barberton Mountain Land, South Africa. Geol Soc Am Bull 91:21 – 26

Eriksson KA (1980b) Transitional sedimentation styles in the Moodies and Fig Tree Groups, Barberton Mountain Land, South Africa: evidence favouring an Archean continental margin. Precambrian Res 12:141 – 160

Eriksson KA (1981) Archean platform-to-trough sedimentation, east Pilbara Block, Australia. Spec Publ Geol Soc Aust 7:235 – 244

Eriksson KA (1982a) Sedimentation patterns in the Barberton Mountain Land, South Africa, and the Pilbara Block, Australia: evidence for Archean rifted continental margins. Tectonophys 81:179 – 193

Eriksson KA (1982b) Archean and early Proterozoic sedimentation styles in the Kaapvaal Province, South Africa and Pilbara Block, Australia. Rev Bras Geosci 12:121 – 131

Folk RL (1968) Petrology of sedimentary rocks. Hemphill's, Austin

Fyle WS (1974) Archean tectonics. Nature 249:338

Gee RD, Baxter JL, Wilde SA, Williams IR (1981) Crustal development in the Archean Yilgarn Block, Western Australia. Spec Publ Geol Soc Aust 7:43 – 56

Glikson AY (1971) Archaean geosynclinal sedimentation near Kalgoorlie, Western Australia. Spec Publ Geol Soc Aust 3:443 – 460

Goodwin AM (1977) Archean volcanism in Superior Province, Canadian Shield. Geol Assoc Can Spec Pap 16:205 – 241

Goodwin AM (1981) Archaean plates and greenstone belts. In: Kröner A (ed) Precambrian plate tectonics. Elsevier, Amsterdam Oxford New York

Graf JL (1977) Rare earth elements as hydrothermal tracers during the formation of massive sulphide deposits in volcanic rocks. Econ Geol 72:527 – 548

Hargraves RB (1976) Precambrian geologic history. Science 193:363 – 371

Hargraves RB (1981) Precambrian tectonic style: a liberal uniformitarian interpretation. In: Kroner A (ed) Precambrian plate tectonics. Elsevier, Amsterdam Oxford New York

Haskin MA, Haskin LA (1966) Rare earths in European shales: a redetermination. Science 154:507 – 509

Henderson JB (1975) Sedimentology of the Archean Yellowknife Supergroup at Yellowknife, District of Mackenzie. Geol Surv Can Bull 246

Henderson JB (1981) Archaean basin evolution in the Slave Province, Canada. In: Kröner A (ed) Precambrian plate tectonics. Elsevier, Amsterdam Oxford New York

Hubert J, Neil PB (1967) Mineral composition and dispersal patterns of deep-sea sands in the western North Atlantic petrologic province. Geol Soc Am Bull 78:749 – 772

Jahn B, Sun S-S (1979) Trace element distribution and isotopic composition of Archean greenstones. Phys Chem Earth 11:597 – 618

Jakeš P, Taylor SR (1974) Excess Eu content in Precambrian rocks and continental evolution. Geochim Cosmochim Acta 38:793 – 795

Jenner GA, Fryer BJ, McLennan SM (1981) Geochemistry of the Archean Yellowknife Supergroup. Geochim Cosmochim Acta 45:1111 – 1129

Kerrich R, Fryer BJ (1979) Archaean precious-metal hydrothermal systems, Dome Mine, Abitibi, Greenstone Belt. II REE and oxygen isotope relations. Can J Earth Sci 16:440 – 458

Kröner A (1982) Archean to early Proterozoic tectonics and crustal evolution: a review. Rev Bras Geosci 12:15 – 31

Kröner A (1984) Evolution, growth and stabilization of the Precambrian lithosphere. Phys Chem Earth 15 (to be published)

Laird MG (1972) Sedimentology of the Greenland Group in the Paparoa Range, west coast, South Island. NZ J Geol Geophys 15:372 – 393

Langmuir CH, Vocke RD, Hanson GN, Hart SR (1978) A general mixing equation with applications to Icelandic basalts. Earth Planet Sci Lett 37:380 – 392

Lowe DR (1980) Archean sedimentation. Ann Rev Earth Planet Sci 8:145 – 167

Lowe DR (1982) Comparative sedimentology of the principal volcanic sequences of Archean greenstone belts in South Africa, Western Australia and Canada: implications for crustal evolution. Precambrian Res 17:1 – 29

Martin J-M, Hogdahl O, Philippot JC (1976) Rare earth element supply to the ocean. J Geophys Res 81:3119 – 3124

Maynard JB, Valloni R, Yu H (1982) Composition of modern deep-sea sands from arc-related basins. Geol Soc Lond Spec Publ 10:551 – 561

McBride EF (1962) Flysh and associated beds of the Martinsburg Formation (Ordovician), central Appalachians. J Sediment Petrol 32:39 – 91

McCulloch MT, Compston W (1981) Sm-Nd age of Kambalda and Konowna greenstones and heterogeneity in the Archaen mantle. Nature 294:322 – 327

McLennan SM (1981) Trace element geochemistry of sedimentary rocks: implications for the composition and evolution of the continental crust. PhD thesis, The Australian National University, Canberra

McLennan SM (1982) On the geochemical evolution of sedimentary rocks. Chem Geol 37:335 – 350

McLennan SM (1984) Petrological characteristics of Archean greywackes. J Sediment Petrol 54 (to be published)

McLennan SM, Taylor SR (1980) Th and U in sedimentary rocks: crustal evolution and sedimentary recycling. Nature 285:621 – 624

McLennan SM, Taylor SR (1982) Geochemical constraints on the growth of the continental crust. J Geol 90:347 – 361

McLennan SM, Nance WB, Taylor SR (1980) Rare earth element-thorium correlations in sedimentary rocks, and the composition of the continental crust. Geochim Cosmochim Acta 44:1833 – 1839

McLennan SM, Taylor SR, Eriksson KA (1983a) Geochemistry of Archean shales from the Pilbara Supergroup, Western Australia. Geochim Cosmochim Acta 47:1211 – 1222

McLennan SM, Taylor SR, Kröner A (1983b) Geochemical evolution of Archean shales from South Africa 1: the Swaziland and Pongola Supergroups. Precambrian Res 22:93 – 124

McLennan SM, Taylor SR, McGregor VR (1984) Geochemistry of Archean metasedimentary rocks from West Greenland. Geochim Cosmochim Acta 48:1 – 13

Morgan P (1984) The thermal structure and thermal evolution of the continental lithosphere. Phys Chem Earth 15 (in press)

Muecke GK, Price C, Sarkar P (1979) Rare-earth element geochemistry of regional metamorphic rocks. Phys Chem Earth 11:449 – 464

Nance WB, Taylor SR (1976) Rare earth element patterns and crustal evolution I: Australian post-Archean sedimentary rocks. Geochim Cosmochim Acta 40:1539 – 1551

Nance WB, Taylor SR (1977) Rare earth element patterns and crustal evolution II: Archean sedimentary rocks from Kalgoorlie, Australia. Geochim Cosmochim Acta 41:225 – 231

Nathan S (1976) Geochemistry of the Greenland Group (early Ordovician), New Zealand, NZ J Geol Geophys 19:683 – 706

Nesbitt HW (1979) Mobility and fractionation of rare earth elements during weathering of a granodiorite. Nature 279:206 – 210

Nutman AP, Bridgwater D (1983) Deposition of Malene supracrustal rocks on an Amîtsoq basement in Outer Ameralik, southern West Greenland. Rapp Grønlands Geol Unders (to be published)

Nutman AP, Dimroth E, Rosing M, Bridgwater D, Allaart J (1984) Stratigraphic and geochemical evidence for the depositional environment of the early Archean Isua supracrustal belt, West Greenland. Precambrian Res (to be published)

O'Nions RK, Pankhurst RJ (1978) Early Archaean rocks and geochemical evolution of the earth's crust. Earth Planet Sci Lett 38:211 – 236

Ojakangas RW (1972) Archean volcanogenic graywackes of the Vermilion District, northeastern Minnesota. Geol Soc Am Bull 83:429 – 442

Pettijohn FJ (1972) The Archean of the Canadian Shield: a résumé. Geol Soc Am Mem 135:131 – 149

Pettijohn FJ (1975) Sedimentary rocks (3rd edition). Harper and Row, New York

Pettijohn FJ, Potter PE, Siever R (1973) Sand and sandstone. Springer Berlin Heidelberg New York

Schwab FL (1975) Framework mineralogy and chemical composition of continental margin-type sandstone. Geology (Boulder) 3:487 – 490

Shaw DM, Bugry R (1966) A review of boron sedimentary geochemistry in relation to new analyses of some North American shales. Can J Earth Sci 3:49 – 63

Sun SS, Nesbitt RW (1978) Petrogenesis of Archean ultrabasic and basic volcanics: evidence from rare earth elements. Contrib Mineral Petrol 65:301 – 325

Tarney J, Windley BF (1981) Marginal basins through geological time. Philos Trans R Soc Lond A 301:217 – 232

Tarney J, Dalziel IWD, de Wit MJ (1976) Marginal basin 'Rocas Verdes' complex from S. Chila: a model for Archaean greenstone belt formation. In: Windley BF (ed) The Early History of the Earth. Wiley, London, pp 131 – 146

Taylor SR (1967) The origin and growth of continents. Tectonophys 4:17 – 34

Taylor SR (1977) Island arc models and the compositon of the continental crust. Am Geophys Union Maurice Ewing Series I:325 – 335

Taylor SR, Hallberg JA (1977) Rare-earth elements in the Marda calc-alkaline suite: an Archean geochemical analogue of Andean-type volcanism. Geochim Cosmochim Acta 41:1125 – 1129

Taylor SR, McLennan SM (1981) The composition and evolution of the continental crust: rare earth element evidence from sedimentary rocks. Philos Trans R Soc Lond A 301:381 – 399

Taylor SR, McLennan SM (1984) The continental crust: its composition and evolution. Blackwells, Oxford (to be published)

Taylor SR, McLennan SM, McCulloch MT (1983a) Geochemistry of loess, continental crustal composition and crustal model ages. Geochim Cosmochim Acta 47:1897 – 1905

Taylor SR, McLennan SM, Arculus RJ, McCulloch MT (1983b) Residual lower continental crustal compositions. Lunar and Planetary Science XIV, Lunar and Planetary Institute, Houston, pp 781 – 782

Walker RG, Pettijohn FJ (1971) Archaean sedimentation: analysis of the Minnitaki Basin, northwestern Ontario, Canada. Geol Soc Am Bull 42:2099 – 2130

Whetten JT, Hawkins JW (1970) Diagenetic origin of graywacke matrix minerals. Sedimentology 15:347 – 361

Wildeman TR, Condie KC (1973) Rare earths in Archean graywackes from Wyoming and from the Fig Tree Group, South Africa. Geochim Cosmochim Acta 37:439 – 453

Spatial and Temporal Variations of Archaean Metallogenic Associations in Terms of Evolution of Granitoid-Greenstone Terrains with Particular Emphasis on the Western Australian Shield

D. I. GROVES and W. D. BATT[1]

Contents

1 Introduction 74
2 Granitoid-Greenstone Terrains 77
3 Spatial Variations in Greenstone Belts 78
3.1 Platform-Phase Greenstones 78
3.2 Rift-Phase Greenstones 81
4 Temporal Variations in Greenstone Belts 83
4.1 Platform-Phase Greenstones 84
4.2 Rift-Phase Greenstones 84
4.3 Summary 85
5 Metallogenic Associations in Granitoid-Greenstone Terrains .. 85
5.1 Older Greenstone Terrains 85
5.2 Younger Greenstone Terrains 87
5.3 Summary of Spatial and Temporal Variations 89
6 Evolution of Granitoid-Greenstone Terrains and Their Metallogenetic Associations 90
6.1 Nature of the Basement 91
6.2 Greenstone Terrain Evolution 93
6.3 Consideration of Tectonic Evolution 95
References 96

Abstract

There is marked spatial and temporal heterogeneity in diversity and intensity of metallogenic associations in Archaean greenstone belts. On the greenstone basin scale, parameters such as intensity of faulting, rapidity of burial, water depth and extent of irruption of komatiitic and felsic magma appear to have controlled the nature and intensity of mineralization. These inter-related parameters apparently depend on the degree of extension during basin development.

Initial development of both older (3.5 – 3.3 Ga) and younger (3.0 to 2.7 Ga) volcanic repositories appears to have occurred on platforms or in shallow basins with zero or negative marginal relief, probably under conditions of low extension. Older platform-phase greenstones formed in very shallow water and the metallogenic associations, including evaporative barite, small Pb-and sulphate-rich volcanogenic massive sulphides and porphyry-style Mo-Cu deposits, reflect the shallow marine to subaerial environments. Younger platform-phase greenstones formed in deeper water basins and have more conventional metallogenic associations, but the volcanogenic massive sulphides, komatiite-associated Ni-Cu

---

1 Geology Department, University of Western Australia, Nedlands, WA 6009, Australia

Archaean Geochemistry (ed. by A. Kröner et al.)
© Springer-Verlag Berlin Heidelberg 1984

deposits and gold mineralization are normally spatially restricted, and the greenstones have a relatively low intensity of mineralization.

Greenstone metallogenesis peaked in the late Archaean (2.8 − 2.7 Ga) in association with the development of major linear rift zones, probably related to increased extension and crustal thinning. Overlap of magmatic, volcanogenic and metamorphogenic mineralization resulted from the eruption of thick sequences of volcanics, including komatiites and felsic rocks, into rapidly subsiding deep water troughs and subsequent metamorphism and deformation of these sequences. The rift phase of greenstone basin development may be represented in the older terrains by more limited, dominantly sediment-filled grabens which are poorly mineralized relative to younger rift zones. This appears the major reason for the temporal contrast in intensity of mineralization, whereas temporal contrasts in the nature of metallogenic associations relate largely to the anomalous, very shallow-water environments at the platform stage in the older greenstone basins.

1 Introduction

Variations in the secular and spatial distribution of Archaean metallogenic associations reflect the tectonic evolution of Archaean terrains. However, these variations are generally poorly documented and seldom emphasized, despite the potential role of mineral deposits in contributing to an understanding of tectonic environments (Hutchinson 1981). Instead, discussion of Archaean metallogenesis and crustal evolution (e.g. Watson 1976; Anhaeusser 1981; Lambert and Groves 1981) generally focuses attention on metallogenic associations characteristic of Archaean granitoid-greenstone terrains and contrasts the volcanic-dominated 'greenstone-style' of mineralization with that of the late Archaean/early Proterozoic epicontinental basins.

The objectives of this paper are to draw attention to the markedly heterogeneous spatial distribution of Archaean mineralization and its peak development in late Archaean times (ca. 2.8 − 2.7 Ga), and to discuss these aspects in terms of evolution of Archaean crust. There are four major types of Archaean terrain, each with a distinctive suite of metallogenic associations (Table 1). However, attention is focused on granitoid-greenstone terrains because (1) they provide a 1 Ga volcanic, sedimentary and tectonic record of the evolution of the early continental crust; (2) they are the most diversely mineralized terrains; and (3) they are areally the most significant. In comparison, high-grade gneiss terrains contain relatively insignificant mineralization and, although preserving an extended tectonothermal history, they generally have a poor volcanic and sedimentary record. While intracratonic basins host some of the world's most economically significant mineral deposits (Table 1), the basins postdate the main craton-forming events and probably herald a new tectonic cycle. Late Archaean basement complexes to early Proterozoic geosynclinal sedimentation are generally poorly understood.

In this paper spatial and temporal variations in greenstone belts are initially discussed to constrain evolutionary models. Length restrictions inhibit exhaus-

Table 1. Contrasts in major features and metallogenic associations of the four major types of Archaean terrain

Major features	High grade gneiss belts	Granitoid-greenstone terrains	Intracratonic basins	Late Archaean complexes
1. Age	ca. 3.8 Ga and younger	Largely ca. 3.5 – 2.7 Ga	ca. 3.0 – 2.4 Ga	ca. 2.6 – 2.4 Ga
2. Lithofacies	Mainly granitic orthogneiss. Supracrustals dominated by shelf and lesser trough sediments intruded by ultramafic, mafic and/or anorthositic bodies. Volcanics rare	Mainly intrusive granitoids and derived gneisses. Supracrustals dominated by early mafic-ultramafic and lesser felsic volcanics and later clastic sediments	Shallow marine to fluvial clastic sediments including significant conglomerate sequences. Subaerial basalt sequences may be developed	Mainly granitic orthogneiss and variable paragneisses derived from sedimentary and volcanic precursors
3. Structure	Widespread early subhorizontal deformation including major recumbent folding and thrusting. Subsequent upright folding. Major dyking between deformation stages. Generally high strain	Local evidence for subhorizontal deformation and transition to gneiss belts. Structural trends dominated by upright folding and diapiric uprise of granitoids. Generally low strain	Relatively mild deformation. Normal faulting common, but folding and reverse faulting also present in parts of the basins. Low strain	Complex deformation involving both subhorizontal structures and upright folding. Generally high strain
4. Metamorphism	Complex polymetamorphic history. Generally granulite facies or amphibolite-granulite transition	Single-stage prograde metamorphism with local retrogression. Generally greenschist facies or below, with zones of amphibolite facies commonly developed adjacent to granitoids	Generally sub-greenschist facies metamorphism	Variable metamorphic grade, generally amphibolite facies

Table 1 (continued)

Major features	High grade gneiss belts	Granitoid-greenstone terrains	Intracratonic basins	Late Archaean complexes
5. Subsequent history	Commonly site of later Proterozoic tectono-thermal mobile belts	Commonly site of late Archaean or early Proterozoic sedimentary basins	Commonly site of Proterozoic or younger sedimentary basins	Commonly basement to early Proterozoic shelf to trough sedimentary basins
6. Metallogenic associations (with major review articles where appropriate)	Terrains poorly mineralized. Generally subeconomic associations include anorthosite-chromite, volcanic- or sedimentary-hosted disseminated Cu, ultramafic-mafic intrusive Ni-Cu, stratiform Au, and pegmatite Sn-Ta (largely related to younger granitoids)	Terrains heterogeneously but extensively mineralized. Major associations include: i) komatiite Ni-Cu ± PGE (Marston et al. 1981). ii) volcanogenic Fe-Cu-Zn massive sulphide association (Franklin et al. 1981) iii) Volcanic- and BIF-hosted Au (Boyle 1979); probably largely metamorphogenic deposits (Kerrich and Fryer 1979) iv) pegmatite Sn-Ta-Li v) layered intrusion chromite-asbestos-talc-magnetite (Anhaeusser 1976b) vi) Algoma-type BIF iron	Terrains contain some of the major sedimentary ore deposits of the world, including: i) quartz-pebble conglomerate Au-U (Pretorius 1981) ii) Superior-type BIF enriched Fe (Bayley and James 1973) Other deposit types include stratiform Mn, disseminated Cu in altered basalts, shale-hosted Au and small Pb-Zn-F deposits	Terrains, themselves, are generally poorly mineralized, although they may host small base-metal deposits and pegmatite Sn-Ta associations. However, they exert an important control on U deposits located at or adjacent to the unconformity with overlying Lower Proterozoic sedimentary sequences — the so-called unconformity-type U deposits (Nash et al. 1981)

tive discussion of critical evidence: Marston et al. (1981) provide additional useful data and discussion. However, largely of the basis of our data from the Western Australian Shield and correlation with information from other terrains, it is suggested that two stages of evolution, an early *platform phase* and a late *rift phase,* produced spatially discrete greenstone sequences of contrasting type in individual terrains. The distribution of metallogenic associations is discussed in terms of this subdivision. Emphasis is placed both on contrasts between associations in platform- and rift-phase greenstones within terrains of similar age, and on variations in associations related to each phase of evolution in terrains of contrasting age.

2 Granitoid-Greenstone Terrains

Most granitoid-greenstone terrains exhibit rather similar volcanic and sedimentary lithofacies, granitoid intrusion histories and deformation sequences extending over a variable interval from <0.1 Ga to 0.5 Ga. These features have been exhaustively described (see Windley 1977), and much emphasis has been placed on the uniformity of greenstone belts and the search for an all-embracing model of their evolution. The common features do have important implications to gross tectonic setting and nature of basement to greenstones. However, it is contrasts between greenstone terrains that are critical to the recognition of evolutionary processes. There is growing evidence for such differences, both between greenstone belts within discrete cratons (Gee et al. 1981) and between greenstone belts of contrasting age (Lowe 1982). These contrasts and their significance are emphasized below as a prelude to discussion of the distribution of Archaean metallogenic associations and formulation of a tectonic and metallogenic model.

It is first important to establish a temporal framework. Pre-3.5 Ga greenstones appear confined to the essentially unmineralized Isua sequences in Greenland and are not discussed further. All other greenstones can be broadly subdivided into two major temporal groups (Windley 1977). *Older greenstones* with supracrustals formed in the interval ca. 3.5 – 3.3 Ga are preserved in the east Pilbara Block, Barberton Mountain Land and restricted parts of Zimbabwe. More widespread *younger greenstones* developed in the interval ca. 3.0 – 2.7 Ga and form much of the Yilgarn Block, Superior Province and Rhodesian Craton and probably much of the Brazilian, Indian and Baltic Shields. There is little evidence of superposition of younger belts on much older granitoid-greenstone terrains in most cratons. It is emphasized that our nomenclature of older and younger greenstones refers to gross differences in age between discrete granitoid-greenstone terrains. They are *not* equivalent to the commonly used terms lower and upper greenstones which refer to subdivisions within individual older and younger greenstone terrains.

3 Spatial Variations in Greenstone Belts

In the following discussion, differences between greenstone belts within individual cratons are identified. Subdivision into platform-phase and rift-phase greenstones (Table 2) is justified. Considerable emphasis is placed on relationships within the Western Australian Shield, where the authors have personal experience. The reconnaissance nature of much of the data essential to formulation of tectonic and metallogenic models makes such models speculative, even in the Western Australian Shield. They are even more speculative for other terrains, but there are strong indications that similar relationships hold in well-documented terrains such as the Canadian Shield.

3.1 Platform-Phase Greenstones

The gross tectonic pattern of many granitoid-greenstone terrains is dominated by more or less equi-spaced, ovoid granitoid batholiths with intervening stellate, grossly synformal greenstone belts (e.g. east Pilbara, Murchison and Southern Cross Provinces of the Yilgarn Block in Western Australia).

The lower, normally volcanic-dominated parts of such sequences are very extensive and, despite evidence for early recumbent folding and thrusting (e.g. Bickle et al. 1980), a coherent stratigraphy has been erected in most regional studies of such terrains (Hickman 1981; Gee et al. 1981). There is commonly a vertical transition from volcanic-dominated sequences to clastic-dominated sedimentary sequences. Sediments within the volcanic sequences comprise largely volcaniclastic sediments, derived almost exclusively from felsic volcanic centres, or orthochemical sediments including BIF. Basement rocks and lower mafic-ultramafic volcanics provided insignificant detritus to the basin. Felsic centres are normally isolated and poorly represented in many terrains, and some vertical sections through lower sequences comprise almost entirely mafic-ultramafic sequences.

Extrusive komatiite sequences are poorly developed in most greenstone belts of this type (e.g. east Pilbara; Murchison and Southern Cross Provinces, Yilgarn Block), although significant sequences are developed in the Barberton Mountain Land. Where present, they are normally the "aluminium-depleted" komatiite type (Nesbit et al. 1979).

The nature of the lower volcanic sequences suggests that they developed in an extensive shallow basin or platform with essentially zero or negative marginal relief (cf. Lowe 1982). They are thus referred to as platform-phase greenstones. Although the depository was volcanically active, and presumably formed in an extensional regime, there is no evidence for widespread fault control on basin development. The resultant tectonic pattern of large equi-spaced granitoid batholiths is consistent with diapiric uprise of underlying granitoids through a greenstone slab of more or less constant thickness. Initial uplift may have instigated widespread clastic sedimentation and basin restriction in some instances, but the major phase of vertical tectonics including granitoid diapirism post-dated at least the early phase of major clastic sedimentation.

Table 2. Major features, emphasizing contrasts, of platform-phase and rift-phase greenstones of older and younger granitoid-greenstone terrains (Major contrasts in metallogenesis are emphasized in point 8)

Older terrains (3.5 – 3.0 Ga)	Younger terrains (3.0 – 2.7 Ga)
Platform-phase greenstones	
1. Maximum exposure of terrain is ca. 50000 km² (east Pilbara)	1. Maximum exposure of terrain is ca. 300000 km² (Murchison and Southern Cross Provinces)
2. No unequivocal basement exposed; sialic basement inferred from indirect evidence	2. Earlier sialic crust exposed in some terrains; sialic crust inferred in others
3. Gross tectonic pattern of more or less equi-spaced granitoid domes with intervening greenstones	3. Gross tectonic pattern of more or less equi-spaced granitoid domes with intervening greenstones
4. Formation interval 300 – 500 Ma; metamorphism at least 300 Ma after earliest volcanics	4. Formation interval 100 – 300 Ma; metamorphism at least 200 Ma after earliest volcanics
5. Coherent volcanic stratigraphy; indirect evidence of more or less constant thickness of greenstone pile	5. Coherent volcanic stratigraphy; indirect evidence of more or less constant thickness of greenstone pile
6. Basalts dominate volcanic sequences; komatiites and felsic volcanics rare	6. Basalts dominate volcanic sequences; komatiites and felsic volcanics rare
7. Very shallow-water depositional environments; sediments include subaerial to shallow-water volcaniclastic sediments, evaporites and accretionary lapilli	7. Variable depositional environments, but very shallow-water basins rare or absent; sediments include shallow-water clastics, turbidites and BIF
8. Anomalous metallogenic associations include evaporative barite, small Pb- and sulphate-rich volcanogenic massive sulphides and porphyry-style Mo-Cu. Fe-sulphides rare. Gold deposits small and dispersed. Komatiite-associated Ni-Cu deposits absent. Mineralization potential low	8. Widespread volcanic-hosted and BIF-hosted gold deposits. Generally relatively small volcanogenic massive Cu-Zn sulphides and komatiite-associated Ni-Cu deposits are widely dispersed. Fe-sulphides common. Mineralization potential higher than older counterparts, but lower than rift-phase greenstones

Table 2 (continued)

Older terrains (3.5 – 3.0 Ga)	Younger terrains (3.0 – 2.7 Ga)
Rift-phase greenstones	
1. Maximum exposure of terrain is ca. 10000 km² (West Pilbara)	1. Maximum exposure is ca. 150000 km² (Norseman-Wiluna Belt)
2. Developed on earlier greenstone belts of platform phase	2. Probably partly developed on earlier platform-phase greenstones and partly on thinned sialic crust
3. Gross tectonic pattern is linear	3. Gross tectonic pattern markedly linear with elongate granitoid domes and strike-slip faults
4. Formation interval unknown	4. Formation interval probably less than 100 Ma.
5. Thick turbidite sequences; complex stratigraphy where volcanics present	5. Complex volcanic stratigraphy; may have elongate zones dominated by felsic volcanics
6. Felsic volcanics and/or komatiites present where volcanics occur	6. Komatiites and/or felsic volcanics widespread; thick volcanic sequences developed
7. Variable sedimentary environments, but dominant deposition in deep troughs; sediments dominated by turbidites	7. Variable depositional environments ranging from subaerial to deep water troughs in axial zones of rift; sediments include turbidites and sulphidic shales/cherts
8. Variable metallogenic associations include Sb- or Bi-rich gold deposits, small volcanogenic massive Cu-Zn sulphides and small gabbroid-associated Ni-Cu deposits. Mineralization potential greater than adjacent platform-phase greenstones, but much lower than younger counterparts	8. Major metallogenic associations; commonly spatial overlap of two or more major mineralization types. Important deposits include volcanogenic massive Cu-Zn sulphides, komatiite-associated Ni-Cu deposits and volcanic-hosted gold deposits. Greatest mineralization potential of all greenstone terrains

3.2 Rift-Phase Greenstones

Within some granitoid-greenstone terrains, there are segments with more linear tectonic patterns, characterized by elongate greenstone belts and intervening granitoid domes; major regional faults commonly cut the greenstone sequences into tectonic slices. They are commonly characterized by a complex stratigraphy of mafic-ultramafic sequences, felsic volcanics and volcaniclastic sediments. Volcaniclastic sediments were deposited both in local terrestrial environments related to volcanic vent construction and faulting and in large submarine fan systems. Clasts include submarine mafic and ultramafic rocks, suggesting uplift and subaerial exposure of these sequences: granitoid clasts may also be locally abundant. Argillaceous sediments are widespread, and there are rapid facies changes across and along depositional basins. The belts commonly have thicker and more extensive extrusive komatiite (normally 'aluminium-undepleted' type) and/or felsic volcanic sequences than the more widespread platform-phase greenstone belts.

Although linear segments of the Abitibi Belt almost certainly represent rift-phase greenstones, the type example of such a terrain is taken here as the stratigraphically complex Norseman-Wiluna Belt of the Yilgarn Block (Fig. 1). This linear zone is characterized by an elongate western segment in which thick extrusive komatiite sequences, intrusive komatiitic dunites and sulphidic shales and cherts are abundant. The presence of distal turbidites and extensive, thick, non-vesicular basalt sequences containing only thin, very distal (Bavinton 1981) sulphidic shale horizons indicate deep water conditions in the central part of the elongate basin, at least in the southern part. The eastern segment of the belt is characterized by subaerial to shallow marine felsic volcanic centres and associated volcaniclastic sediment wedges. The flanking, more extensive greenstone belts have a contrasting lithofacies dominated by basaltic lavas with only minor extrusive komatiites and isolated felsic volcanic centres (Fig. 1). BIF is abundant in these greenstone sequences, but is generally absent from the Norseman-Wiluna Belt.

These features suggest that the greenstones formed in tectonically more active basins than the platform-phase greenstones. As in the latter, felsic volcanic centres formed significant relief, but submarine mafic-ultramafic sequences and granitoids were exposed, presumably by faulting and/or isostatic readjustment. Rapid facies changes support the concept of fault-bounded basins, and sedimentary facies suggest that parts of these basins were deep troughs. The occurrence of bimodal basalt-rhyolite volcanism in some belts provides supporting evidence for a tensional tectonic setting (McGeehan and MacLean 1980). It is likely that the greenstones formed in an actively extensional, fault-bounded basin. For this reason, they are termed rift-phase greenstones. The present linear pattern of the terrains is largely the result of imposed strike-slip faults, but evidence from such diverse sources as gross lithofacies variation and orientation of mineral deposits (e.g. Kambalda Ni-Cu ores) suggests an original linear aspect (Groves 1982). The gross linear tectonic pattern can be explained in terms of vertical tectonics with late diapiric uplift of granitoids controlled by edge effects related to the boundary faults between greenstones and basement (Archibald and Bettenay 1977).

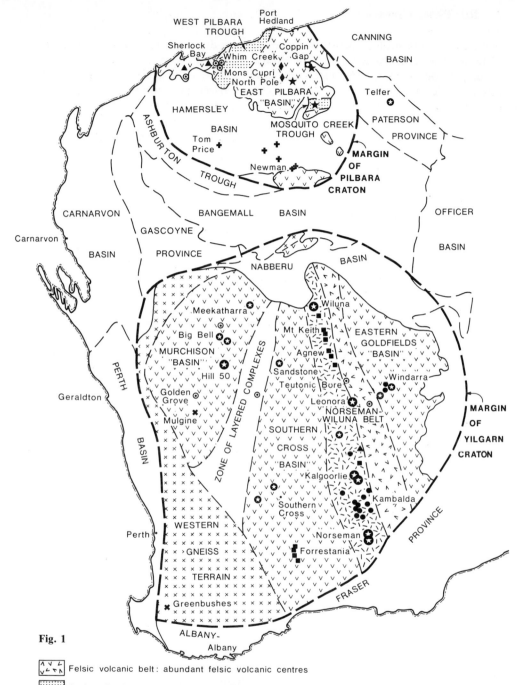

Fig. 1

	Felsic volcanic belt: abundant felsic volcanic centres
	Graben basin sequence: clastic sediments -mainly turbidites
	Rift basin sequence: basalt, komatiites, sulphidic shales, epiclastic sediments, BIF absent
	Platform/extensive basin sequence: mainly basalts, BIF or shallow water cherty sediments

The precise temporal relationship between platform-phase and rift-phase greenstones within the same craton is not completely clear. There is normally insufficient superior geochronological information (U-Pb in zircons, Sm-Nd whole-rock isochrons) for greenstones to precisely define their ages. Available data from the Norseman-Wiluna Belt, however, suggest that it may have evolved over a much shorter period (ca 0.1 Ga) than many platform-phase greenstone belts. In addition, available isotopic data firmly establishes an age of ca. 2.8 Ga for this belt (e.g. McCulloch and Compston 1981; Cooper and Dong 1983), whereas reconnaissance Sm-Nd isotopic studies suggest that at least some greenstones were formed at ca 3.0 Ga in the adjacent Murchison and Southern Cross Provinces (e.g. Fletcher et al. 1982). Possible stratigraphic confirmation that the rifts are superimposed on earlier basins is provided by the distribution of BIFs. These represent major lithostratigraphic units in the platform-phase greenstones, commonly marking the boundary betwen early basalt-dominated sequences and younger, commonly sediment-dominated sequences (e.g. Gee 1979). In the Norseman-Wiluna Belt, however, they only occur at the lowest stratigraphic levels at the southern end of the belt (e.g. Gemuts and Theron 1975), suggesting that the bulk of rift-related volcanism and sedimentation may have post-dated BIF development. Similarly, in the Pilbara Block there is a close temporal relationship between widespread BIF deposition and the appearance of thick trough sedimentary sequences (Eriksson 1981). It is possible that the appearance of widespread BIF in the more stable portions of greenstone basins reflects an increase in hydrothermal circulation of Fe-rich fluids related to the initial stages of rifting.

4 Temporal Variations in Greenstone Belts

Recent sedimentological studies have highlighted important differences between older (3.5 – 3.3 Ga) greenstone belts and younger (3.0 – 2.7 Ga) counterparts. Before discussing these differences (summarized in Table 2), it must be emphasized that younger terrains are far more widespread and extensive than older terrains, so that it is difficult to establish if the exposed segments of the latter are truly representative of greenstone belts formed at that time. The close similarities of the best-documented older greenstones in the east Pilbara and Barberton

---

✪	Gold deposit > 50t Au	⊙	Volcanogenic Cu-Zn deposit > 10,000t Cu+Zn
◐	Gold deposit > 10t Au	▢	Porphyry-style Cu-Mo deposit > 10,000t Cu+Mo
★	Pilbara gold deposits > 2t Au	✖	Tin-tantalum or tungsten deposit > 10,000t W or Sn-Ta
■	IDA nickel deposit > 10,000t Ni	♦	Major barite deposit
●	VPA nickel deposit > 10,000t Ni	✚	Iron ore deposit > 1000 Mt Fe
▲	Other nickel deposits > 10,000t Ni		

Fig. 1. Regional map of the Western Australian Shield showing the major tectonic units and major ore deposits of the Archaean Pilbara and Yilgarn Blocks. The distribution of platform-phase and rift-phase greenstones is somewhat schematic, and the distribution of sequences with abundant felsic volcanics is generalized

ment of evaporative gypsum and reworking of these deposits in shallow water (Dunlop and Groves 1978; Reimer 1980). In addition, barite is an important component of small, uneconomic base-metal deposits that have significant Pb and Ag and low concentrations of Fe-sulphides (e.g. Big Stubby and Yandicoogina, east Pilbara). In the east Pilbara there are several examples of stockwork Cu-Mo and porphyry-style Mo-Cu deposits associated with probable subvolcanic granitoid intrusives (Barley 1982). There are no significant komatiite-associated Ni-Cu deposits in these terrains.

It is difficult to make generalizations concerning gold mineralization. There has been very low gold production from the relatively extensive greenstone belts of the east Pilbara. In contrast, the Barberton Mountain Land has a very high gold production (Anhaeusser 1976a), but much of the gold is geographically restricted, is associated spatially with early (now refolded) faults, and deposits are largely hosted in the lower parts of the upper sedimentary sequences. This may represent a unique style of mineralization in the older greenstone terrains and be atypical of the Kaapvaal Craton in general; trace element studies of pyrites from the younger Witwatersrand gold-bearing conglomerates (Hallbauer and Kable 1982), for example, suggest that mineralization of the Barberton type was not a major contributor to these gold deposits.

The absence of economic komatiite-associated Ni-Cu or volcanogenic Cu-Zn massive sulphide deposits reflects the relative lack of peridotitic komatiites and the widespread shallow-water depositional environment characteristic of the early platform phase.

The absence of komatiites may be a direct consequence of the low rate (and total) extension in development of platform-phase greenstone basins. Under such conditions, fractionation of komatiitic magmas in intra- or sub-crustal magma chambers and the eruption of basaltic magmas rather than komatiites is probably favoured (e.g. Nisbet 1982). The shallow deposits and environment also militated against the deposition of sulphidic sediments, the potential source of sulphur for komatiite-associated deposits (e.g. Lesher et al. 1984).

Lack of synvolcanic deep water basins inhibited development of hot water plumes at depths below which boiling would be prevalent, and massive sulphides could be formed and preserved (cf. Solomon and Walshe 1979). The small, relatively Cu-poor deposits that are present probably represent lower temperature, epithermal-style mineralization formed in a shallow-water environment. The widespread occurrence of evaporative sulphates, the more restricted occurrence of stratiform barite and the association of barite with the small volcanogenic massive sulphide deposits all point to a relatively oxidized shallow hydrosphere (Lambert and Groves 1981). The relatively Fe-poor nature of all these deposit types suggests that at the volcanic stage this hydrosphere was above the level of available ferrous iron reservoirs or was isolated from deeper basins in which such reservoirs were present.

The occurrence of porphyry-style Mo-Cu mineralization is consistent with the subaerial nature of the felsic volcanism in the older terrains (Barley 1982). Stockwork Cu-Mo mineralization possibly represents the subvolcanic equivalent of massive sulphide deposits in shallow-water envionments, where boiling of ore fluids took place below the water interface.

Lack of large volcanic-hosted metamorphogenic gold deposits such as those that typify the younger greenstone belts may be due to (1) the restricted occurrence of exposed rocks within the metamorphic window (greenschist to low amphibolite facies) for gold deposition (Groves et al. 1984), (2) the lack of high temperature (high sulphur) komatiitic volcanics to act as suitable source rocks (Keays 1984); (3) volatilization of sulphur during shallow eruption and consequent lack of extractable sulphide-hosted gold for subsequent metamorphic leaching; and/ or (4) early leaching of available gold from potential source rocks during extensive synvolcanic alteration. Such intense alteration is characteristic of subaerial to shallow-water volcanic sequences where multi-stage hydrothermal flushing and horizontal channelling of fluids is common (e.g. Spooner and Fyfe 1973; Henley and Ellis 1983).

Rift-phase greenstones are poorly developed in older terrains, where they are perhaps more appropriately termed graben-phase supracrustals. A characteristic association of Sb-, Bi-, and/or As-rich gold deposits occurs within trough sediments of the Mosquito Creek Beds in the east Pilbara, and more typical Archaean pyritic massive base-metal sulphides lacking sulphates are associated with a felsic volcanic belt in the West Pilbara (Fig. 1), although there is some debate as to the contemporaneity of the volcanics and trough sediments (Fitton et al. 1975; Horwitz 1979; Hickman 1980). The deeper water environments are clearly important for formation and preservation of volcanogenic sulphides and retention of gold in potential source rocks. Despite this, the diversity and intensity of mineralization is lower than that shown by rift-phase greenstones in younger terrains, and the majority of deposits are uneconomic.

5.2 Younger Greenstone Terrains

As in the older terrains, younger platform-phase greenstones show a diversity of metallogenic associations. Individual metallogenic associations may show geographic concentrations, for example, several gold provinces and the Forrestania Ni-Cu deposits in the Yilgarn Block (Fig. 1), whereas volcanogenic base-metal deposits tend to be scattered throughout the belts. There are few areas where there is spatial overlap of two or more important associations.

In accord with their deeper water depositional basins than in older counterparts, the greenstones contain no significant sulphate-bearing deposits, and porphyry-style Mo-Cu deposits are rare. Small enriched iron ore deposits are associated with BIFs towards the top of volcanic-dominated sequences.

The major mineralization developed is an epigenetic gold association. This is widespread, but individual deposits cluster, commonly with one or two large deposits surrounded by a number of much smaller deposits. Both volcanic-hosted and BIF-hosted gold deposits are well-developed, particularly in mafic volcanic-dominated sequences. BIFs are commonly preferred host rocks; for example, they provide almost half the gold production from the Murchison and Southern Cross Provinces of the Yilgarn Block despite their low total volume. In the Western Australian Shield gold concentration in the younger platform-phase greenstones (ca. 16 kg Au/km^2) is considerably higher than that of older counterparts in the east Pilbara (<1 kg Au/km^2). In both terrains preferred gold source-rocks (komatiites – Keays 1984) are rare, and it is possibly the lower intensity of syn-volcanic alteration of source rocks in the deeper water Yilgarn basins that is the critical factor in their higher mineralization potential. Further generalizations on the relative gold content of platform phases in younger and older terrains are re-strained by the Barberton Mountain Land with an anomalously high concentra-tion of ca. 50 kg Au/km^{2}of greenstone. However, the limited extent of the ter-rain and the uncertainty of the tectonic setting of the gold-hosting sedimentary sequences make the significance of this value uncertain: selected equivalent areas of younger terrains would have even higher concentrations. The Kaapvaal and Rhodesian Cratons also appear to be particularly enriched in gold: Zimbabwe greenstone belts in toto, dominated areally by younger sequences, have a concen-tration in excess of 70 kg Au/km^2 of greenstone based on data provided by An-haeusser (1976a).

Rift-phase greenstones in younger greenstone belts are intensively mineralized and typified by an abundance of major volcanic-hosted gold deposits and either volcanogenic massive Fe-Cu-Zn sulphide deposits or komatiite-associated Fe-Ni-Cu deposits: there is an antipathetic relationship between base-metal deposit types. For example, parts of the Abitibi Belt are typified by abundant volcano-genic massive sulphide deposits, commonly related to specific stratigraphic inter-vals within thick mafic-felsic volcanic piles (e.g. Franklin et al. 1981). Small por-phyry-type Mo-Cu deposits are also present (e.g. Ayres et al. 1982). In these ter-rains, komatiite-associated Ni-Cu deposits are generally small and/or low grade (Green and Naldrett 1981). In contrast, the Norseman-Wiluna Belt contains abundant high-grade komatiite-associated Ni-Cu deposits (Marston et al. 1981), but rare and geographically discrete volcanogenic massive sulphide deposits. Over 90% of pre-mining reserves of nickel in the Western Australian Shield are concentrated in the Norseman-Wiluna Belt (Ross and Travis 1981). Contrasts be-tween the Abitibi Belt and Norseman-Wiluna Belt may, at least in part, be due to the stratigraphic level of greenstones exposed in the two belts, although the sub-aerial nature of felsic centres in the Norseman-Wiluna Belt (Hallberg 1980) is clearly a contributing factor.

Gold deposits are well-developed in rift-phase greenstones, and major gold centres from such settings tend to be an order of magnitude more productive than those in platform associations. Volcanic-hosted vein or lode deposits are most widespread, and in many provinces metabasalt- or metadolerite-hosted de-posits are dominant (Woodall 1979). On the regional scale gold deposits may be concentrated at specific stratigraphic levels (Hodgson and McGeehan 1982), and/or in zones of major faulting and uplift, commonly typified by extensive carbonate alteration, in the axial region of the rifts (Fig. 1; Groves et al. 1984). In the Yilgarn Block gold mineralization is more intense in the Norseman-Wiluna Belt (ca. 32 kg Au/km^2 of greenstones) than in flanking zones (ca. 16 kg Au/km^2).

The spatial overlap of major ore deposits of different metallogenic associa-tions and with markedly contrasting genesis provides important insights into the regional tectonic controls of metallogenesis. In the Norseman-Wiluna Belt, for example, the major volcanic komatiite-associated Ni deposits occupy the central part of the belt and appear related to extensive eruption of thick sequences of komatiites in the more rapidly subsiding, deeper-water axial zone of the rift during the major extensional stage of greenstone belt evolution (Groves 1982). This is consistent with the models of Nisbet (1982). The presence of non-vesicular lavas and sulphidic shales or chert with a very low clastic component in these zones is consistent with a deep-water setting. The abundance of Fe-sulphides in these zones and possible syn-early rifting oxide-facies BIF in flanking zones combined with the general absence of sulphates suggests relatively reducing basins open to the influx of ferrous iron during the formation of younger rift-phase greenstone volcanic sequences. These would also be ideal conditions for the formation, and importantly preservation, of volcanogenic massive sulphide deposits in areas where there was coincident felsic volcanism (e.g. Abitibi Belt).

Several of the major gold producers of the Yilgarn Block occur in the same axial zone along the same structural high occupied by the major nickel deposits (Fig. 1). Their occurrence is probably related to generation of thick sequences of relatively unaltered, sulphide-bearing mafic/ultramafic volcanic rocks, which subsequently represented ideal source rocks (Keays 1984) for metamorphic gold-bearing fluids (Groves 1982).

5.3 Summary of Spatial and Temporal Variations

1. *Older platform-phase greenstones* are poorly mineralized. Relatively small sulphate-rich sedimentary and volcanogenic deposits and porphyry-style Mo-Cu deposits reflect the subaerial to shallow water depositary and a relatively oxidized shallow hydrosphere. Lack of Ni-Cu deposits reflects the general poor development of komatiites (perhaps more specifically "aluminium-undepleted" types) due to low total extension and/or low extension rate, and lack of very large gold deposits may also reflect limited irruption of high temperature komatiitic magmas and/or early intense alteration and removal of gold from the subaerial to shallow-water volcanic pile.

2. *Rift (graben)-phase greenstones* are poorly expressed and poorly mineralized in older greenstone terrains where they are sediment-dominated. However, if flanking volcanic zones are co-genetic, as is possible in the West Pilbara Block, they may contain a diversity of more typical Archaean metallogenic associations due to the deeper water depositories. However, individual volcanogenic deposits are generally relatively small.

3. *Platform-phase greenstone* in younger terrains are better mineralized than their older equivalents due to the greater chance of formation and preservation in deeper-water basins, but significant base-metal deposits are normally isolated and generally poorly developed. Gold deposits are widespread, and BIF-hosted deposits are particularly important.

4. *Rift-phase greenstones* in younger terrains contain the greatest concentration and diversity of important metallogenic associations. The great majority of greenstone-hosted, komatiite-associated Ni-Cu deposits are concentrated in such sequences; data from Ross and Travis (1981) suggest that at least 85% of global

resources of such deposits occur in these zones and that over 60% are confined to the Norseman-Wiluna Belt alone. The majority of Cu-Zn resources in Archaean volcanogenic massive sulphides are concentrated in the Abitibi Belt (e.g. Franklin et al. 1981). Similarly, many of the large gold deposits occur in such belts. Six of the ten largest deposits in the Western Australian Shield (including the four largest) occur in the Norseman-Wiluna Belt and 26 of the 33 largest producers (over 10^6 oz Au) in the Canadian Shield occur in the Abitibi Belt (Hodgson and McGeehan 1982). The most important, inter-related parameters that lead to the high mineralization potential of the rift-phase greenstones appear to be:

a) high total extension (and extension rates?) that favour eruption of komatiites rather than basalts only (Nisbet 1982): komatiites are the source of magmatic Ni-Cu deposits and represent potentially enriched Au sources.
b) deeper water environments that favour formation and preservation of volcanogenic massive sulphide deposits, the deposition of sulphidic shales as potential sulphur sources for komatiite-associated Ni-Cu deposits (Lesher et al. 1984) and less intensive hydrothermal alteration (Henley and Ellis 1983).

6 Evolution of Granitoid-Greenstone Terrains and Their Metallogenic Associations

The two most important features of the heterogeneous time-space distribution of metallogenic associations in Archaean granitoid-greenstone terrains are:

1. the existence of discrete zones or belts of mineralization with a preferential concentration of a diverse range of mineral deposits (Fig. 1), and
2. the peak development of such restricted belts in the period 2.8 – 2.7 Ga ago (Fig. 2).

These features are further examined in terms of tectonic evolution of platform- and rift-phase greenstones and implications for early crustal development. The schematic model outlined in Fig. 3 is largely based on contrasts between, and within, the Pilbara and Yilgarn Blocks of the Western Australian Shield. Available information suggests it has a wider application, although the relative importance of platform- versus rift-phase greenstones may vary widely between cratons: for example, the Rhodesian craton appears dominated by platform-phase greenstones in the period 3.0 – 2.7 Ga, whereas the Superior Province appears to have a higher proportion of 2.8 – 2.7 Ga rift-phase greenstones.

6.1 Nature of the Basement

An initial constraint on any model is the nature of basement, if any, on which greenstones formed, and this has been one of the major debates of Archaean geology. It has been argued that temporal differences in greenstones relate to changes in tectonic setting from ensimatic to ensialic (Lowe 1982), but the overall

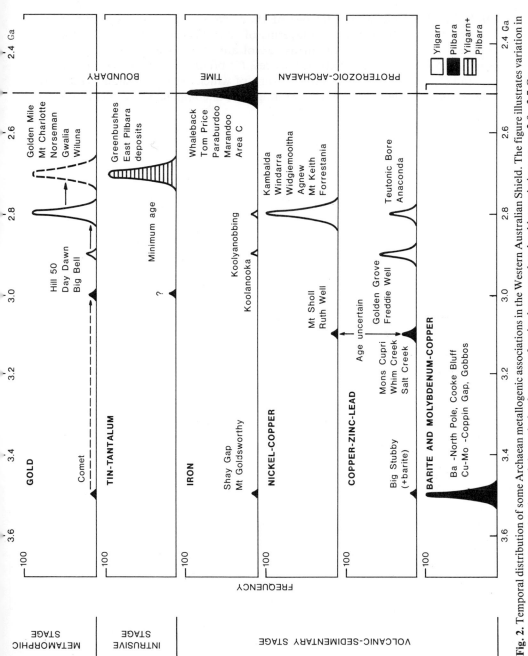

Fig. 2. Temporal distribution of some Archaean metallogenic associations in the Western Australian Shield. The figure illustrates variation in the nature of associations with time and the peak development of major base-metal and gold associations at ca. 2.8–2.7 Ga

similarity of lithofacies and history of the terrains suggest a common setting. Features such as (1) multiple interlayering of volcanics and sediments; (2) lack of constant vertical zonation of intrusive/volcanic lithologies and metamorphic grade in lower mafic-ultramafic sequences; (3) presence of calc-alkaline volcanics; and (4) abundance of granitoids, some of which are synchronous with early greenstones (Bickle et al. 1983) are among several compelling arguments against

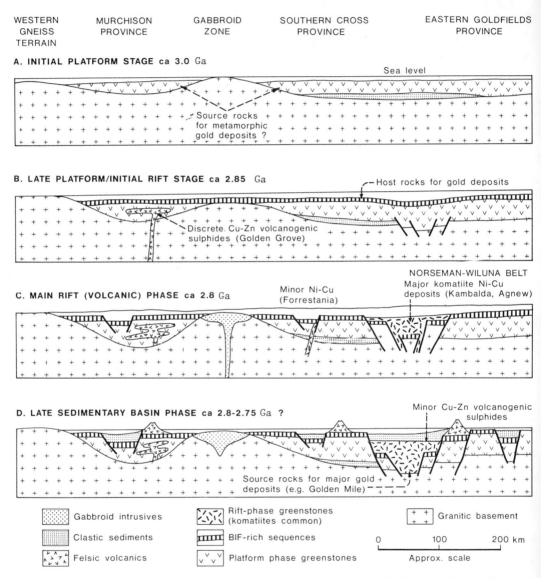

Fig. 3. History of development of greenstone basins through platform- and rift-phases illustrated by reference to the Yilgarn Block. The sequence of events must be considered tentative due to the reconnaissance nature of available geochronology. The vertical scale is exaggerated to allow illustration of relationships within the greenstone belts

greenstones representing oceanic crust. (*For alternative view see Smith et al., this Vol., eds.*) Unequivocal evidence for sialic basement is provided for some sequences by the occurrence of supracrustal sequences in younger greenstones unconformably overlying older granitic gneisses (e.g. Bickle et al. 1975). In most terrains, however, such evidence is lacking, and indirect evidence such as (1) the rare occurrence of probable granitoid-derived basal sediments (Gee et al. 1981); (2) the nature of crustal profiles (Archibald et al. 1981); (3) the presence of calc-alkaline volcanics with geochemical characteristics similar to those erupted through continental crust (Bickle et al. 1983); and (4) the hiatus between volcanism and metamorphism that is suggestive of a thermal buffer such as continental crust, provides equivocal evidence of sialic basement.

For these reasons it is assumed that greenstones formed on sialic crust in the model outlined below (Fig. 3).

6.2 Greenstone Terrain Evolution

The apparent broad lateral continuity of basalt-dominated lower volcanic sequences and the lack of orogenic sediments suggest that platform-phase greenstones evolved as extensive anorogenic basins or platforms with zero or negative marginal relief (cf. Lowe 1982) in an extensional tectonic regime. A model involving early regional arching and fracturing of a relatively rigid pre-existing sialic crust over divergent zones of mantle convection cells is considered most appropriate (cf. Hargraves 1981). Basaltic volcanism and more restricted felsic volcanism produced volcanic sequences up to several kilometres thick (Fig. 3a); peridotitic komatiites were generally poorly represented.

In older terrains (3.5 – 3.3 Ga) intervolcanic sediments indicate that deposition occurred more or less at base level (defined as the hydrosphere-atmosphere interface) with rate of deposition maintaining pace with rate of subsidence. In younger terrains (3.0 – 2.7 Ga) platform-phase greenstones are similar to their older counterparts, except that deposition appears to have been at greater average water depths, suggesting either greater initial subsidence or subsidence rates greater than those of deposition. The major tectonic implication is that the amount of relative crustal thinning was greater under the younger terrains than the older, which may reflect a secular change in the degree of fractionation and thermal regime in the younger crust.

The difference in depositional environments has a major influence on the relative importance of certain metallogenic associations within the platform phases of younger and older terrains; for example, volcanogenic Cu-Zn massive sulphides are preferentially developed and preserved in deeper-water environments of the younger terrains, and high level subvolcanic Cu-Mo deposits are more common in the subaerial to shallow-water environments of the older terrains. Sulphate deposits are restricted to the older terrains.

Synvolcanic metallogenic associations are poorly developed in both older and younger platform-phase terrains. Possible exceptions include the intrusive dunite-associated deposits at Forrestania in the Yilgarn Block (Fig. 1) which occur within a narrow linear zone along the extension of which volcanic- and BIF-hosted gold deposits also appear preferentially concentrated. The relationship of such zones to the platform-phase greenstones is uncertain: they may be related to superimposed rift-phase events. Similarly, in Zimbabwe low grade komatiite-associated Ni-Cu deposits are preferentially developed in the youngest greenstone sequences (Williams 1979) stratigraphically above thick BIF units.

The end of the platform-phase in both younger and older terrains is marked by widespread deposition of BIFs which may be due to increased hydrothermal activity related to the initial stages of rifting, heralding the onset of a new phase of greenstone evolution (Fig. 3c).

The rift-phase greenstones have a different expression in granitoid-greenstone terrains of different ages (Fig. 4). However, in both cases evidence suggests de-

RIFT OR GRABEN PHASE

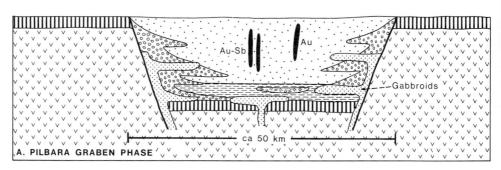

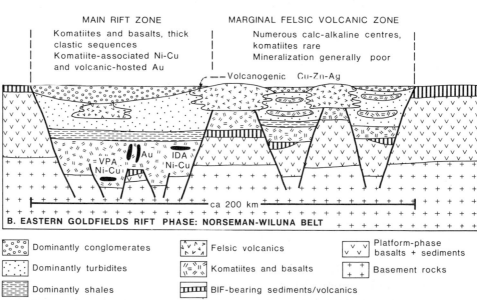

Dominantly conglomerates	Felsic volcanics	Platform-phase basalts + sediments
Dominantly turbidites	Komatiites and basalts	Basement rocks
Dominantly shales	BIF-bearing sediments/volcanics	

Fig. 4. Representation of the contrast between rift (graben)-phase greenstones in older and younger granitoid-greenstone terrains illustrated with reference to the Pilbara and Yilgarn Blocks of the Western Australian Shield. The vertical scale is exaggerated to allow illustration of relationships within the greenstone belts

velopment of linear basins or grabens and syntectonic volcanism and sedimentation in response to renewed or greater extension over an evolving crust already modified by development of the platform-phase greenstones (Fig. 3c, d).

In older granitoid-greenstone terrains the depositories are graben-like and dominated by trough sediments (turbidites and proximal submarine-fan deposits). The volcanic component is small and restricted to the lower part of the stratigraphy. The main metallogenic association is sediment-hosted $Au \pm Sb \pm As$ lodes, the intensity of which may reflect the availability of appropriate source rocks (komatiites, sulphidic interflow sediments) in the underlying pre-rift stratigraphy and the subsequent tectono-thermal history.

In younger granitoid-greenstone terrains, rift-phase greenstones appear to reflect greater extension and crustal thinning than in older terrains. The thinned crust and extensional faulting has been important in promoting irruption of thick piles of komatiites (e.g. Nisbet 1982) with associated Ni-Cu mineralization in lower parts of the stratigraphic sequence. Felsic volcanic centres may also have developed along major fracture zones with rapid lateral facies changes into volcaniclastic turbidite sequences. Associated rapid subsidence produced deeper water basins suitable for generation and preservation of volcanogenic massive sulphide deposits and conducive to rapid burial and retention of potentially extractable gold in volcanic rocks representing the source of metamorphic gold-bearing ore fluids. At the time of rifting basins of deposition appear to have been open with respect to reservoirs of ferrous iron such that discontinuous Fe-rich sulphidic sediments were deposited in the more active, strongly mineralized rift zones.

Contemporaneous volcanism and sedimentation may have continued within platform-phase greenstones outside the major rift zones, with increasing restriction of basins high in the sequence, perhaps in part related to faulting in flanking zones and in part to initiation of diapiric uplift of sialic basement. Significant base-metal mineralization appears lacking from such sequences.

Narrow linear belts of clastic sediment appear to be superimposed on rift- and platform-phase greenstones in both younger and older terrains. Such basins, where well-studied, appear to develop in oblique-slip tectonic regimes (B. Krapez, in prep.) and lack endogenous metallogenic associations.

6.3 Consideration of Tectonic Evolution

The broader questions of the precise tectonic setting of the greenstone belts and whether there are modern analogues are beyond the scope of this paper. These and related problems concerning whether deformation was due to internal processes or was imposed from outside (e.g. by continental collision) do not affect conclusions relating to the relationships between metallogenic associations and recognized phases of greenstone evolution discussed above. A major enigma that arises from our model is that the older greenstones apparently developed under more stable conditions than younger greenstones, contrary to theoretical considerations that tectonic activity would have been more vigorous in earlier times (e.g. Burke et al. 1976). The observed evolutionary trend may, however, relate more to changes in the thickness, degree of differentiation and thermal state of the crust on which greenstones formed than to changes in tectonic regime.

Irrespective of these considerations, the major conclusion of our paper is that initiation of major extension and crustal thinning with the development of rapidly subsiding, volcanically active, deep-water rift zones at ca. 2.8 Ga ago was the major control on the marked peak in Archaean greenstone belt metallogenesis (Fig. 2).

Acknowledgements. The concepts put forward in this paper have developed from data collected by numerous Geological Survey, CSIRO, university and mining company geologists in Western Australia. We are particularly indebted to postgraduate students at the University of Western Australia, especially M. E. Barley, R. Buick, J. S. R. Dunlop and C. M. Lesher, and to R. D. Gee and R. J. Marston for useful discussions over a period of several years. The paper was improved by critical comments on an earlier manuscript by G. N. Phillips.

References

Anhaeusser CR (1976a) The nature and distribution of Archaean gold mineralization in southern Africa. Mineral Sci Engng 8:46 – 84

Anhaeusser CR (1976b) Archaean metallogeny in southern Africa. Econ Geol. 71:16 – 43

Anhaeusser CR (1981) The relation of mineral deposits to early crustal evolution. Econ Geol 75th Anniv Vol: 42 – 62

Archibald NJ, Bettenay LF (1977) Indirect evidence for tectonic reactivation of a pre-greenstone sialic basement in Western Australia. Earth Planet Sci Lett 33:370 – 378

Archibald NJ, Bettenay LF, Bickle MJ, Groves DI (1981) Evolution of Archaean crust in the eastern Goldfields Province of the Yilgarn Block, Western Australia. Geol Soc Aust Spec Publ 7:491 – 504

Ayres LD, Averill SA, Wolfe AJ (1982) An Archean molydenite occurrence of possible porphyry type at Setting Net Lake, Northwestern Ontario, Canada. Econ Geol 77:1105 – 1119

Barley ME (1982) Porphyry-style mineralization associated with early Archaean calc-alkaline igneous activity, eastern Pilbara, Western Australia. Econ Geol 77:1230 – 1235

Barley ME, Dunlop JSR, Glover JE, Groves DI (1979) Sedimentary evidence for an Archaean shallow-water volcanic-sedimentary facies, eastern Pilbara Block, Western Australia. Earth Planet Sci Lett 43:74 – 84

Bavinton OA (1981) The nature of sulfidic metasediments at Kambalda and their broad relationships with associated ultramafic rocks and nickel ores. Econ Geol 76:1606 – 1628

Bayley RW, James HL (1973) Precambrian iron formations of the United States. Econ Geol 68:934 – 959

Bickle MJ, Bettany LF, Barley ME, Chapman HJ, Groves DI, Campbell IH, de Laeter JR (1983) A 3500 Ma plutonic and volcanic calc-alkaline province in the Archaean East Pilbara Block. Contrib Mineral Petrol 84:25 – 35

Bickle MJ, Bettenay LF, Boulter CA, Groves DI, Morant P (1980) Horizontal tectonic interactions of an Archaean gneiss belt and greenstones, Pilbara Block, Western Australia. Geology (Boulder) 8:525 – 529

Bickle MJ, Martin A, Nisbet EG (1975) Basaltic and peridotitic komatiites and stromatolites above a basal unconformity in the Belingwe greenstone belt, Rhodesia. Earth Planet Sci Lett 27:155 – 162

Boyle RW (1979) The geochemistry of gold and its deposits. Geol Surv Can Bull 280:584

Buick R, Dunlop JSR, Groves DI (1981) Stromatolite recognition in ancient rocks: an appraisal of irregularly laminated structures in an early Archaean chert-barite unit from North Pole, Western Australia. Alcheringa 5:161 – 181

Burke K, Dewey JF, Kidd WSF (1976) Dominance of horizontal movements, arc and microcontinent collision during the late permobile regime. In: Windley BF (ed) The early history of the Earth. Wiley, New York, pp 113 – 129

Cooper JA, Dong YB (1983) Zircon age data from a greenstone of the Yilgarn Block, Australia. Mid-Proterozoic heating or uplift. Contrib Mineral Petrol 82:397 – 402

Dunlop JSR, Buick R (1981) Archaean epiclastic sediments derived from mafic volcanics, North Pole, Pilbara Block, Western Australia. Geol Soc Aust Spec Publ 7:225 – 234

Dunlop JSR, Groves DI (1978) Sedimentary barite of the Barberton Mountain Land: a brief review. Geol Dept Extension Service, Univ West Aust Publ 2:39 – 44

Eriksson KA (1981) Archaean platform-to-trough sedimentation, east Pilbara Block, Australia. Geol Soc Aust Spec Publ 7:235 – 244

Fitton MJ, Horwitz RC, Sylvester G (1975) Stratigraphy of the early Precambrian in the west Pilbara, Western Australia. CSIRO Aust Min Res Lab Rep FP11

Fletcher IR, Rosman KJR, Trendall AF, de Laeter JR (1982) Variability of E_{Nd}^i in greenstone belts in the Archaean of Western Australia. 5th Internat Conf Geochron Cosmochron Isotope Geol Nikko Japan (Abstract)

Franklin JM, Sangster DM, Lydon JW (1981) Volcanic-associated massive sulfide deposits. Econ Geol 25th Anniv Vol: 485 – 627

Gee RD (1979) Explanatory notes on the Southern Cross 1:250,000 geological sheet, Western Australia. Rec Geol Surv West Aust 1979/5

Gee RD, Baxter JL, Wilde SA, Williams IR (1981) Crustal development in the Archaean Yilgarn Block, Western Australia. Geol Soc Aust Spec Publ 7:43 – 56

Gemuts I, Theron A (1975) The Archaean between Coolgardie and Norseman – stratigraphy and mineralization. In: Knight CL (ed), Economic geology of Australia and Papua New Guinea, I Metals. Australas Inst Min Metall, pp 66 – 74

Green AH, Naldrett AJ (1981) The Langmuir volcanic peridotite-associated nickel deposits: Canadian equivalents of the Western Australian occurrences. Econ Geol 76:1503 – 1523

Groves DI (1982) The Archaean and earliest Proterozoic evolution and metallogeny of Australia. Rev Bras Geocien 12:135 – 148

Groves DI, Phillips GN, Ho SE, Henderson CA, Clark ME, Woad GM (1984) Controls on distribution of Archaean hydrothermal gold deposits in Western Australia. In: Foster RP (ed) Gold '82: The Geology, Geochemistry and Genesis of Gold deposits. AA Balkema Rotterdam, pp 689 – 712

Hallbauer DK, Kable EJD (1982) Fluid inclusions and trace element content of quartz and pyrite pebbles from Witwatersrand conglomerates: Their significance with respect to the genesis of primary deposits. In: Amstutz GC et al. (eds) Ore genesis – the state of the art. Springer, Berlin Heidelberg New York, pp 742 – 752

Hallberg JA (1980) Archaean geology of the Leonora-Laverton area. Excursion guide northeast Yilgarn. 2nd Int Archaean Symp Perth 1980:37 p

Hargraves RB (1981) Precambrian tectonic style: a liberal uniformitarian interpretation. In: Kröner A (ed) Precambrian plate tectonics. Elsevier, Amsterdam, pp 21 – 56

Henderson JB (1975) Archaean stromatolites in the northern Slave Province. Northwest Territories, Canada. Can J Earth Sci 12:1619 – 1630

Henley RW, Ellis AJ (1983) Geothermal systems ancient and modern: A geochemical review. Earth Sci Rev 19:1 – 50

Hickman AH (1980) Excursion guide Archaean geology of the Pilbara Block. 2nd Int Archaean Symp Perth 1980 55 p

Hickman AH (1981) Crustal evolution of the Pilbara Block, Western Australia. Geol Soc Aust Spec Publ 7:57 – 70

Hodgson CJ, McGeehan PJ (1982) A review of the geological characteristics of "gold-only" deposits in the Superior Province of the Canadian Shield. Can Inst Min Metall Spec Vol 24:211 – 229

Horwitz RC (1979) The Whim Creek Group: a discussion. J R Soc West Aust 61:67 – 72

Hutchinson RW (1981) Metallogenic evolution and Precambrian tectonics. In: Kroner A (ed) Precambrian plate tectonics. Elsevier, Amsterdam, pp 733 – 760

Keays RR (1984) Archaean gold deposits and their source rocks: the upper mantle connection. In: Foster RP (ed) Gold '82: The geology, geochemistry and genesis of gold deposits. AA Balkema Rotterdam, pp 17 – 51

Kerrich R, Fryer BJ (1979) Archaean precious-metal hydrothermal systems, Dome Mine, Abitibi Greenstone Belt II. REE and oxygen isotope relations. Can J Earth Sci 16:440 – 458

Lambert IB, Groves DI (1981) Early Earth evolution and metallogeny. In: Wolf KH (ed) Handbook of strata-bound and stratiform ore deposits. Elsevier, Amsterdam 8:339 – 447

Lesher CM, Arndt NT, Groves DI (1984) Genesis of komatiite-associated nickel sulphide deposits at Kambalda: a distal volcanic-assimilation model. In: Buchanan DL, Jones ML (eds) Sulphide deposits in mafic and ultramafic rocks. Inst Min Metal Lond Spec Publ (to be published)

Lowe DR (1980) Stromatolites 3400-Myd old from the Archean of Western Australia. Nature 284:441 – 443

Lowe DR (1982) Comparative sedimentology of the principal volcanic sequences of Archean green-stone belts in South Africa, Western Australia and Canada: implications for crustal evolution. Precambrian Res 17:1 – 29

Lowe DR, Knauth LP (1977) Sedimentology of the Onverwacht Group (3.4 billion years), Transvaal, South Africa, and its bearing on the characteristics and evolution of the early Earth. J Geol 85:699 – 723

Lowe DR, Knauth LP (1978) The oldest marine carbonate ooids reinterpreted as volcanic accre-tionary lapilli, Onverwacht Group, South Africa. J Sediment Petrol 49:664 – 666

Marston RJ, Groves DI, Hudson DR, Ross JR (1981) Nickel sulfide deposits in Western Australia: a review. Econ Geol 76:1330 – 1363

McCulloch MT, Compston W (1981) Sm-Nd age of Kambalda and Kanowna greenstones and hetero-geneity in the mantle. Nature 294:322 – 327

McGeehan PJ, MacLean WH (1980) Tholeiitic basalt-rhyolite magmatism and massive sulphide de-posits at Matagami, Quebec. Nature 283:153 – 157

Nash JT, Granger HC, Adams SS (1981) Geology and concepts of genesis of important types of uranium deposits. Econ Geol 25th Anniv Vol: 63 – 116

Nesbitt RW, Shen-Su Sun, Purvis AC (1979) Komatiites: geochemistry and genesis. Can Mineral 17:165 – 186

Nisbet EG (1982) The tectonic setting and petrogenesis of komatiites. In: Arndt NT, Nisbet EG (eds) Komatiites. George Allen and Unwin, London, pp 501 – 520

Pretorius DA (1981) Gold and uranium in quartz-pebble conglomerates. Econ Geol 25th Anniv Vol: 117 – 138

Reimer TO (1980) Archaean sedimentary baryte deposits of the Swaziland Supergroup (Barberton Mountain Land, South Africa). Precambrian Res 12:393 – 410

Ross JR, Travis GA (1981) The nickel sulfide deposits of Western Australia in global perspective. Econ Geol 76:1291 – 1329

Smith HS, O'Neil JR, Erlank AJ (1984) Oxygen isotope compositions of minerals and rocks and chemical alteration patterns in pillow lavas from the Barberton greenstone belt, South Africa. This volume, 115 – 137

Solomon M, Walshe JL (1979) The formation of massive sulfide deposits on the sea floor. Econ Geol 74:797 – 813

Spooner ETC, Fyfe WA (1973) Sub-sea floor metamorphism, heat and mass transfer. Contrib Mineral Petrol 42:287 – 304

Stanistreet IG, de Wit MJ, Fripp REP (1981) Do graded units of accretionary spheroids in the Barberton Greenstone Belt indicate Archaean deep water environment. Nature 293:280 – 284

Walter MR, Buick R, Dunlop JSR (1980) Stromatolites 3400 – 3500 Myd old from the North Pole area, Western Australia. Nature 284:443 – 445

Watson J (1976) Mineralization in Archaean provinces. In: Windley BF (ed) The early history of the Earth. Wiley, London, pp 443:5454

Williams DAC (1979) The association of some nickel sulfide deposits with komatiitic volcanism in Rhodesia. Can Mineral 17:337 – 350

Windley BF (1977) The evolving continents. Wiley, London, 385 p

Woodall R (1979) Gold – Australia and the world. Geol Dept Extension Service Univ West Aust Publ 3:1 – 34

Magma Mixing in Komatiitic Lavas from Munro Township, Ontario

N. T. Arndt[1] and R. W. Nesbitt[2]

Contents

1 Introduction 100
2 Location of Samples 101
3 Analytical Methods 105
4 Results 105
5 Compositions of Erupted Liquids 107
6 Main Problems in the Genesis of Munro Komatiitic Lavas 108
7 Textural Features Consistent with Magma Mixing 110
8 The Extent of Magma Mixing in the Munro Volcanic Succession 113
References 113

Abstract

Komatiites at Munro Township, northeast Ontario, show greater LREE depletion and have lower ratios of highly to moderately incompatible elements (e.g. Ti/Sc) than associated komatiitic basalts. These differences indicate that the two magma types are not related to one another by low pressure fractional crystallization: they formed either from mantle sources with slightly different compositions, or from the same source under different conditions of partial melting or high pressure fractionation.

In some situations these two magma types have mixed together to form hybrid magmas. The best example is Fred's Flow, a thick mafic-ultramafic layered unit that has chemical characteristics intermediate between those of the komatiites and komatiitic basalts. Textural evidence for mixing is found in the flow top breccia which contains two types of fragment, one with komatiitic composition and the second with basaltic composition. Particularly significant are augite phenocrysts in the breccia, which have compositions indicating that they could not have crystallized from the liquid that formed the bulk of Fred's Flow.

Magma mixing may also have played a role in the formation of komatiitic basaltic flows with acicular pyroxene textures and komatiites which contain anomalously Fo-rich olivine xenocrysts or unusually high concentrations of incompatible trace elements.

1 Max-Planck-Institut für Chemie, Postfach 3060, 6500 Mainz, FRG
2 Department of Geology, The University, Southampton SO9 5NH, England

Archaean Geochemistry (ed. by A. Kröner et al.)
© Springer-Verlag Berlin Heidelberg 1984

1 Introduction

The petrology and geochemistry of the komatiites and basalts of Munro Township, Ontario, have been described in detail by Pyke et al. (1973), Arndt et al. (1977), Arndt (1977), Arth et al. (1977), Sun and Nesbitt (1978), Whitford and Arndt (1977), Basaltic Volcanism Study Project (1981) and Arndt and Nesbitt (1982). In the last paper, detailed analyses of the basalts allowed a preliminary interpretation to be made the petrological processes that had influenced the compositions of these rocks. It was shown that there existed, within a relatively thin (700 m) volcanic pile, representatives of magmas from at least three chemically distinct mantle sources. The principal characteristics of these magmas are given in Fig. 1 and Table 1. Two of the magma types, the Fe-rich tholeiites and olivine porphyries, come from distinctive mantle sources, one very rich in FeO and in-

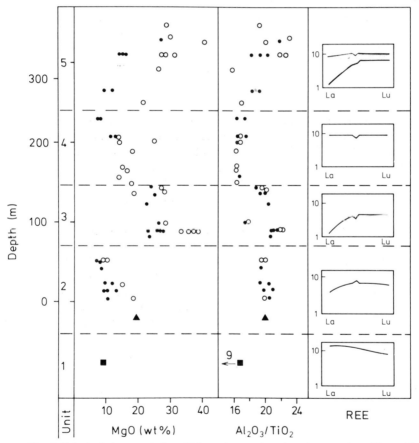

Fig. 1. Principal characteristics of Munro lavas, plotted against stratigraphic position. ● Lavas representative of erupted liquid compositions; ○ cumulate and evolved lavas; ▲ Fred's Flow erupted liquid; ■ erupted liquid of Theo's Flow, a thick, layered Fe-rich tholeiite flow. Note that the stratigraphic depth scale applies only to the komatiitic lavas, not to Fred's Flow and underlying tholeiites. Data from Arndt and Nesbitt (1982), this study, and Arndt (unpubl.)

Table 1. Element and oxide ratios for the main magma types in Munro Township

	Unit 1 Fe-rich tholeiites	Unit 2 Komatiitic basalts	Fred's Flow	Unit 3 Komatiites	Unit 4 Olivine porphyries	Unit 5 Komatiites, komatiitic basalts	Chondrite
MgO	5 – 14	7 – 15	16 – 19	22 – 28	7 – 15	7 – 28	–
Al_2O_3/TiO_2	7 – 9	20	20	20.5	17	16 – 22	20
Ti/Zr	95	120	125 – 130	120 – 125	87	95 – 125	120
Zr/Y	3 – 4	2	2	2	2.5	1.7 – 2.0	–
(Sc/Ti) × 1000	3 – 4	10 – 11	11 – 12	12 – 14	7 – 8	10 – 12	1.3
$(Sm/Ce)_N$	0.9 – 1.0	1.31 – 1.34	1.34 – 1.5	1.87 – 1.94	0.9 – 1.1	2.09	1
$(Yb/Gd)_N$	0.7 – 1.0	0.9 – 1.0	0.9 – 1.0	1.0	0.9 – 1.0	0.9 – 1.0	1

compatible elements, and the other with low FeO and high Ni, Cr and incompatibles (Table 1). These magmas are probably unrelated to the komatiites and komatiitic basalts that are the main subject of this paper, and little further mention will be made of them. The komatiites and komatiitic basalts share similar chemical characteristics and were believed to have come from a third source.

At the time the last paper was written (Arndt and Nesbitt 1982), few trace element data for komatiites from the northern part of the Township were available, and the relationship between komatiites and the analyzed basalts (all of which came from the northern part of the Township) could only be inferred using previously published data from the komatiites of Pyke Hill in the centre of the Township. Because of lack of intervening outcrop, correlation between the two areas was uncertain. On the basis of the available information it was suggested that the komatiitic basalts were derived from komatiite parent magmas by low pressure fractional crystallization.

In this paper we present new analyses of the komatiites from the north of the Township and additional analyses of komatiitic basalts. We then discuss the petrogenesis of these rock types, using field observations and textural data to deduce the compositions that the liquids had at the time they were extruded before they were affected by post-eruptive fractionation. Using this information, we then show that two different magmas were erupted, an ultramafic magma with about 25% MgO and a basaltic magma with about 11% MgO. The data suggest that these magmas were not related to one another by low pressure fractionation, and that mixing between these magmas played a small but significant role in the formation of these rocks.

2 Location of Samples

Figure 1 is a modified version of the stratigraphic profile that has appeared in a number of earlier papers (e.g. Arth et al. 1977; Arndt and Nesbitt 1982). The figure shows the relationship between the various units that make up the stratigraphic sequence in the northern part of Munro Township and the main geochemical characteristics of each.

Table 2a. Major elements in Munro lavas (analyses normalized to 100% on a water-free basis)

	Komatiites							Komatiitic basalts						
	C68	C18	M620	C17	M625	M610	C37	C84A	C87A	C9	C85A	C80	C126	C125
SiO_2	45.56[c]	44.17	45.60	44.99	45.40	46.95	45.56	47.30	49.60	51.00	51.60	50.77	51.02	50.63
TiO_2	0.33	0.35	0.40	0.45	0.45	0.52	0.31	0.55	0.64	0.68	0.65	0.75	0.61	0.63
Al_2O_3	7.48	7.98	7.95	9.26	9.30	10.10	7.45	11.30	13.00	13.16	12.90	13.94	12.20	13.20
Fe_2O_3[a]	13.02	13.54	12.66	13.33	12.70	12.70	12.06	14.00	13.10	12.80	12.70	14.00	11.60	11.30
MnO	0.24	0.23	0.20	0.23	0.19	0.20	0.21	0.23	0.19	0.21	0.19	0.23	0.19	0.21
MgO	27.24	26.99	25.00	23.70	22.40	18.90	26.82	15.20	10.60	9.85	9.50	7.28	12.20	9.80
CaO	6.34	5.81	7.60	7.54	9.03	9.50	7.10	9.20	10.90	10.10	10.30	9.43	8.50	10.20
Na_2O	0.11	0.09	0.01	0.10	0.01	0.76	0.20	1.80	1.80	1.37	1.79	3.56	3.39	3.54
K_2O	0.11	0.06	0.02	0.04	0.05	0.01	0.07	0.13	0.06	0.69	0.18	0.07	0.09	0.15
P_2O_5	0.02	0.02		0.04			0.03			0.06		0.07	0.07	0.06
LOI[b]	7.31	7.61		6.62			7.89	3.59		2.24		2.37	2.55	1.94
Cr	1322[c]	2944	2900	3210	2806	2046	2285	1100	553	530	595	173	1089	595
Ni	1200	1200	1220	909	808	690	1199	276	130	89	85	68	285	171

	Olivine porphyries								Fred's Flow			
	M614	M621	M612	M622	M613	M615	C76	C31	C11	C123	C55	C6
SiO_2	47.00	47.80	49.10	50.00	48.70	50.90	49.29	52.05	45.85	50.1	46.68	50.54
TiO_2	0.52	0.64	0.63	0.60	0.68	0.72	0.71	0.88	0.49	0.42	0.55	0.82
Al_2O_3	8.70	10.40	10.30	9.90	11.50	12.00	12.20	14.40	9.36	8.39	10.84	14.08
Fe_2O_3[a]	12.00	12.80	11.60	10.90	11.70	11.50	10.59	11.20	13.07	11.78	13.64	13.90
MnO	0.16	0.20	0.17	0.16	0.19	0.17	0.17	0.17	0.24	0.24	0.20	0.22
MgO	24.70	17.90	17.70	16.60	15.00	13.90	13.01	7.50	19.26	14.71	16.35	7.27
CaO	6.40	8.20	8.40	9.80	10.10	7.60	11.07	10.50	10.85	13.57	10.23	9.51
Na_2O	0.01	1.60	1.63	1.50	1.82	2.90	2.60	3.09	0.18	0.31	1.16	3.22
K_2O		0.02					0.03	0.09	0.03	0.03	0.06	0.27
P_2O_5							0.08	0.09	0.03	0.04	0.05	0.07
LOI[b]							3.59	2.76	4.07	3.59	4.34	2.04
Cr	2585	1890	1750	1750	1302	1360	1746	294	3000	2058	3270	149
Ni	1230	804	765	765	465	574	509	78	758	461	481	46

[a] Total iron as Fe_2O_3; [b] Loss on ignition; [c] Major elements in wt %, trace elements in ppm
Samples M610 to 625 and C84A to C87A analyzed at Max-Planck-Institut für Chemie; others analyzed at University of Adelaide.

Sample Descriptions

Komatiites

C68, C18, M620, C17, M625, M610: Komatiites from Unit 3. All samples are spinifex-textured. Samples M620 and M625 have fine random spinifex textures; the others have medium to coarse plate spinifex textures. In all samples but C68 olivine is entirely replaced by chlorite and/or serpentine; augite and chromite are fresh; and the glass is replaced by fine-grained hydrous minerals. In C68 about 20% of olivine has escaped alteration.
C37: Komatiite from Unit 5. Texturally similar to C18 but more highly altered. In C37 augite as well as olivine is replaced by chlorite and amphibole.

Komatiitic Basalts

C84A, C87A, C9, C85A, C80: Komatiitic basalts from Unit 2. Samples C9 and C80 are evolved augite porphyritic basalts. The others contain randomly oriented augite needles in a matrix of finer augite and altered glass. C84A contains about 20% small equant olivine phenocryst (now chloritized) in addition to acicular pyroxenes.
C126, C125: Komatiitic basalts from Unit 5. C125 contains spinel-shaped augite grains in a matrix of augite, plagioclase and iron oxides. C126 is pyroxene spinifex-textured. In both samples augite is fresh but other phases are altered.

Olivine Porphyries

C76, M612 – 615, M621 – 622: Olivine porphyritic basalts, Unit 4. Equant solid to skeletal olivine phenocrysts in a matrix of skeletal olivine, fine chromite grains and altered glass. All samples contain appreciable solid, presumably accumulative olivine and probably do not represent liquid compositions. In all samples olivine is completely chloritized.
C31: Evolved basalt, Unit 4. Augite phenocrysts in an augite, plagioclase, iron oxides matrix. Pyroxene is fresh but other phases are altered.

Fred's Flow

C11: Olivine porphyritic lava from 2 m below flow top. Olivine is chloritized.
C123: Fine-grained hyaloclastite breccia, 50 cm below flow top. All primary phases altered. Further description in text.
C55: Random olivine spinifex texture. Olivine completely chloritized.
C6: Gabbroic textured lava from 25 m below flow top. Augite, plagioclase, minor quartz and iron oxides.

Table 2b. Trace elements in Munro lavas

	Komatiites				Komatiitic basalts			
	C68	C18	C17	C37	C9	C80	C126	C125
MgO	27.24	26.99	23.70	26.82	9.85	7.28	12.20	9.80
Rb	6.2	4.5	2.3	0.90	15.5	2.8	1.8	3.0
Sr	10.7	6.3	5.3	23.4	112	83.0	73.0	161
Ba	7.0	6.0	8.0	10.0	82.0	41.0	102	142
Zr	16.1	18.1	21.3	13.7	33.0	43.0	38.0	37.0
Nb	0.30	0.60	1.2	0.50	1.5	1.8	1.8	1.4
Y	8.1	7.3	9.6	7.8	16.0	20.0	17.0	18.0
Sc	25.1	28.0	34.0	26.1	43.0	41.0	38.0	41.0
V	147	154	190	141	259	270	237	269
Cr		2944	3210	2285	530	173	1089	595
Ni	1332	1200	909	1199	89.0	68.0	285	171
La					1.47	2.07	1.70	1.88
Ce		1.73	1.85	1.16	4.54	5.81	4.88	5.27
Nd		1.80	2.03	1.39	3.74	4.94	4.17	4.48
Sm		0.76	0.85	0.57	1.37	1.78	1.53	1.67
Eu		0.25	0.28	0.26	0.61	0.67	0.55	0.63
Gd		1.20	1.32	0.96	2.13	2.60	2.35	2.50
Dy		1.46	1.66		2.61	3.20	2.78	3.15
Er			1.12	0.84	1.67	2.07	1.79	2.03
Yb		0.94	1.08	0.78	1.63	2.04	1.73	1.94
Lu		0.14			0.24	0.31	0.26	0.29

	Olivine porphyries			Fred's Flow			
	M612	C76	C31	C11	C123	C55	C6
MgO	17.70	13.01	7.50	19.26	14.71	16.35	7.27
Rb		1.0	2.7	2.5	2.0	3.0	6.5
Sr		43.0	98.0	4.4	16.0	53.0	98.0
Ba		12.0	72.0	13.0	11.0	24.0	63.0
Zr		48.0	63.0	22.7	20.0	26.0	41.0
Nb		1.1	3.8	1.5	1.5	0.20	2.7
Y		19.0	24.4	12.5	11.0	14.0	20.0
Sc		36.0	36.0	29.0	28.0	36.0	39.0
V		225	259	175	160.0	217	263
Cr		1746	294	3000	2058	3270	149
Ni		509	78.0	758	461	481	46.0
La		2.49			0.80		
Ce	6.70	6.36	11.40	2.79	2.38	3.51	5.00
Nd	4.92	5.14	7.36	2.77	2.20	3.18	4.24
Sm	1.57	1.76	2.25	1.05	0.84	1.22	1.60
Eu	0.48	0.57	0.74	0.41	0.46	0.44	0.63
Gd	2.15	2.51	2.98	1.55	1.26	1.55	2.76
Dy	2.63	2.97	3.49	1.89	1.54	2.31	3.28
Er	1.69	1.90	2.22	1.20	0.99	1.48	2.05
Yb	1.64	1.86	2.32	1.13	0.95	1.42	1.94
Lu	0.25	0.28	0.40		0.143		

The composition of Fred's Flow, a thick, layered komatiitic unit at the base on Unit 2, is represented by the average composition of the flow top breccia and uppermost spinifex rocks. The stratigraphic section had previously been divided into a number of 'cycles' within which the chemical composition seemed to change in a systematic manner, usually from more to less mafic. The results of Arndt and Nesbitt (1982) showed, however, that some cycles contained more than one magma type and that the apparently systematic compositional variation, from ultramafic to mafic, was largely fortuitous. In this paper we use the term unit rather than cycle and divide the stratigraphic section into five units, numbered from the base upward. In this system Unit 1 corresponds to the tholeiitic unit of Arndt and Nesbitt (1982), Unit 2 to Cycle 1, Unit 3 to the lower part of Cycle II, Unit 4 to the upper part of Cycle II, and Unit 5 to Cycle III.

Figure 1 contains analyses of 18 new samples, these being komatiites in Units 3 and 5 and basalts in Unit 5.

3 Analytical Methods

The major and trace element compositions of most samples were determined at the University of Adelaide using XRF techniques described by Nesbitt and Stanley (1980). Certain samples were also analyzed by XRF at the Australian National University using the method of Norrish and Chappel (1967). The newly collected samples M610 − 625 were analyzed for major elements at the Max-Planck-Institut für Chemie (MPI) in Mainz using the XRF method of Palme and Jagoutz (1977). Rare earth elements were analyzed by isotope dilution at MPI with the method described by White and Patchett (1984); other trace element concentrations were measured using INAA with the method of Wänke et al. (1977). The new analyses and their sources are listed in Table 2.

4 Results

The new analytical data confirm many of the results of the earlier study (Arndt and Nesbitt 1982). These results are summarized in the introduction and in Table 1; for more detailed information the reader is referred to the earlier publication. Important observations that can be drawn from the new data are:

a) The komatiites of Units 3 and 5 have essentially the same geochemical characteristics as the komatiites of Pyke Hill. Both have strongly depleted LREE (Fig. 2a) and low concentrations of other incompatible elements, but roughly chondritic ratios of elements such as Al, Ti, Zr, and HREE.

b) The komatiitic basalts of Units 2 and 5 have chemical characteristics that are broadly similar to the komatiites, i.e. depleted LREE, $Al_2O_3/TiO_2 = 20$, $Ti/Zr = 120$. However, the extent of LREE depletion in the basalts is significantly less than that in the komatiites ($(Sm/Ce)_N = 0.7$ cf. 0.5; Fig. 2b), and Sc/Ti ratios are higher ($90 - 100$ cf. $70 - 80$).

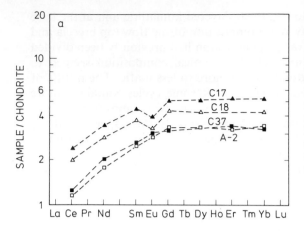

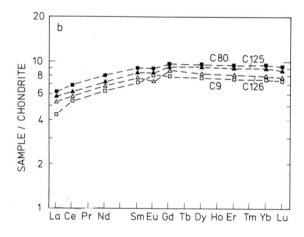

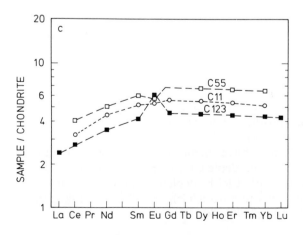

Fig. 2a – c. Rare earth element concentrations, normalized against chondritic values, for three magma types: **a** komatiites of Unit 3 (*C17* and *C18*) and Unit 5 (*C37*). *A-2* is an analysis of a komatiite from Pyke Hill (from Arth et al. 1977); **b** komatiitic basalts of Unit 2 (*C9* and *C80*) and Unit 5 (*C125* and *C126*); **c** flow top breccias and spinifex lavas from Fred's Flow. Analyses of *C11* and *C55* are from Whitford and Arndt (1977); (see Table 2 caption for further description of the samples)

c) The lava of Fred's Flow has a composition intermediate between that of the komatiites and komatiitic basalts ($(Sm/Ce)_N = 0.6$; $Sc/Ti = 90$).

5 Compositions of Erupted Liquids

Much of the compositional variation within the lavas is probably caused by post-eruption processes. These include crystal settling which forms cumulates in the lower parts of flows, in situ crystal growth and accumulation which produces the parallel spinifex textures that commonly contain a cumulus component, and within-flow fractional crystallization which produces more evolved liquids in the flow centres (Donaldson 1982; Lesher et al. 1983). These processes operate most efficiently in the more magnesian, low viscosity liquids such as those that erupted at the start of each magmatic pulse at Munro Township. The preferential formation of cumulate rocks enriched in mafic minerals in the lower part of each unit has the effect of exaggerating the extent of compositional variation within each unit.

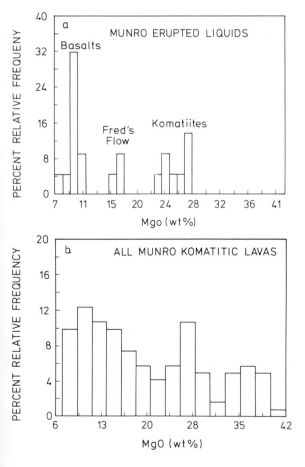

Fig. 3a, b. Histogram showing frequency of (a) lavas representative of erupted liquids and (b) all Munro lavas

If field data and textural criteria are used to eliminate both the cumulate rocks formed by crystal settling or crystal growth and the locally differentiated portions of lava flows, a far more restricted range of compositions is revealed. The compositions, as shown in Figs. 1 and 3 and Table 1, fall into two relatively narrow groups: komatiitic basalts with 7 – 12% MgO and komatiites with 23 – 27% MgO. These compositions probably represent the compositions that the magmas had at the time they were erupted. The only major exception is Fred's Flow in which the initial liquid had 16 – 18% MgO, intermediate between that of the komatiites and komatiitic basalts.

6 Main Problems in the Genesis of Munro Komatiitic Lavas

In the foregoing discussion two main problems were raised. The first is to explain why the lavas should have erupted with relatively restricted compositions, and particularly why there should be such a large gap between the compositions of komatiites and basalts. The second is to explain what was the origin of the two main lava types: whether the basalts were derived from parental komatiite magmas or whether the two were independently derived primary magmas.

Part of the explanation may lie in the ideas proposed by Nisbet and Chinner (1981) for komatiitic lavas of the Ruth Well area in Western Australia. These authors suggest that komatiites (25 – 28% MgO) are primary magmas which erupted essentially unmodified from their mantle source, whereas komatiitic basalts (8 – 11% MgO) are products of fractionation in high level magma chambers. Following the model of Sparks et al. (1980) and Stolper and Walker (1980), they suggest that when a magma chamber exists, it traps the dense primary komatiite magmas; only the fractionation products of the komatiites, which have basaltic compositions, are likely to erupt. The primary komatiite magmas reach the surface only when a magma chamber did not exist. Nisbet and Chinner (1981) show that the most abundant basaltic composition at Ruth Well corresponds to the minimum density for basaltic compositions, and propose this as a factor enhancing the probability of eruption of basalts with 8 – 10% MgO.

Much the same applies to the Munro lavas: the komatiites probably are primary, and the composition of the most commonly basalts coincides with the density minimum. Where the situation differs is in the relationship between the two magma types. In Fig. 4 ratios of $(Sm/Ce)_N$, $(Sm/Nd)_N$, and Sc/Ti are plotted against MgO contents. Each of the trace elements plotted is highly incompatible, and the ratios should not be changed by olivine fractionation. Experimental studies of komatiitic magmas have shown that at 1 atm only olivine and minor chromite crystallize in liquids with MgO contents between 30% and 9% (Arndt 1976). Clinopyroxene fractionation, which could change the ratios, would be expected only in less magnesian liquids. As can be seen from Figs. 4 and 5, the basalts have ratios distinctly different from those of the komatiites, effectly eliminating the possibility that the two magma types are related by low pressure fractionation. The magmas could have been independently derived by melting of sources with different compositions (differing degrees of initial depletion), or they could have

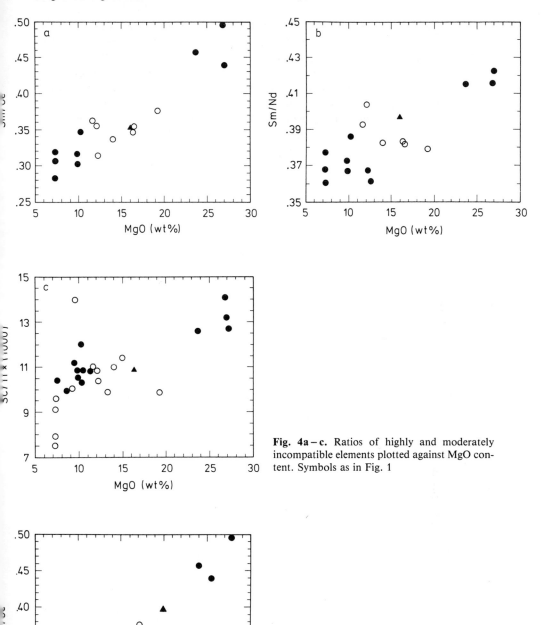

Fig. 4a–c. Ratios of highly and moderately incompatible elements plotted against MgO content. Symbols as in Fig. 1

Fig. 5. Sm/Ce vs Sc/Ti. Symbols as in Fig. 1

formed from the same depleted source, with some pyroxene being left as a residual phase or fractionating at high pressures, during the formation of the basalts.

The magma that erupted to form Fred's Flow, as represented by the uppermost breccia and random spinifex rocks, has $(Sm/Ce)_N$, $(Sm/Nd)_N$, and Sc/Ti ratios intermediate between those of the komatiite and basalt. As mentioned before the bulk composition also lies between those of the two more common rock types. This magma could have been produced from a source of intermediate composition or by melting with somewhat less residual pyroxene than in the basalt case; but a third explanation exists, namely that this magma is a mixture of the two other types.

7 Textural and Chemical Features Consistent with Magma Mixing

All magmas at Munro are relatively magnesian and erupted at high temperatures. They would have had low viscosities and, if brought into contact with one another, would have mixed efficiently leaving no trace of their original composite nature. Mixtures of glasses with contrasting compositions, which provide clear evidence of mixing in more viscous magmas, would only be preserved under unusual conditions. Most of the textural and mineralogical evidence of mixing between komatiitic magmas is of a more subtle form, as is described below. Some of these observations support mixing between comagmatic magmas with similar compositions, but other data suggest mixing between unrelated magmas such as the types that formed the Fred's Flow lava. The relevant observations are:

a) Many komatiitic basalts with MgO contents between 10 and 7% have augite porphyritic textures. According to Walker et al. (1979) this texture results from mixing of liquids with compositions on the curved olivine-clinopyroxene cotectic such that the resultant mixed liquid composition falls in the pyroxene field.

b) More magnesian basalts (10 – 12% MgO) contain randomly oriented or aligned pyroxene needles (Arndt et al. 1977). This texture and mineralogy suggests crystallization from nucleus-free, superheated liquids such as would be produced by mixing between liquids in the olivine field and liquids on the olivine-pyroxene-plagioclase cotectic. This situation is illustrated in Fig. 6. It should be pointed out at this stage that many samples with aligned pyroxenes and no olivine (the string beef pyroxene spinifex textures) have MgO contents up to 16%. This led Campbell and Arndt (1982) to suggest that these pyroxenes had crystallized metastably from a liquid in the olivine primary phase field. It now seems more likely that these rocks do not represent liquid compositions but form by accumulation of pigeonite and augite during the growth of the spinifex textures, and that the apparently premature pyroxene crystallization can be largely explained by the pyroxene accumulation.

c) The basal parts of many komatiite flows contain crystals of olivine that are too forsteritic to have crystallized from the liquid that formed the bulk of the flow. One example is given in Fig. 11 of Arndt et al. (1977). In the flow M2, olivines at the base of the B_4 zone have the composition Fo_{94-95}, too magnesian to have

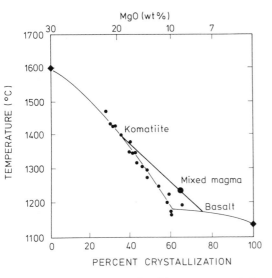

formed from the liquid represented by the spinifex layer. This liquid had 25 – 26% MgO and would have crystallized olivine no more magnesian than $Fo_{93.5}$ (assuming $KD_{Mg-Fe} = 0.30$ and no Fe^{3+} in the liquid). Other, clearer examples are found in komatiitic basalt flows from Gilmour Island, Hudson Bay, where the erupted liquids had no more than 16% MgO, and the olivine xenocrysts have the composition Fo_{92-93} (Arndt unpubl. data). These observations suggest mixing between komatiitic liquids with MgO contents between 15 and 30%.

d) The flow top breccia of Fred's Flow is made up of small fragments which have a variety of textures: some are microspinifex, others are composed entirely of chloritized glass, and others are porphyritic and contain small (∼0.1 mm) phenocrysts of olivine and augite. The pyroxene phenocrysts have compositions similar to cumulus augite grains in lower parts of the flow (Table 3), indicating that they crystallized at low pressure and were not formed during an earlier episode of high-pressure crystallization. Similar augite grains are found in the basal border zone of the flow. Whole rock analyses of the flow top breccia show between 15 and 19% MgO (samples C123, and C11, Table 2), similar to that of the underlying olivine spinifex-textured part of the flow. Individual fragments in the breccia, as analyzed with defocussed-beam microprobe, have MgO contents between 16 and 20% (Table 3). In the case of the olivine spinifex-textured fragments the MgO contents are consistent with their olivine-rich mineralogy. In the case of the more glassy augite prophyritic fragments the glass is completely chloritized, a process that leads to loss of SiO_2 and CaO and gain of MgO. The original MgO content of glass in these fragments probably was lower, in the range of 8 – 10% MgO, which would be consistent with the crystallization of augite (Arndt 1976; Kinzler and Grove 1983).

As explained before, augite can crystallize at low pressures only in liquids with less than about 9% MgO. The presence of augite phenocrysts in the breccia therefore provides clear evidence that the lava that initially erupted in Fred's

Table 3. Compositions of phenocrysts and altered glasses in Fred's Flow breccia and in underlying cumulates

	1	2	3	4
SiO_2	53.4	51.9	52.5	48.8
TiO_2	nd	0.37	0.20	0.75
Al_2O_3	1.59	4.6	2.53	11.4
Cr_2O_3	0.44	0.23	0.81	nd
FeO(t)	6.7	7.7	5.91	13.1
MnO	0.18	0.25	0.15	nd
MgO	18.9	17.8	18.4	14.0
CaO	18.2	17.3	19.7	10.2
Na_2O	nd	0.14	0.11	1.89
K_2O	nd	nd	nd	nd
(H_2O)				(9.8)
	99.35	100.2	100.5	100.0
Mg/(Mg + Fe)	0.83	0.81	0.85	0.66

1 Augite phenocryst in basaltic fragment in flow top breccia.
2 Augite mantle of pyroxene needle in spinifex-textured lava.
3 Cumulus augite from cpx-opx cumulate in lower part of flow.
4 Defocussed beam microprobe analysis of relatively unaltered glassy portion of fragment. Analyses
  4 and 5 are recalculated to 100% water-free.

Analyses 1 and 4 carried out with an ARL-SEMQ electron microprobe with KEVEX 5100 energy dispersive spectrometer at Max-Planck-Institute in Mainz under the following operating conditions: 15 KV accelerating voltage, 5.0 nanoamps sample current on pure copper, beam size between 5 and 100 μm. Calibration samples are a range of pure metals, simple oxides and silicates. Analyses 2 and 3 are from Arndt and Fleet (1979).

Flow consisted of two liquids, one with a composition approaching that of the primary komatiite, and another with a composition similar to that of the basalts.

e) Kinzler and Grove (1983) have suggested that mixing also took place at a later stage in the crystallization of Fred's Flow. They explain orthopyroxene cumulates by mixing between a low density evolved liquid which accumulates beneath the spinifex layer and a denser, less evolved liquid in the lower part of the flow. Further evidence of internal mixing is found in the compositions of the most evolved gabbroic-textured parts of the flow (e.g. sample C5). These have abundances of incompatible elements and ratios of highly to moderately incompatible elements (e.g. Sm/Nd) that are consistently too high to be explained by simple fractional crystallization. Computer modelling of the crystallization indicates that the amount of olivine, pyroxene and plagioclase fractionation necessary to give the bulk composition of sample C5 would produce the following compositions: Ti = 0.77%, Zr = 37 ppm and Sm/Nd = 0.375. These values compare with TiO_2 = 0.82%, Zr = 44 ppm, and Sm/Nd = 0.368 in the rock itself. The discrepancy can be explained by repeated mixing in the lava flow between low density evolved liquid and incoming less evolved liquid, in much the same way as is advocated by O'Hara (1977) in the case of larger intrusive magma chambers.

8 The Extent of Magma Mixing in the Munro Volcanic Succession

As stated earlier the Munro lavas erupted as a number of discrete magma types, each with its own distinctive chemical characteristics. These magmas appear to have been derived from different mantle sources and to have erupted largely independently of one another. Most of the evidence of magma mixing described above applies to co-magmatic magmas of the same magma type within the same unit, such as the komatiitic basalts of Unit 2 or the komatiites of Unit 3. Throughout the entire volcanic succession there are only a few cases of mixing between the different magma types.

These are as follows: (a) at the top of Unit 3 magma of the Ti-enriched, olivine porphyritic Unit 4 type appears to have mixed with the komatiite magma, resulting in a slight but progressive decrease in Al_2O_3/TiO_2 ratios in the uppermost komatiites (Fig. 1). (b) In Unit 5 the lavas show a range of chemical characteristics that spans the range between komatiite and Ti-enriched olivine porphyry, the magmas in the underlying two units. In this unit magmas with compositions approaching those of the two end members, as well as lavas with intermediate compositions, erupted in a more or less sporadic manner. It appears that in this case magmas from two sources erupted simultaneously, in some cases mixing to form hybrid magmas. (c) As described above, Fred's Flow was formed from a mixture of komatiite like that of Unit 3 and komatiitic basalt of Unit 2.

Acknowledgements. We acknowledge analytical assistance or complete analyses provided by B. W. Chappel at the Australian National University, J. Stanley at the University of Adelaide, and B. Spettel, H. Palme, and A. Burghele at the Max-Planck-Institut für Chemie in Mainz. Earlier versions of the paper were reviewed by A. W. Hofmann, T. L. Grove, J. Longhi, S. S. Sun, and E. Stolper.

References

Arndt NT (1976) Melting relations of ultramafic lavas (komatiites) at one atmosphere and high pressure. Carnegie Inst Wash Year Book 75:555 – 562

Arndt NT (1977) Thick, layered peridotite-gabbro lava flows in Munro Township, Ontario. Can J Earth Sci 14:2620 – 2637

Arndt NT, Fleet ME (1979) Stable and metastable pyroxene crystallization in layered komatiitic lava flows. Am Mineral 64:856 – 864

Arndt NT, Naldrett AJ, Pyke DR (1977) Komatiitic and iron-rich tholeiitic lavas in Munro Township, northeast Ontario. J Petrol 18:319 – 369

Arndt NT, Nesbitt RW (1982) Geochemistry of Munro Township basalts. In: Arndt NT and Nisbet EG (eds) Komatiites. George Allen and Unwin, London, pp 309 – 330

Arth JG, Arndt NT, Naldrett AJ (1977) Genesis of Archean komatiites – trace element evidence from Munro Township, northeast Ontario. Geology (Boulder) 5:590 – 594

Basaltic Volcanism Study Project (1981) Basaltic volcanism in the terrestrial planets. Pergamon, New York, 1286 pp

Campbell IH, Arndt NT (1982) Pyroxene accumulation in spinifex rocks. Geol Mag 119:605 – 610

Donaldson CH (1982) Spinifex-textured komatiites: a review of textures, mineral compositions and layering. In: Arndt NT, Nisbet EG (eds) Komatiites. George Allen and Unwin, London, pp 211 – 244

Green DH, Nicholls IA, Viljoen MJ, Viljoen RP (1975) Experimental demonstration of the existence of peridotitic liquids in earliest Archean magmatism. Geology (Boulder) 3:11 – 15

Kinzler RJ, Grove TL (1983) Crystallization and differentiation of Archean komatiite lavas from northeast Ontario: phase equilibrium and kinetic studies. Am Mineral (to be published)

Lesher CM, Arndt NT, Groves DI (1983) Emplacement of nickel sulphide deposits at Kambalda: A distal volcanic model: Buchanan DL, Jones MJ (eds) Sulphide Deposits in Mafic and Ultramafic Rocks. Inst Mining Metall Spec Issue (to be published)

Nesbitt RW, Stanley J (1980) Compilation of analytical geochemistry reports. Centre for Precambrian Research, Research Report 3, University of Adelaide.

Nisbet EG, Chinner GA (1981) Controls on the eruption of mafic and ultramafic lavas. Ruth Well Cu-Ni prospect, West Pilbara. Econ Geol 76:1729–1735

Norrish K, Chappel BW (1967) X-ray fluorescence spectrography. In: Zussman J (ed) Physical Methods in Determinative Mineralogy. pp 161–214

O'Hara MJ (1977) Geochemical evolution during fractional crystallization of a periodically refilled magma chamber. Nature 266:503–507

Palme C, Jagoutz E (1977) Application of the fundamental parameter method for the determination of major and trace elements on fused geological samples with X-ray fluorescence spectrometry. An Chem 49:717–722

Pyke DR, Naldrett AJ, Eckstrand OR (1973) Archean ultramafic flows in Munro Township, Ontario. Geol Soc Am Bull 84:955–978

Sparks RSJ, Meyer P, Sigurdsson H (1980) Density variation amongst mid-ocean ridge basalts: Implications for magma mixing and the scarcity of primitive lavas. Earth Planet Sci Lett 46:419–430

Stolper EM, Walker D (1980) Melt density and the average composition of basalt. Contrib Mineral Petrol 74:7–12

Sun SS, Nesbitt RW (1978) Petrogenesis of Archean ultrabasic and basic volcanics: evidence from rare earth elements. Contrib Mineral Petrol 65:301–325

Walker D, Shibata T, Delong SE (1979) Abyssal tholeiites from the Oceanographer fracture zone, II. Phase equilibria and mixing. Contrib Mineral Petrol 70:111–125

Wänke H, Kruse H, Palme H, Spettel B (1977) Instrumental neutron activation analysis of lunar samples and the identification of primary matter in the lunar highlands. J Radioanal Chem 38:363–378

White WM, Patchett J (1984) Hf-Nd-Sr isotopes and incompatible-element abundances in island arcs: implications for magma origin and crust-mantle evolution. Earth Planet Sci Lett (in press) 67:167–185

Whitford DF, Arndt NT (1977) Rare earth elements in a thick, layered komatiite lava flow. Carnegie Inst Wash Year Book 76:594–597

Oxygen Isotope Compositions of Minerals and Rocks and Chemical Alteration Patterns in Pillow Lavas from the Barberton Greenstone Belt, South Africa

H. S. Smith[1], J. R. O'Neil[2] and A. J. Erlank[1]

Contents

1 Introduction 116
2 Sample Selection and Analytical Methods 117
3 Isotopic Composition of Minerals 119
4 Komatiitic Basalt Pillow Lavas 122
5 Komatiite Lava Flows 127
6 Discussion 133
References 135

Abstract

Oxygen isotope, major and trace element data have been obtained on a suite of rocks and minerals from the Barberton greenstone belt, South Africa. Analyses of relict orthopyroxene grains separated from a cumulate ultramafic rock indicate that $\delta^{18}O$ values of the komatiite magmas were ~5.7. This value is similar to that of modern ocean floor basalts and indicates that source mantle material of the Archaean komatiites had the same isotopic composition as mantle material from which modern ocean ridge basalts are derived.

Oxygen isotope compositions of secondary minerals (antigorite and chrysotile) are similar to those of serpentine minerals from modern altered oceanic rocks. Similarly, low magnesium komatiitic meta-basalts (8 – 10% MgO) from Barberton have a similar range and mean of $\delta^{18}O$ values to those of greenschist facies meta-basalts on the modern ocean floor.

Variations in the major and trace element contents between margins and cores of komatiitic basalt pillows show that the elements analysed can be divided into two groups. In the first group which includes Si, Ti, Al, Mg, P, Nb, Zr, Y, Co, V, Sc, and probably Cr there are no variations within analytical error. In the second group differences between margins and cores of pillows vary from barely detectable up to a few hundred percent. This group consists of the following elements arranged in order of increasing variability: – Fe(II), Mn, Ca, Ni, Ga, Fe(III), Na, Sr, K, S, Rb, Ba, Zn, and Cu. The patterns of variation of Fe(II) and S are particularly significant as they are similar to patterns observed in altered pillow lavas from the ocean floor and indicate that the Barberton pillows were subject to seawater alteration before being metamorphosed. The isotopic data and alteration patterns in the pillows provide strong evidence that the processes

1 Geochemistry Department, University of Cape Town, Rondebosch 7700, South Africa
2 U.S. Geological Survey, Menlo Park, CA 94301, USA

Archaean Geochemistry (ed. by A. Kröner et al.)
© Springer-Verlag Berlin Heidelberg 1984

that occurred in the Archaean rocks were similar to those occurring today during alteration and metamorphism of submarine lavas.

Detailed modelling of $\delta^{18}O$ variations in komatiite lava flows indicates that magmatic or primitive waters ($\delta^{18}O$ = 5 to 7) could have been the major fluids responsible for the hydration of the lavas at temperatures of 240 – 450 °C. Such an interpretation is consistent with the greenschist facies mineralogy of the Barberton rocks. However, waters with $\delta^{18}O$ values of around 0 (Archaean seawater?) may have exchanged with the komatiites at temperatures as low as 100 °C. Mixtures of Archaean seawater and deep-seated water could have exchanged with the komatiites at intermediate temperatures (100 – 450 °C). Specifically a mixture of 75% seawater ($\delta^{18}O$ = 0) and 25% deep-seated water could have given rise to the komatiite isotopic compositions if equilibration occurred at ~130 °C. The available evidence is consistent with the interpretation that Archaean seawater had a $\delta^{18}O$ of around zero, but further work is required to establish a reliable estimate of this value. Oxygen isotope compositions of rocks and minerals from the Komati Formation and the chemical alteration patterns observed in the pillow lavas are compatible with an ocean floor environment for extrusion, alteration and metamorphism of these ancient volcanic rocks.

1 Introduction

The Barberton greenstone belt represents some of the oldest crustal material preserved on the Kaapvaal Craton, South Africa, and consists of volcanic and sedimentary rocks that have been metamorphosed to greenschist facies grade (Anhaeusser et al. 1968). The lower ~15 km of the stratigraphy is known as the Onverwacht Group and consists predominantly of metavolcanic rocks of komatiitic and basaltic composition (Viljoen and Viljoen 1969a, b, c; Williams and Furnell 1979). These metavolcanic rocks have yielded a Sm-Nd isochron age of 3540 ± 30 Ma (Hamilton et al. 1979) which is considered to be the age of formation of the Onverwacht Group. A Rb-Sr age of 3420 ± 200 Ma has been determined on different density fractions of a komatiitic basalt (Jahn and Shih 1974; Jahn et al. 1982) and is interpreted as the time of low grade metamorphism of the Komati Formation. Barton (1981) reported a Rb-Sr age of 3430 ± 135 Ma for the intrusive Threespruit pluton. The errors of these age determinations are not sufficiently small to resolve time differences between the events but do indicate that extrusion and metamorphism of the Onverwacht lavas and their intrusion by tonalitic magmas occurred over a relatively short time span.

The origin of this ancient greenstone belt remains enigmatic partly due to the structural complexities of the belt (De Wit 1982) and to the controversial nature of the granite-greenstone contacts (Anhaeusser 1973; Hunter 1974). Clearly, a subaqueous environment existed during much of the volcanic activity that formed Barberton and other Archaean greenstone belts to account for the common occurrence of thick pillow lava sequences throughout most of the volcanic pile. Numerous origins have been proposed for the formation of greenstone belts such as volcanism induced by meteorite impact (Green 1972), ocean floor volcanism similar to that occurring at modern ocean ridges (Anhaeusser 1973, 1975) and back arc basin volcanism (Weaver and Tarney 1979; Smith and Erlank 1982). (*For alternative view see also Groves and Batt, this Vol., eds.*)

In this work we have attempted to establish the source of the water that caused the extensive development of hydrated minerals in the Onverwacht lavas using oxygen isotope data from rocks and minerals and chemical alteration patterns observed in the pillow lavas. The models proposed above for the origin of the greenstone belt imply that Archaean seawater played a major role in the alteration and metamorphism of the lavas. In addition, magmatic waters that probably emanated from the granitic magmas may have been involved during metamorphism of the lavas as the plutons were emplaced. With respect to the latter possibility Viljoen and Viljoen (1969a) have shown that the metamorphic aureole associated with granitic intrusions is limited to a narrow belt of amphibolite facies rocks along the contacts of the intrusions.

The nomenclature for komatiitic lavas used here is from Smith and Erlank (1982). Lavas are considered to be komatiites if they are non-cumulate rocks containing >24% MgO (volatile free) and komatiitic basalts if they have lower MgO contents of 8 – 24% (volatile free) and are believed to have been derived from komatiites by processes of differentiation. Only rocks that show good preservation of primary igneous textures or structures can be used for classification purposes.

The chemical and isotopic composition of a series of komatiitic basalt pillow lavas have been used to investigate the effects of alteration on different portions of the pillow structures and for comparison with alteration trends observed in modern submarine basalts. Most of the whole rock samples were obtained close to the type section of the Komati Formation (Viljoen and Viljoen 1969b) and represent areas of typical greenschist facies metamorphism. The oxygen isotope data are used to estimate the isotopic composition of the waters responsible for the pervasive hydration of the volcanic rocks. Komatiite flows are particularly suited to this type of investigation as the secondary mineral assemblage (amphibole, antigorite, magnctite, and chlortie) is relatively simple and, consequently, the isotopic compositions are more easily modelled than in the basaltic rock types.

2 Sample Selection and Analytical Methods

Three suites of samples were selected for analysis and are derived mainly from the Komati Formation of the Onverwacht Group. The first suite consists of coarse grained ultramafic rocks from the Stolzburg, Koedoe and Msauli ultramafic bodies (Viljoen and Viljoen 1970) from which primary and secondary minerals were separated and analysed for their oxygen isotopic compositions. The second suite consists of a series of komatiitic basalt pillow lavas from a pillowed horizon in the Komati Formation (Fig. 1). Oxygen isotope compositions as well as major and trace element data were obtained on margin and core samples of the pillows. Oxygen isotope compositions were also obtained for a third suite of samples from komatiite lava flows in the Komati Formation for which major

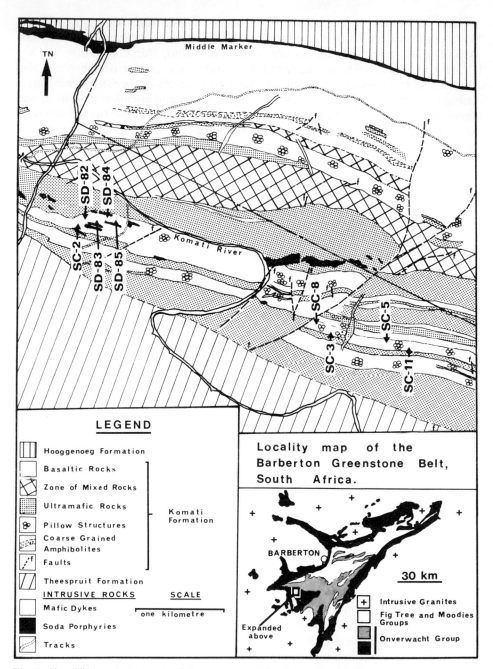

Fig. 1. Simplified geological map of the type area of the Komati Formation, Barberton greenstone belt, taken from Viljoen and Viljoen (1969b). The location of the komatiitic basalt pillow lavas sampled is shown

and trace element data have been reported elsewhere (Smith and Erlank 1982; Viljoen et al. 1983).

The major and trace element data were obtained at Cape Town using precise X-ray fluorescence analysis as described in Willis et al. (1971, 1972). The estimated analytical errors calculated from the counting strategy employed are given in Table 2. Mineral separations were performed on -80 to $+120\#$ powder using standard magnetic and heavy liquid separation procedures. All mineral separates were $>95\%$ pure. CO_2 and H_2O^+ were determined by gas chromatography using a Hewlet Packard 185B CHN Analyser and $20-30$ mg ($-300\#$) of sample powder pre-dried at $110°C$. The following standards were used for the H_2O calibration: PCC-1, T-1, AGV-1, BCR-1, DTS-1, NIM-L, NIM-G, NIM-N, NIM-D, NIM-P and NIM-S with the concentrations given in Abbey (1975). For the CO_2 calibration synthetic quartz-calcium carbonate mixes and the above NIM standards were used. Multiple runs of a metamorphosed basalt sample gave $2.82\% \pm 0.03$ (1σ) H_2O and $1.10\% \pm 0.07$ (1σ) CO_2. FeO was determined by standard wet chemical methods.

Oxygen was extracted from the rocks and minerals for isotopic analysis using the BrF_5 extraction method described by Clayton and Mayeda (1963) and reaction temperatures of 600 to $650°C$. All ^{18}O analyses are reported in the standard δ notation relative to SMOW. Duplicate analyses were made on most samples, and the average error was typically less than $\pm 0.1\%$ for the minerals and $\pm 0.4\%$ for the whole rock samples. Seventeen analyses of the NBS-28 quartz standard carried out during the isotopic analysis of the samples yielded an average $\delta^{18}O$ of $9.63\% \pm 0.14$ (1σ).

3 Isotopic Composition of Minerals

Of the relict igneous minerals that are preserved in the Barberton rocks only orthopyroxene could be extracted in sufficient amounts for isotopic analysis. The orthopyroxene comes from the Stolzburg Body in the Komati Formation. This body and other similar differentiated ultramafic bodies are believed to represent penecontemperaneous intrusions of komatiitic magma (Viljoen and Viljoen 1970) which had a bulk composition similar to the komatiite lava flows. The lower sections of the Stolzburg body consists of alternating layers of dunite (now altered to serpentinite) and pyroxenite. In places these pyroxenite layers are virtually unaffected by alteration and fresh orthopyroxene and chromite crystals are preserved.

The orthopyroxene (enstatite) was separated from one of the pyroxenite layers (compositions reported in Smith and Erlank 1982) and the oxygen isotopic composition is given in Table 1. From the mineral-melt data given in Kyser et al. (1981) a crude estimate of the isotopic composition of the komatiitic magmas from which these orthopyroxene layers crystallized can be obtained. The temperature of crystallization of the orthopyroxene crystals is expected to be significantly lower than the $1650°C$ liquidus temperature of komatiite lava flows (Green et al. 1975) as a thick (tens of metres) layer of dunite had already accumulated

Table 1. Oxygen isotope compositions of primary and secondary minerals from the Onverwacht Group rocks

Sample No.	Mineral	Rock type and location	$\delta^{18}O_{SMOW}$
SD-76	Enstatite	Pyroxenite, Stolzburg Body	6.2
HSS-777	Antigorite	Serpentinized dunite, Stolzburg Body	3.6
STOLZ-S	Antigorite	Serpentinite, Stolzburg Body	4.5
MSAULI-S	Antigorite	Serpentinized dunite, Msauli Asbestos Mine	5.8
HK-1	Antigorite	Serpentinized dunite, Koedoe Body	1.7
HK-8	Antigorite	Serpentinized dunite, Koedoe Body	4.3
20-J	Antigorite	Ultramafic rock, Hooggenoeg Formation	7.1
SSV-1	Chrysotile	Serpentine vein, Stolzburg Body	3.1
STOLZ-C	Chrysotile	Serpentinite, Sterkspruit Mine, Stolzburg Body	3.6
MSAULI-S	Chrysotile	Serpentinized dunite, Msauli Asbestos Mine	6.0

before the onset of pyroxene crystallization in the Stolzburg body. The orthopyroxene crystallization temperature is somewhat arbitrarily assumed to be 1500 °C or lower, and from $\Delta_{basalt-pyroxene}$ value of ~ −0.5 at 1450 °C (Kyser et al. 1981) the magma is estimated to have had a $\delta^{18}O$ of ~5.7. This value of 5.7 is typical of mantle-derived materials such as peridotite xenoliths in kimberlite (e. g. Sheppard and Dawson 1975; Kyser et al 1983) and fresh ocean floor basalts (Muehlenbachs and Clayton 1972a; Kyser and O'Neil 1978) and indicates that there has been no change in the oxygen isotopic composition of the upper mantle over the last 3500 Ma.

Chromites were also extracted from the ultramafic bodies and lava flows, but due to the resistant nature of this mineral to the BrF_5 oxidant, very poor oxygen yields were obtained and thus the data are not reported.

The oxygen isotope data for the secondary minerals analysed are given in Table 1 and consist of the serpentine minerals antigorite and chrysotile. The $\delta^{18}O$ values of six antigorites and three chrysotiles range from 1.7 to 7.1 and 3.1 to 6.1, respectively. As no coexisting minerals could be analysed the temperature of serpentinization is not known, but the large spread in the data indicates that a range of temperatures and/or fluids of different isotopic compositions may have been involved. Wenner and Taylor (1971) have estimated isotopic temperatures of serpentinization and show that antigorite can form over the temperature range of 220 to 460 °C. Even assuming that the Barberton material represents serpentine formed over this temperature range in order to account for the spread in the antigorite isotopic compositions they must have equilibrated with fluids which varied in $\delta^{18}O$ by at least ~2‰ and possibly as much as 9‰.

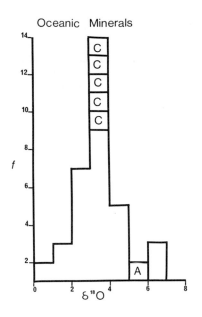

Oceanic Minerals

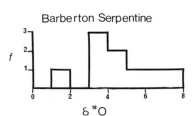

Barberton Serpentine

Fig. 2. Histograms of the oxygen isotopic compositions of modern oceanic serpentine, chlorite *C* and actinolite *A* minerals (data taken from Wenner and Taylor, 1971, 1973; Stakes and O'Neil 1982) and Barberton antigorite and chrysotile minerals

In Fig. 2 the $\delta^{18}O$ values of the Barberton serpentines have been compared to those of modern ocean floor serpentines. The Barberton data are very similar in range and distribution to oceanic serpentines which suggests that similar processes operated in Archaean rocks to form serpentine as is occurring in the modern ocean. Wenner and Taylor (1973) have shown that ocean floor serpentine minerals could have formed in equilibrium with a mixture of 25% magmatic water and 75% seawater at temperatures of 140 – 300 °C. The similarity of the processes of formation of the Barberton and modern serpentine minerals implied by the distributions in Fig. 2 leads to the intriguing possibility that Archaean seawater had an oxygen isotopic composition similar to that of modern seawater. Although no D/H determinations were made on these minerals Wenner and Taylor (1974) reported two values from Barberton serpentinites which have δD_{SMOW} values of – 61 and – 64, at the light end of the range of values of – 35 to – 68 given by Wenner and Taylor (1971) for modern ocean floor serpentines. Detailed investigation of the isotopic compositions of cherts from the Barberton greenstone belt has led Knauth and Lowe (1978) and Perry (1978) to suggest that either the seawater from which the cherts were precipitated was hot (70 – 100 °C) and had a $\delta^{18}O$ of ~0‰ or the temperature was lower and the isotopic composition was different (– 15 to – 18‰) from modern seawater (Perry 1967; Perry and Tan 1972). These possibilities are considered in more detail in a later section, but based on our serpentine data we note that waters with a range of isotopic compositions (2 to 9‰) must have been involved in the formation of the Barberton antigorites.

4 Komatiitic Basalt Pillow Lavas

Numerous well-exposed pillow horizons occur in the Onverwacht Group, and one of these horizons consisting of low-Mg komatiitic basalt pillows was sampled (Fig. 1). The pillows range in diameter from $70-100$ cm and are rimmed by a ~ 2 cm thick chilled margin. Thin cracks can be seen extending from the margins up to 2 cm into the interior of the pillows, while a few large cracks penetrate into the cores of the pillows. These fracture systems may represent relict radial cooling cracks as described in younger pillow lavas by Moore et al. (1971) and Moore (1975).

In thin section the chilled margins consist predominantly of brown crypto-crystalline material (mainly tremolite-actinolite and chlorite as identified by X-ray diffraction) and minor microphenocryst phases consisting of clinopyroxene, plagioclase and glomeroporphyritic spinel octahedra. The interiors of the pillows consist of a groundmass of fan-shaped spherulites of actinolite with occasional phenocrysts of clinopyroxene. Chlorite, quartz, epidote and minor calcite are also present in the groundmass of all the pillow cores along with glomeropor-phyritic spinel microphenocrysts. The pyroxene and plagioclase micropheno-crysts are generally altered with pyroxene being pseudomorphed by actinolite and chlorite with variable amounts of epidote, quartz and calcite, while the plagioclase crystals are usually altered to albite with minor quartz, clinozoisite, sericite and calcite.

Major and trace element data for the pillow samples are given in Table 2 along with estimated analytical errors, and the oxygen isotope data are reported in Table 3. Examination of the data in Table 2 shows that there are no variations in the concentrations of Ti, P, Nb, Zr, Y, Co, V, and Sc within analytical error $(\pm 2\sigma)$. As this group of elements is homogeneously distributed within and be-tween pillows they are considered to have remained immobile during the altera-tion processes. Ti, Nb, Zr, Y, Co, and P have previously been shown to remain immobile during low grade metamorphic and seawater alteration processes (Cann 1970; Pearce and Cann 1973; Hart et al. 1974; Humphris and Thompson 1978a, b; Floyd and Winchester 1975). Only small changes are observed in the concentrations of Si, Al, and Mg between some of the pillow pairs, and in gener-al these elements are also considered to have remained immobile. Small but sig-nificant changes in the Cr content within pillows could be accounted for by minor variations in the content of spinel microphenocrysts and need not be due to alteration effects.

The most obvious indications of chemical alteration of these pillows are the H_2O^+ and CO_2 contents. No systematic trend can be seen in the H_2O^+ content between margins and cores of the pillows. However, based on thin section exami-nation and the generally high H_2O^+ values it is clear that none of the pillow sam-ples is pristine unaltered material.

The CO_2 contents of the pillow pairs are low and generally very similar. From this observation it follows that the large and inconsistent variations in the Ca contents within pillows (see Table 2) cannot be related to formation of variable amounts of calcite in the samples. Neither can the distribution of Ca be account-ed for by fractionation of phenocryst phases (clinopyroxene or plagioclase) as

Table 2. Major and trace element compositions and inter-element ratios of komatiitic basalt pillow lavas from the Komati Formation

	Western Pillows					
	SC-2		SD-82	SD-83	SD-84	SD-85
	Margin	Core	Core	Core	Core	Core
SiO_2	49.50	51.26	52.14	50.72	51.40	52.27
TiO_2	0.71	0.73	0.75	0.74	0.72	0.74
Al_2O_3	10.38	10.88	10.97	10.73	10.67	10.91
Fe_2O_3	2.86	2.21	2.22	2.57	2.47	1.93
FeO	9.02	8.32	8.51	8.74	8.45	8.56
MnO	0.20	0.17	0.17	0.19	0.17	0.17
MgO	10.11	9.79	9.10	9.80	9.65	9.25
CaO	12.64	10.38	10.21	11.83	11.36	10.14
Na_2O	1.23	2.17	2.90	1.70	1.93	2.73
K_2O	0.41	0.37	0.10	0.12	0.24	0.25
P_2O_5	0.10	0.10	0.11	0.11	0.10	0.10
H_2O^+	3.08	2.84	2.54	2.64	2.80	2.43
H_2O^-	0.03	0.08	0.05	0.04	0.04	0.04
CO_2	0.40	0.45	0.21	0.22	0.44	0.25
Total	100.47	99.75	99.97	100.14	100.45	99.77
S(ppm)	149	189	469	269	203	300
Nb	2.4	3.6	3.1	3.7	3.1	2.2
Zr	60	62	62	61	61	63
Y	18.5	18.2	17.3	17.6	17.4	17.1
Sr	265	232	156	203	220	175
Rb	9.4	7.9	<2	<2	5.0	5.2
Zn	102	92	139	384	81	77
Cu	67	172	217	559	115	149
Ni	209	220	155	184	179	197
Co	61	62	57	57	57	59
Cr	622	662	567	618	635	663
V	192	192	191	193	192	196
Ba	113	106	21	32	48	59
Ga	14.1	11.5	12.0	12.2	12.8	11.9
Sc	26.3	26.8	28.0	27.5	26.8	27.4
Ratios						
Si/Al	4.2	4.2	4.2	4.2	4.3	4.2
Si/Ti	54	55	54	54	56	55
Al/Ti	13	13	13	13	13	13
Ca/Ti	21.2	16.9	16.2	19.2	18.8	16.3
Fe^{3+}/Ti	4.7	3.5	3.4	4.1	4.0	3.0
Ti/Zr	71	71	73	72	71	71

H. S. Smith, J. R. O'Neil and A. J. Erlank

Table 2 (continued)

	Eastern pillows									
	SC-8		SC-3			SC-5		SC-11		
	Margin	Core	Margin	Inter.	Core	Margin	Core	Margin	Core	Error [a]
SiO_2	52.43	52.06	52.55	53.01	52.37	52.65	52.62	53.23	53.98	0.50
TiO_2	0.77	0.76	0.76	0.78	0.77	0.77	0.79	0.76	0.76	0.02
Al_2O_3	11.23	11.13	10.64	11.37	11.84	11.63	11.65	11.65	11.61	0.25
Fe_2O_3	1.55	1.51	1.67	1.51	1.67	1.81	1.72	1.67	1.72	0.20
FeO	9.54	9.06	9.66	9.35	8.90	9.50	9.39	8.77	8.10	0.05
MnO	0.18	0.18	0.22	0.18	0.17	0.18	0.17	0.18	0.17	0.01
MgO	8.69	9.52	8.95	9.37	9.18	9.31	8.87	8.76	8.36	0.25
CaO	10.64	10.76	11.00	9.84	9.83	9.67	10.18	10.35	11.08	0.10
Na_2O	2.44	2.10	2.36	2.73	2.59	2.68	2.56	2.96	3.02	0.05
K_2O	0.21	0.50	0.22	0.22	0.24	0.29	0.28	0.12	0.12	0.01
P_2O_5	0.10	0.10	0.10	0.10	0.10	0.10	0.11	0.10	0.10	0.006
H_2O^+	1.76	2.28	2.42	1.65	1.81	1.91	1.89	2.00	1.51	0.07
H_2O^-	0.04	0.07	0.05	0.05	0.04	0.06	0.04	0.04	0.05	–
CO_2	0.21	0.24	0.39	0.26	0.22	0.18	0.15	0.30	0.30	0.14
Total	99.79	99.82	100.98	100.41	99.73	100.74	100.41	100.89	100.87	
S (ppm)	191	144	182	337	348	166	523	264	357	4.0
Nb	3.2	3.3	3.7	3.4	3.0	3.5	3.6	3.3	3.4	0.6
Zr	65	64	65	66	65	66	66	65	64	1.4
Y	17.8	18.3	17.6	17.9	18.2	18.5	18.8	18.0	18.1	1.2
Sr	203	211	149	141	158	159	137	149	129	1.8
Rb	4.0	13.6	5.1	4.5	7.0	7.5	6.9	1.7	1.9	0.5
Zn	108	127	89	80	102	89	82	116	77	1.0
Cu	160	135	88	121	199	100	170	162	126	1.4
Ni	194	196	178	190	201	174	160	178	167	2.0
Co	59	60	59	60	59	60	61	56	59	2.4
Cr	695	653	633	646	643	613	582	623	639	4.0
V	201	193	204	195	195	199	204	192	196	2.6
Ba	44	58	31	24	29	61	40	19	14	1.0
Ga	11.0	11.2	10.2	11.4	11.3	11.7	12.8	9.6	11.0	0.2
Sc	29.6	28.7	28.5	28.3	29.1	28.8	28.4	28.5	28.6	0.5
Ratios										
Si/Al	4.1	4.1	4.4	4.1	3.9	4.0	4.0	4.0	4.1	
Si/Ti	53	54	54	53	53	54	52	55	56	
Al/Ti	13	13	12	13	14	13	13	14	14	
Ca/Ti	16.5	17.0	17.3	15.1	15.3	15.0	15.3	16.3	17.4	
Fe^{3+}/Ti	2.4	2.3	2.6	2.3	2.5	2.8	2.5	2.6	2.7	
Ti/Zr	71	71	70	71	71	70	72	70	71	

[a] Analytical error given as 2 standard deviations

Table 3. Oxygen isotope compositions of komatiitic basalt pillow lavas from the Komati Formation

Sample No.	Type	$\delta^{18}O_{SMOW}$
SC-13	Core	7.3
SC-11M	Margin	6.5
SC-11C	Core	6.6
SC-5M	Margin	7.3
SC-5C	Core	6.9
SC-3M	Margin	6.7
SC-3I	Intermediate	6.9
SC-3C	Core	6.7
SC-8M	Margin	6.8
SC-8C	Core	6.8
SD-85	Core	6.4
SD-84	Core	6.0
SD-83	Core	6.2
SD-82	Core	6.3
SC-2M	Margin	5.9
SC-2C	Core	6.4

there are no similar changes in the Mg or Al contents. The high Fe(III), Ca, Sr, and Ga contents in a sample such as SC-2M is probably related to the appearance of epidote grains (clinozoisite) in the devitrified matrix of this sample. Epidote in low grade metamorphic rocks has been shown to accumulate Ca, Sr, and to a lesser extent Ga (Melson and van Andel 1966; Smith 1968; Smith and Smith 1976; Condie et al. 1977).

The effects of epidote formation on the composition of the pillows are illustrated (Fig. 3) by the variations of the Fe(III)/Ti and Ca/Ti ratios. The linear trend obtained for most of the western pillows is interpreted as a mixing line and shows that Fe(III) and Ca have been added to these samples in a constant ratio. The average Ca/Fe(III) ratio computed for the five samples that plot along the trend is 4.7 and is essentially identical to that of clinozoisite [4.5 assuming all Fe as Fe(III)] analysed in the SC-11 pillow core (Smith 1980). The Ca and Fe(III) added to the western pillows have therefore probably been incorporated into clinozoisite during low grade metamorphism.

There is a regular decrease of Fe(II) in the cores of the pillows relative to the margins, and a similar trend is observed for Mn. The loss of iron from the cores of recent marine pillows has been recorded by Scott and Hajash (1976), and this feature in the Barberton pillows may be a relict seawater alteration effect.

The K, Rb, and Ba contents of the pillows are also variable, but the distribution patterns of the three elements are similar in core-margin pairs. These changes are considered to reflect the effects of low temperature alteration and/or low grade metamorphism on the pillow lavas. For the remaining elements Na, Zn, Cu, Ni, and S there are large and variable changes between cores and margins of the Barberton pillows. S and Cu have similar distribution patterns in

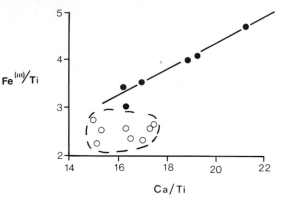

Fig. 3. Plot of Fe(III)/Ti against Ca/Ti for the komatiitic basalt pillow lavas. The western pillows are plotted as *dots* and the eastern pillows as *circles*. The linear array formed by the western pillows is considered to be due to the addition of Fe(III) and Ca to form additional epidote in these samples

the pillow pairs (see Table 2) and are lower in all the margins than cores except in the SC-8 pillow. Consistently lower S contents in the margins of recent ocean floor pillows have been recorded by Hart et al. (1974), and this similar trend in the Barberton pillows may be a relict effect of seawater alteration. The chemical alteration patterns observed in the Komati Formation pillow lavas are therefore compatible with the interpretation that these lavas were altered and metamorphosed in a submarine environment.

There are no systematic trends in the $\delta^{18}O$ values of the different portions of the pillow lavas, and in fact margins and cores have similar isotopic compositions indicating that the hydrothermal fluids equilibrated with the entire pillow. As these komatiitic basalt magmas probably had isotopic compositions similar to those of the komatiite magmas, the effects of alteration on the pillows has been to increase the $\delta^{18}O$ values from 5.7 to 6.6 ± 0.6. These latter values are typical of greenschist facies metabasalts from the ocean floor (e.g. Muehlenbachs and Clayton 1972b; Stakes and O'Neil 1982), and in Fig. 4 the isotopic compositions of ocean floor metabasalts are compared to the Barberton pillow values. Although the submarine metabasalts show a greater spread than the Barberton pillows the two distributions and modes are very similar. These data indicate that similar processes may have occurred in the two environments.

The similarity of these two contrasted environments must also take into account the ranges of temperatures involved and the range of isotopic compositions of the hydrothermal fluids that interacted with the lavas. As greenschist facies rocks are being considered in both cases, the temperature requirement is hardly surprising, but the similarity of the fluids leads to the same possibility that was noted for the serpentine data, namely that the Archaean ocean water had a $\delta^{18}O$ value of around 0. The above chemical and isotopic data therefore show that at least two distinguishable events have affected the pillow lavas, one being seawater alteration which gave rise to trends such as those observed for Fe(II) and S in the pillow pairs, and the other being greenschist facies metamorphism. While every effort has been made to sample only the freshest material available the effects of more recent events such as interaction with modern groundwaters has not been evaluated and for the purposes of the present interpretation will be assumed to have been negligible. Nevertheless, if analogies with the modern ocean ridge processes are valid as indicated by Fig. 4, the Archaean seawater probably con-

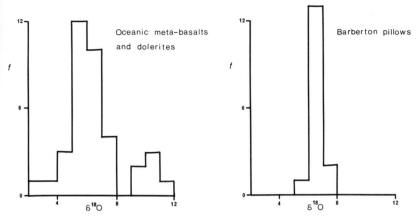

Fig. 4. Histograms of the oxygen isotopic compositions of greenschist facies rocks from modern oceanic crust (data compiled in Sheppard 1980) and the komatiitic basalts from the Barberton greenstone belt

tributed to a large extent to the hydrothermal fluids associated with metamorphism and, consequently, by analogy had a $\delta^{18}O$ value close to 0.

5 Komatiite Lava Flows

The $\delta^{18}O$ values for the komatiites analysed are given in Table 4. Petrographic descriptions and major and trace element data are given in Viljoen et al. (1983) for Richard's Flow and in Smith and Erlank (1982) for other komatiite samples. Richard's Flow is a thin (33 cm thick) komatiite flow occurring in the type section of the Komati Formation. It has excellent textural preservation as well as all the classical spinifex textured units observed in thicker komatiite flows (Pyke et al. 1973; Smith et al. 1980). Although all the primary olivine blades and phenocrysts have been altered to serpentine and magnetite, detailed modelling of the chemical variations in the flow section shows that the flow differentiated by olivine accumulation towards the base (Viljoen et al. 1983).

Variations in $\delta^{18}O$ with depth in Richard's Flow (Fig. 5) show that the flow has a relatively uniform isotopic composition regardless of chemical and mineralogical variations that occur in this section. This is in contrast to the findings of Beaty and Taylor (1982) who noted a close correspondence between mineralogy and $\delta^{18}O$ values in a section through a komatiite flow in Munro Township. Because the Munro Township rocks were subjected to a metamorphic grade (Arndt et al. 1977) that was lower than that of the Barberton material, different secondary minerals have developed in these two flow sections, and this, combined with the lower temperatures of metamorphism in the Munro Township rocks, could account for the differences observed in the oxygen isotopic compositions from the two flows.

Table 4. Oxygen isotope compositions of komatiites (All samples are from the Komati Formation unless otherwise indicated; samples MDF-1 to MDF-10 are from Richard's Flow)

Sample No.	Rock type and location	$\delta^{18}O_{SMOW}$
MDF-1	Flow top	5.5
MDF-2	Spinefex A_{2R} zone	6.1
MDF-3	Spinefex A_{2P} zone	5.7
MDF-4	Spinifex A_{2P} zone	5.9
MDF-5	Spinifex A_{2P} zone	6.0
MDF-6	B1 zone	6.0
MDF-7	B2/B4 zone	5.8
MDF-8	B2/B4 zone	5.5
MDF-9	B2/B4 zone	6.3
MDF-10	Basal Chilled Margin	6.0
HSS-523	Pillow lava	5.7
MF-5	B2/B4 zone, Morris's Flow	5.8
HSS-1	Porphyritic komatiite	4.9
HSS-31	Porphyritic komatiite	3.6
HSS-33	Porphyritic komatiite	4.8
VS-11/80	Porphyritic komatiite	5.2
R-13	Porphyritic komatiite	4.8
20-J	Porphyritic komatiite Hooggenoeg Formation	7.5
HSS-92	Porphyritic komatiite	4.8
HSS-93	Porphyritic komatiite	5.5
HSS-95	Spinifex textured komatiite	5.2
HSS-105	Serpentinite, Sandspruit Formation	3.3

The komatiites from the Komati Formation are remarkably uniform in oxygen isotopic composition with the majority of samples having $\delta^{18}O$ values in the range of 4.8 to 6.3. This is surprising in view of the wide range in chemical compositions (24 to 45% MgO on a volatile free basis) and associated range of mineral proportions.

Three samples 20-J, HSS-31 and HSS-105 (Table 4) do not fall in this range of isotopic compositions. 20-J is from the lower portion of the Hooggenoeg Formation and in thin section consists almost entirely of antigorite with minor amounts of chlorite and very minor amounts of magnetite and chromite. The relict igneous texture indicates that antigorite has replaced olivine although the individual serpentine pseudomorphs lack the characteristic dusting of secondary magnetite along their margins, and in fact this sample lacks the abundant magnetite that usually occurs in serpentinites from the Onverwacht Group. As magnetite usually has a relatively low $\delta^{18}O$ value (Taylor and Epstein 1962) the low content of magnetite in this sample partly accounts for its heavier isotopic composition compared to the other serpentinized ultramafic samples. On the other hand, sample HSS-31 contains occasional thin magnetite rich veins, and the lighter ^{18}O composition of this sample is possibly related to the magnetite in these veins.

Sample HSS-105 is from the Sandspruit Formation and occurs very close to a granite contact. The isotopic composition of this sample probably resulted

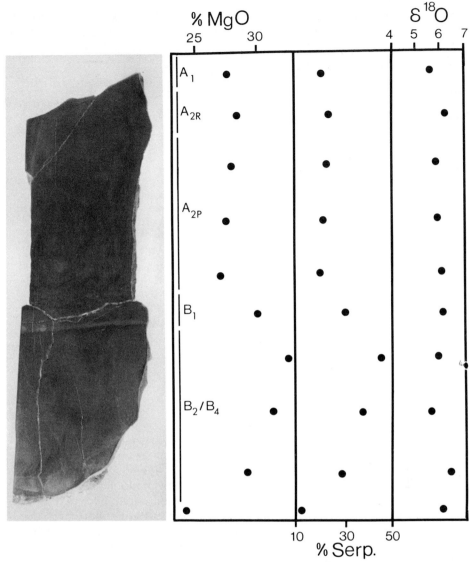

Fig. 5. Polished slab of a section of Richard's Flow (33 cm thick) (Viljoen et al. 1983). The designations of the different textural zones are from Smith et al. (1980). Variations in mineralogy, chemical and isotopic composition with position in the flow section are illustrated

from higher temperature reactions than those of the typical greenschist facies rocks in the Komati Formation because of its close proximity to the intrusive contact. From the mineral proportions in Table 6 and using the mineral-water fractionation equations discussed below it can be calculated that if this sample equilibrated with a large amount of magmatic water ($\delta^{18}O = 5$ to 7) at $\sim 470\,^\circ C$ near the upper limit of serpentine stability at 3 kb pressure (Scarfe and Wyllie 1967) it would have a $\delta^{18}O$ value between 2.9 and 4.9. The measured oxygen composition of HSS-105 is 3.3‰ and suggests that this sample may in fact have equilibrated with magmatic waters from the intrusive pluton.

Table 5. Mineral compositions and example of mix calculation. Mineral data are from Smith (1980)

Input data

Oxide	Antigorite	Magnetite	Actinolite	Chlorite
SiO_2	41.25	0.00	55.26	31.44
Al_2O_3	0.54	0.00	2.23	17.62
FeO	1.42	100.00	6.82	0.00
MnO	0.07	0.00	0.31	0.00
MgO	41.84	0.00	20.41	37.64
CaO	0.02	0.00	12.07	0.01
H_2O^+	13.68	0.00	1.81	13.19

Output data

Oxide	MDF-1 measured composition	MDF-1 calculated composition	Difference
SiO_2	44.70	44.68	-0.02
Al_2O_3	3.86	3.87	0.01
FeO	11.87	11.87	0.00
MnO	0.17	0.19	0.02
MgO	25.66	25.82	0.16
CaO	6.99	6.88	-0.11
H_2O^+	6.23	5.80	-0.43
		Sum of Squares of Differences =	0.23

Vector	Coefficient	Standard deviation
Antigorite	0.2122	0.0173
Magnetite	0.0768	0.0028
Actinolite	0.5697	0.0094
Chlorite	0.1412	0.0157
Totals	1.0000	0.0253

The mineralogy of the Barberton komatiites is now predominantly serpentine (antigorite), chlorite (clinochlore), amphibole (actinolite-tremolite) and magnetite. The proportions of each of these minerals in the komatiites have been calculated using the least squares mixing method (Bryan et al. 1969) and the mineral compositions given in Table 5. Only the major constituents SiO_2, Al_2O_3, FeO, MnO, MgO, CaO, and H_2O^+ (or LOI) of the rocks and minerals have been used in the mixing calculations to provide a semi-quantitative estimate of the metamorphic mineral proportions. The results of a typical mixing calculation are given in Table 5. Mineral percentages calculated for the rocks are given in Table 6 along with the sum of squares of differences (SSD) obtained for each of the mixing calculations. For the majority of the samples the SSD values are low (<2) and indicate that the major element compositions of the rocks are adequately accounted for by mixtures of these four minerals. For samples with SSD values >2 (not reported) this is not the case, and additional mineral phases should be considered. These latter samples are discounted from further discussion.

The magnetite contents calculated for all the komatiites do not vary much (Table 6), and if the magnetites are all of similar isotopic composition they would

Table 6. Mineral proportions calculated for the komatiites

Sample No.	Antigorite	Magnetite	Actinolite	Chlorite	SSD
MDF-1	21	8	57	14	0.2
MDF-2	24	8	53	15	0.5
MDF-3	23	8	55	14	0.4
MDF-4	21	8	56	15	0.7
MDF-5	20	8	58	13	0.6
MDF-6	30	6	50	14	1.2
MDF-7	44	8	38	10	1.7
MDF-8	37	8	43	12	1.0
MDF-9	28	7	52	12	0.6
MF-5	13	7	64	15	0.7
HSS-523	29	8	49	14	1.2
HSS-33	49	9	26	15	1.9
HSS-1	49	6	31	14	0.9
HSS-95	13	9	60	18	1.1
HSS-93	37	7	40	17	1.0
HSS-92	31	8	45	17	1.0
20-J	82	5	0	12	0.1
HSS-105	46	6	37	12	0.1

not contribute to isotopic variations in the whole rock values. Wenner and Taylor (1971) and Stakes and O'Neil (1982) have argued that the fractionation of ^{18}O between fluid and chlorite is similar to that between fluid and serpentine, and as such these minerals will have similar isotopic compositions if equilibrated with the same fluid. In terms of isotopic variations they can be treated together. In Fig. 6 the $\delta^{18}O$ values of the komatiites have been plotted against their amphibole and serpentine + chlorite contents. As can be seen from this diagram there is no systematic change in the $\delta^{18}O$ values with large variations in the proportions of these silicate minerals. This indicates that the amphibole (actinolite-tremolite) must have a similar isotopic composition to serpentine and chlorite in the komatiites and that, in essence, all three silicate minerals can be treated together when considering the isotopic composition of waters that may have equilibrated with them.

In order to determine the isotopic composition of the fluids that caused the pervasive hydration using the equation given in Taylor (1978), the water/rock ratio must be estimated. This ratio is assumed to be large because over the whole areal extent of the Barberton greenstone belt and other greenstone remnants in South Africa the volcanic rocks have been extensively hydrated. In addition, pillow lavas occur throughout the Onverwacht stratigraphy and are in turn overlain by the Fig Tree Group which consists in part of deep water sediments (De Wit 1982). Consequently a water/rock ratio of $\geqslant 1$ has been assumed. The isotopic compositions of the komatiites in equilibrium with different waters have been calculated over a range of temperatures using the mineral contents given in Table 6, the magnetite-water fractionation equation of Bottinga and Javoy (1973) and the chlorite-water fractionation equation of Wenner and Taylor (1971). Based on the interpretation of Fig. 6 as discussed above the chlorite-water equation has been used for serpentine-water and amphibole-water as well.

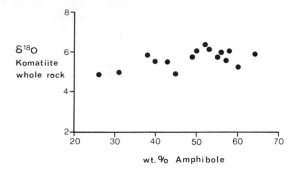

Fig. 6. Illustration of small and unsystematic variations of the komatiite whole rock samples although large variations in the mineral proportions occur in these rocks. The inference made from these trends is that the fractionation of ^{18}O between fluid and serpentine (or chlorite) is similar to that between fluid and actinolite-tremolite

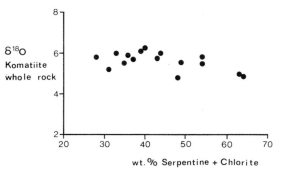

Three isotopically distinct waters have been considered. The first, with $\delta^{18}O$ values of 5 to 7, could represent magmatic water from the granitic intrusions or deep-seated water from the mantle. The second, with $\delta^{18}O$ value of 0, represents our best estimate, although poorly constrained, of Archaean seawater as indicated by the serpentine, metabasalt and chert data. The third water has a $\delta^{18}O$ value of -15 and is the alternative isotopic composition suggested for Archaean seawater by Perry and Tan (1972). The range of possible isotopic compositions of komatiites from the Komati Formation that would be in equilibrium with these waters over the temperature range 0 to 500 °C is shown in Fig. 7.

Clearly the isotopically light seawater suggested by Perry and Tan (1972) can be disregarded as it could not have been responsible for the hydration of the komatiites at any geologically reasonable temperature. The magmatic or deep-seated waters could have caused the hydration of the komatiites at temperatures of 240 to 450 °C. These temperatures are consistent with the greenschist facies mineral assemblage, and the implied source of the water from either the intrusive plutons or mantle is geologically reasonable.

Alternatively, seawater with a $\delta^{18}O$ value of 0 could also have equilibrated with the komatiites at ~100 °C. This interpretation is consistent with the model of hot Archaean seawater proposed by Knauth and Lowe (1978) and Perry (1978). The lowest 'isotopic' temperature of formation for antigorite is 220 °C (Wenner and Taylor 1971), and the low equilibration temperatures implied by this model therefore could not be the alteration event in which antigorite was formed but could reflect the exchange of the hydrated rocks with hydrothermal fluids after the peak of metamorphism.

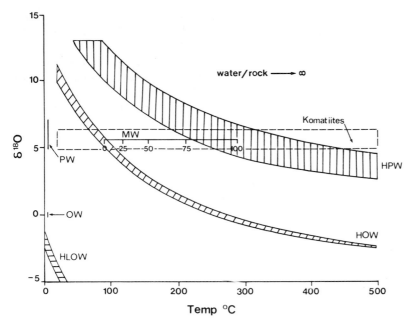

Fig. 7. Modelling of isotopic composition vs temperature for waters that may have equilibrated with the komatiite rocks. *OW* ocean water $\delta^{18}O = 0$; *PW* primary water $\delta^{18}O = 5$ to 7; *MW* mixed water; *HOW* field of altered komatiite compositions hydrated by OW; *HPW* field of altered komatiite compositions hydrated by PW; *HLOW* field of altered komatiite compositions hydrated by isotopically light water ($\delta^{18}O = -15$).

The isotopically light water could not have equilibrated with the komatiites at geologically reasonable temperatures. The ocean water and primary water and mixtures of these waters could have equilibrated with the greenschist facies minerals in the komatiites. See text for further discussion

The participation of waters of mixed origin in the alteration of the lavas is difficult to evaluate, as possible end member compositions are not yet well constrained. However, if Archaean seawater had $\delta^{18}O$ close to 0 then a 75/25 mixture of seawater/magmatic water as suggested by Wenner and Taylor (1973) for the modern ocean environment could have equilibrated with the komatiites at ~130 °C (Fig. 7).

6 Discussion

On a global scale oxygen isotope compositions of meta-lavas from the younger 2800 – 2600 Ma (Gee et al. 1981) Yilgarn block granite-greenstone terrain by Hoefs and Binns (1978) show that for the best preserved material the isotopic compositions of the komatiites and komatiitic basalts are similar to those of the equivalent Barberton rocks. This similarity shows that hydrothermal fluids of uniform isotopic composition existed at 3500 Ma and ~2700 Ma ago. A possible source of water for these fluids is deep-seated water from the mantle. As mantle derived water would be in equilibrium with mantle minerals at high temperatures it would be expected to maintain a constant isotopic composition throughout the geological record. Thus, if water from this source gave rise to the hydrothermal fluids responsible for the alteration of greenstone belt lavas it would account for both the uniformity of isotopic compositions of the meta-lavas and existence of these fluids during a large part of the history of the Earth.

The participation of Archaean seawater is more difficult to evaluate because if it had a similar isotopic composition to modern seawater the processes controlling the isotopic composition of Archaean oceans would involve such factors as the volume of ocean water, the proportions of continental and oceanic crust, the rate of production of new crust and the weathering and sedimentation rates. None of these factors is well constrained for the Archaean. Nevertheless, the existence of Archaean oceans with a $\delta^{18}O$ value of 0 remains a possibility. Beaty and Taylor (1982) have demonstrated that the Munro Township flows could also have equilibrated with ocean water of $\delta^{18}O = 0$ and Costa et al. (1980) have shown that sedimentary talc in Archaean greenstone belts could have formed by the reaction of wall rock with heated ocean water which had an oxygen isotope composition similar to that of the modern ocean.

Gregory and Taylor (1981) have argued that hydrothermal interaction of oceanic crust and seawater results is no net change of the ^{18}O content of average oceanic crust or seawater and that such buffering of the isotopic composition of seawater would have occurred in the past, provided world wide spreading rates exceeded 1 km^2/yr. They further suggest that this conclusion can be extended as far back in time as the Archaean and that ocean water has probably had a constant $\delta^{18}O$ value of -1 to $+1$ during most of Earth history. These contentions are entirely compatible with the ^{18}O data obtained in this work from the 3500 Ma old Barberton greenstone belt. If the isotopic data for the komatiitic pillows (average of ten pillows is 6.6‰) and komatiites (average of Richard's Flow and other komatiites in Table 4 except 20-J and HSS-105 is 5.2‰) are representative of the lower formations of the Onverwacht, then the bulk $\delta^{18}O$ value of this Archaean crustal material is 6.2 (assuming it consists of 70% komatiitic basalt and 30% komatiite). Gregory and Taylor (1981) have calculated the steady state bulk fractionation of ^{18}O between oceans and mid-ocean ridge basalt magmas to be 6.1 ± 0.3. The value calculated above for the Archaean crust is in excellent agreement with this value and suggests that the overall Archaean crustal temperature was similar to that prevailing at present.

Based on the simple modelling discussed in this work, the isotopic composition of the komatiite lavas could have been established by equilibration with deep-seated waters at relatively high temperatures (240 – 450 °C), mixtures of deep-seated water and water with $\delta^{18}O = 0$ at intermediate temperatures (240 – 100 °C) or Archaean seawater at ~100 °C. Either of the latter two possibilities are consistent with the oxygen isotope compositions of serpentine minerals and pillow lavas and the chemical alteration patterns in the pillows which indicates that Archaean seawater had a significant influence during the alteration and metamorphism of these materials. It is therefore concluded that from the data presented for the Komati Formation lavas an ocean floor origin for this volcanic pile is consistent with the data, and that magmatic waters from the intrusive plutons had little effect on the isotopic composition of the typical greenschist facies meta-lavas except immediately adjacent to the intrusive contacts.

Acknowledgements. This work was carried out in part with financial support from the CSIR and the University of Cape Town Staff Research Fund. The authors are indebted to J. P. Willis for providing the Ga data and assistance with the XRF analyses and to D. Wenner and N. Ikin for their comments on various drafts of this paper. The staff and students of Trailer 14 are thanked for providing guidance and other help to H. S. S. during his stay in Menlo Park.

References

Abbey S (1975) Studies in "Standard Samples" of silicate rocks and minerals. In: Part 4: 1974 ed. "Usable" Values. Geol Surv Pap 74–41, p 1

Anhaeusser CR (1973) The evolution of the early Precambrian crust of southern Africa. Philos Trans R Soc Lond A 273:359

Anhaeusser CR (1975) Precambrian tectonic environments. Ann Rev Earth Planet Sci 3:31

Anhaeusser CR, Roering C, Viljoen M, Viljoen RP (1968) The Barberton Mountain Land: a model of the elements and evolution of an Archaean fold belt. Trans Geol Soc S Afr 71 (annex):225

Arndt NT, Naldrett AJ, Pyke DR (1977) Komatiitic and iron-rich tholeiitic lavas of Munro Township, northeast Ontario. J Petrol 18:319

Barton JM Jr (1981) The pattern of Archaean crustal evolution in Southern Africa as deduced from the evolution of the Limpopo Mobile Belt and the Barberton granite-greenstone terrain. In: Glover JE, Groves DI (eds) Archaean geology: Second International Symposium, Perth, 1980. Spec Publ Geol Soc Aust 7:21

Beaty DW, Taylor HP Jr (1982) The oxygen isotope geochemistry of komatiites: evidence for water-rock interaction. In: Arndt NT, Nisbet EG (eds) Komatiites. Allen, London, p 267

Bottinga Y, Javoy M (1973) Comments on oxygen isotopic geothermometry. Earth Planet Sci Lett 20:250

Bryan W, Finger L, Chayes F (1969) Estimating proportions in petrogenic mixing equations by least squares approximation. Science 163:926

Cann JR (1970) Rb, Sr, Y, Zr and Nb in some ocean floor basaltic rocks. Earth Planet Sci Lett 10:7

Clayton RN, Mayeda TK (1963) The used of bromine pentafluoride in the extraction of oxygen from oxides and silicates for isotopic analysis. Geochim Cosmochim Acta 27:43

Condie KC, Viljoen MJ, Kable EJD (1977) Effects of alteration on element distributions in Archaean tholiites from the Barberton greenstone belt, South Africa. Contrib Mineral Petrol 64:75

Costa UR, Fyfe WS, Kerrich R, Nesbitt HW (1980) Archaean hydrothermal talc evidence for high ocean temperatures. Chem Geol 30:341

De Wit MJ (1982) Gliding and overthrust nappe tectonics in the Barberton greenstone belt. J Struc Geol 4:117

Floyd PA, Winchester JA (1975) Magma type and tectonic setting discrimination using immobile elements. Earth Planet Sci Lett 27:211

Gee RD, Baxter JL, Wilde SA, Williams IR (1981) Crustal evolution of the Pilbra Block, Western Australia. In: Glover JE, Groves DI (eds) Archaean geology: Second International Symposium, Perth, 1980. Spec Publ Geol Soc Aust 7:43

Green DH (1972) Archaean greenstone belts may include terrestrial equivalents of lunar maria? Earth Planet Sci Lett 15:263

Green DH, Nicholls IA, Viljoen MJ, Viljoen RP (1975) Experimental demonstration of the existence of peridotitic liquids in earliest Archaean magmatism. Geology 3:11

Gregory RT, Taylor HP Jr (1981) An Oxygen Isotope Profile in a Section of Cretaceous Oceanic Crust, Samail Ophiolite, Oman: Evidence for $\delta^{18}O$ buffering of the oceans by deep (>5 km) seawater-hydrothermal circulation at mid-ocean ridges. J Geophys Res 86:2737

Groves DI, Batt WD (1984) Spatial and temporal variations of Archaean metallogenic associations in terms of evolution of granitoid-greenstone terrains with particular emphasis on the Western Australian shield. Chap. 4, this Vol., 73–98

Hamilton PJ, Evensen NM, O'Nions RK, Smith HS, Erlank AJ (1979) Sm-Nd dating of Onverwacht Group volcanics, Southern Africa. Nature 279:298

Hart SR, Erlank AJ, Kable EJD (1974) Sea floor basalt alteration: Some chemical and Sr isotopic effects. Contrib Mineral Petrol 44:219

Hoefs J, Binns RA (1978) Oxygen-isotope compositions in Archaean rocks from Western Australia, with special reference to komatiites. Geol Surv Open File Rep 78–101, p 180

Humphris SE, Thompson G (1978a) Hydrothermal alteration of oceanic basalts by sea water. Geochim Cosmochim Acta 42:107

Humphris SE, Thompson G (1978b) Trace element mobility during hydrothermal alteration of oceanic basalts. Geochim Cosmochim Acta 42:127

Hunter DR (1974) Crustal development in the Kaapvaal craton. 1 The Archaean. Precambrian Res 1:259

Jahn B-M, Gruau G, Glikson AY (1982) Komatiites of the Onverwacht Group, S. Africa: REE geochemistry, Sm/Nd age and mantle evolution. Contrib Mineral Petrol 80:25

Jahn B-M, Shih CY (1974) On the age of the Onverwacht Group, Swaziland Sequence, South Africa. Geochim Cosmochim Acta 39:1679

Knauth LP, Lowe DR (1978) Oxygen isotope geochemistry of cherts from the Onverwacht Group (3.4 billion years), Transvaal, South Africa, with implications for secular variations in the isotopic composition of cherts. Earth Planet Sci Lett 41:209

Kyser TK, O'Neil JR (1978) Oxygen isotope relations among oceanic tholeiites, alkali basalts, and ultramafic nodules. Geol Surv Open File Rep 78–701, p 237

Kyser TK, O'Neil JR, Carmichael ISE (1981) Oxygen isotope thermometry of basic lavas and mantle nodules. Contrib Mineral Petrol 77:11

Kyser TK, O'Neil JR, Carmichael ISE (1983) Genetic relationships among basic lavas and ultramafic nodules: Evidence from oxygen isotope compositions. Contrib Mineral Petrol 81:88

Melson WG, van Andel TH (1966) Metamorphism in the Mid-Atlantic Ridge, 22° N latitude. Mar Geol 4:165

Moore JG (1975) Mechanism of formation of pillow lava. Amer Scientist 63:269

Moore JG, Cristofolini R, Lo Giudice A (1971) Development of pillows on the submarine extension of recent lava flows, Mount Etna, Sicily. US Geol Surv Prof Pap 750-c, p 89

Muehlenbachs K, Clayton RN (1972a) Oxygen isotope studies of fresh and weathered submarine basalts. Can J Earth Sci 9:172

Muehlenbachs K, Clayton RN (1972b) Oxygen isotope geochemistry of submarine greenstones. Can J Earth Sci 9:471

Pearce JA, Cann JR (1973) Tectonic setting of basic volcanic rocks determined using trace element analyses. Earth Planet Sci Lett 19:290

Perry EC Jr (1967) The oxygen isotope chemistry of ancient cherts. Earth Planet Sci Lett 3:62

Perry EC Jr (1978) The oxygen isotope composition of 3,800 m.y. old metamorphosed chert and iron formation from Isukasia, West Greenland. J Geol 86:223

Perry EC Jr, Tan C (1972) Significance of oxygen and carbon isotope variations in early precambrian cherts and carbonate rocks of Southern Africa. Bull Geol Soc Am 83:647

Pyke DR, Naldrett AJ, Eckstrand OR (1973) Archaean ultramafic flows in Munro Township, Ontario. Geol Soc Am Bull 84:955

Scarfe CM, Wyllie PJ (1967) Experimental redetermination of the upper stability limit of serpentine up to 3 kbar pressure. Trans Am Geophys Union 48:225

Scott RB, Hajash A Jr (1976) Initial submarine alteration of basaltic pillow lavas: a microprobe study. Am J Sci 276:480

Sheppard MF (1980) Isotopic evidence for the origins of water during metamorphic processes in oceanic crust and ophiolite complexes. Colloques Internationaux du CNRS No. 272 – Association Mafiques Ultra-Mafiques Dans Les Orogenes, p 135

Sheppard SMF, Dawson JB (1975) Hydrogen, carbon and oxygen isotope studies of megacryst and matrix minerals from Lesothan and South African Kimberlites. In: Ahrens LH, Dawson JB, Duncan AR, Erlank AJ (eds) Physics and chemistry of the Earth. 9. Pergamon, Oxford, p 747

Smith HS (1980) Aspects of the geochemistry of Onverwacht Group lavas from the Barberton greenstone belt. Ph D Thesis, University of Cape Town, South Africa. pp 237

Smith HS, Erlank AJ (1982) Geochemistry and petrogenesis of komatiites from the Barberton greenstone belt, South Africa. In: Arndt NT, Nisbet EG (eds) Komatiites. Allen, London, p 347

Smith HS, Erlank AJ, Duncan AR (1980) Geochemistry of some ultramafic komatiite lava flows from the Barberton Mountain Land, South Africa. Precambrian Res 11:399

Smith RE (1968) Redistribution of major elements in the alteration of some basic lavas during burial metamorphism. J Petrol 9:191

Smith RE, Smith SE (1976) Comments on the use of Ti, Zr, Y, Sr, K, P and Nb in classification of basaltic magmas. Earth Planet Sci Lett 32:144

Stakes DS, O'Neil JR (1982) Mineralogy and stable isotope geochemistry of hydrothermally altered oceanic rocks. Earth Planet Sci Lett 57:285

Taylor HP Jr (1978) Oxygen and hydrogen isotope studies of plutonic granitic rocks. Earth Planet Sci Lett 38:177

Taylor HP Jr, Epstein S (1962) Relationship between O^{18}/O^{16} ratios in coexisting minerals of igneous and metamorphic rocks. Part 2. Application to petrologic problems. Geol Soc Am Bull 73:675

Viljoen RP, Viljoen MJ (1969a) The effects of metamorphism and serpentinization on the volcanic and associated rocks of the Barberton region. Geol Soc S Afr Spec Publ 2:29

Viljoen MJ, Viljoen RP (1969b) The geology and geochemistry of the Lower Ultramafic Unit of the Onverwacht Group and a proposed new class of igneous rock. Geol Soc S Afr Spec Publ 2:55

Viljoen MJ, Viljoen RP (1969c) Evidence for the existence of mobile extrusive peridotitic magma from the Komati Formation of the Onverwacht Group. Geol Soc S Afr Spec Publ 2:87

Viljoen RP, Viljoen MJ (1970) The geology and geochemistry of layered ultramafic bodies of the Kaapmuiden area, Barberton Mountain Land. Geol Soc S Afr Spec Publ 1:661

Viljoen MJ, Viljoen RP, Smith HS, Erlank AJ (1983) Geological, textural and geochemical features of komatiitic flows from the Komati Formation. Geol Soc S Afr Spec Publ 9:1

Weaver BL, Tarney J (1979) Thermal aspects of komatiite generation and greenstone belt models. Nature 299:689

Wenner DB, Taylor HP Jr (1971) Temperature of serpentinization of ultramafic rocks based on O^{18}/O^{16} fractionation between coexisting serpentine and magnetite. Contrib Mineral Petrol 32:165

Wenner DB, Taylor HP Jr (1973) Oxygen and hydrogen isotope studies of the serpentinization of ultramafic rocks in oceanic environments and continental ophiolite complexes. Am J Sci 273:207

Wenner DB, Taylor HP Jr (1974) D/H and O^{18}/O^{16} studies of serpentinization of ultramafic rocks. Geochim Cosmochim Acta 38:1255

Williams DAC, Furnell RG (1979) Reassessment of part of the Barberton type area, South Africa. Precambrian Res 9:325

Willis JP, Ahrens LH, Danchin RV, Erlank AJ, Gurney JJ, Hofmeyer PK, McCarthy TS, Orren MJ (1971) Some interelement relationships between lunar rocks and fines and stony meteorites. In: Levinson ED (ed) Proceedings of the second lunar science conference. 2. MIT Press, Cambridge, Mass, p 1123

Willis JP, Erlank AJ, Gurney JJ, Theil RH, Ahrens LH (1972) Major, minor and trace element data for some Apollo 11, 12, 14 and 15 samples. In: Heymann LD (ed) Proceedings of the third lunar science conference. 2. MIT Press, Cambridge, Mass, p 1269

Petrology and Geochemistry of Layered Ultramafic to Mafic Complexes from the Archaean Craton of Karnataka, Southern India

C. SRIKANTAPPA[1,2], P. K. HÖRMANN[1] and M. RAITH[1,3]

Contents

1 Introduction	139
1.1 Analytical Procedure	140
1.2 Regional Geological Setting	140
1.3 Structure of the Sargur Region	142
2 Field Relations and Petrography of the Ultramafic and Mafic Rocks	144
2.1 Ultramafic Unit	144
2.2 Mafic Unit	145
2.3 Late Gabbros	145
2.4 Metamorphism	146
3 Geochemistry	147
3.1 Major Elements	147
3.2 Trace Elements	151
3.3 Distribution of the Rare Earth Elements (REE)	153
4 Discussion and Conclusions	155
References	157

Abstract

The Archaean ultramafic to mafic rocks emplaced into the Sargur supracrustal series (>3.0 Ga) in the Karnataka craton of southern India vary in composition from spinel dunite, harzburgite, bronzite peridotite, pyroxenite to gabbro and gabbroic anorthosite. Relic igneous features like layering, cumulus and poikilitic textures, chromite seams with "way up" stratigraphy and occasional cryptic layering indicate an origin by a process of gravitational differentiation in a stable crustal environment. The rock units are cut by late gabbro dykes.

The complexes show a predominantly tholeiitic trend in the AFM diagram. Major as well as trace element data differ significantly from those of komatiites. The element and oxide ratios TiO_2/Al_2O_3, TiO_2/CaO, Ti/Zr, Ti/Y, Y/Zr and Sc/Zr are non-chondritic. The REE distribution patterns of anorthositic gabbros show an enrichment by a factor of $5-15$ compared to chondrites and LREE/HREE ratios of $1-3$. The patterns display weak positive Eu anomalies.

1 Mineralogisch-Petrographisches Institut, Universität Kiel, Olshausenstraße 40 – 60, 2300 Kiel, FRG
2 Present address: Department of Geology, University of Mysore, Manasa Gangotri, Mysore – 570006, India
3 Present address: Mineralogisch-Petrologisches Institut, Universität Bonn, Poppelsdorfer Schloß, 5300 Bonn, FRG

Archaean Geochemistry (ed. by A. Kröner et al.)
© Springer-Verlag Berlin Heidelberg 1984

The Ce_N/Yb_N ratios correspond closely to those of Archaean layered anorthositic gabbros from Canada and West Greenland. The late gabbros exhibit fractionated REE patterns enriched by a factor of about 40 compared to chondrites.

The stratiform igneous bodies are comparable with the layered meta-igneous complexes of the Lewisian in NW Scotland and the Limpopo belt of southern Africa. The nature of the igneous complexes and the asscociated sediments favours an intracrustal development of the Sargur Group rather than the formation by rifting and ocean opening processes. Conditions of granulite facies metamorphism ($700 \pm 50\,°C$; 9 ± 1 kb) indicate a minimum crustal thickness of about 35 km at ca. 2.6 Ga ago.

1 Introduction

The study of Archaean ultramafic and mafic complexes has received much attention in recent years as these rocks represent important components of many greenstone belts and of supracrustal series within high-grade gneiss terranes (Windley and Smith 1976; Glickson 1977; Windley et al. 1981). These complexes now occur as lenses, sheets and disrupted layers, tectonically interleaved, deformed and metamorphosed together with the associated supracrustal rocks. Despite a strong tectonic and metamorphic overprint, the original igneous stratigraphy, the process of generation and differentiation of the magmas and their mode of emplacement may be elucidated in many cases by careful investigation of the field relations, petrography and geochemistry of the complexes.

Apart from the layered igneous complexes, a distinctive group of ultramafic to mafic rocks, the komatiite-tholeiite series, occurs in many Archaean greenstone belts, either forming the basal members (Viljoen and Viljoen 1969; Arndt et al. 1977; Blais et al. 1978) or overlying the sedimentary units (Bickle et al. 1975; Henderson 1981; Kröner et al. 1981). Komatiites were thought to be restricted to Archaean greenstone belts, indicating unusually high geothermal gradients in the upper mantle before 2.5 Ga ago. The recognition of Proterozoic and Palaeogene komatiites (Gansser et al. 1979), however, indicated that such conditions may not have been unique to the Archaean. In highly deformed and metamorphosed Archaean terranes, the distinction between komatiites and layered igneous complexes has major implications for models of crustal evolution.

In the Archaean Karnataka craton of southern India, deformed and metamorphosed ultramafic to mafic complexes occur within several belts of supracrustal rocks, e.g., the Nuggihalli, Holenarsipur and Sargur schist belts (Swami Nath et al. 1974).

Viswanathan (1974), Naqvi and Hussain (1979), Hussain et al. (1982) and Jafri et al. (1982) suggest that the ultramafic to mafic rocks in the Nuggihalli and Holenarsipur belts represent komatiitic volcanic rocks that form the basal members of the individual schist belts. Fine-grained anorthositic rocks associated with the ultramafic complexes were compared to lunar anorthosites and interpreted as primitive anorthositic basalt lavas (Drury et al. 1978; Naqvi and Hussain 1979).

Based on detailed geological and structural work in these belts, Swami Nath et al. (1974), Janardhan et al. (1978) and Srikantappa (1979) have shown however, that the ultramafic to mafic complexes are stratiform layered intrusives and are not komatiite metalavas. The fine-grained anorthosites of the Holenarsipur belt are now regarded as cumulates from basic magmas, their fine-grained texture being attributed to intense shear deformation (Bhaskara Rao and Veerabhadrappa 1982).

These contrasting views regarding the origin of the ultramafic to mafic complexes led to the postulation of different models for the evolution of the Archaean crust in southern India. Formation of the greenstone belts by a process of initial rifting and eventual ocean opening was proposed by Naqvi (1976), in contrast to an intracrustal development implied by Swami Nath et al. (1974).

In order to identify the true nature of the complexes, the ultramafic to mafic rocks from the high-grade Sargur schist belt in the Karnataka craton have been investigated. The discussion of their petrogenesis is based on field observations and on detailed petrographic and geochemical data. The significance of these complexes in the Sargur high-grade terrane is discussed with regard to the possible nature of early continental crust in southern India. The P-T conditions of the high-grade phase of metamorphism that affected these rocks are evaluated by geothermobarometry on garnet-pyroxene bearing mafic rocks.

1.1 Analytical Procedure

Major element compositions were determined by colorimetry, atomic absorption spectrometry (AAS) and X-ray fluorescence spectrometry (XRS). References are given in Hörmann et al. (1973). Rb was separated from the rocks by an Amberlite IR 120 ion exchange method (Hörmann and Fulert 1980) and measured subsequently by AAS. The elements Sr, Ba, Ni, Cr, V, Sc, Y, and Zr were measured by ICP spectrometry. The REE were isolated by means of ion exchange chromatography with DOWEX 50WX8 (Broekaert and Hörmann 1981) and measured by ICP spectrometry.

Precision of the major element analyses is better than 1%. The standard deviation of the Rb measurements is 10%, those of Sr and Ba 1%. The standard deviation of the Ni, Cr, V, Sc, Y, and Zr measurements is 10% for contents lowert han 10 ppm, 5% at the 10 ppm level and smaller than 1% for contents higher than 100 ppm. The precision of the REE measurements is better than 5%, except for Ho where it is about 10%.

Compositional data of minerals reported in this paper are based on microprobe analyses carried out on a wavelength dispersive microprobe (Siemens) at the Mineralogical Institute, University of Kiel. Detailed analyses are reported in Srikantappa et al. (1984).

1.2 Regional Geological Setting

The Archaean Karnataka craton of southern India is a classical example of a greenstone-granite terrane which grades progressively into a high-grade granu-

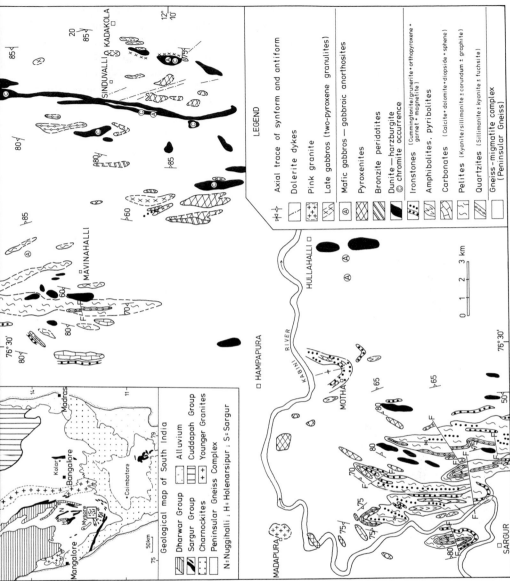

Fig. 1. Geological map of the Sargur schist belt after Srikantappa (1979) and Viswanatha and Ramakrishnan (1981)

lite-gneiss terrane from north to south (Fig. 1). There are two groups of supra-crustal sequences, the older Sargur Group (>3.0 Ga) and the younger Dharwar Group (2.9 – 2.5 Ga) (cf. Glickson 1982). They occur as belts and as isolated enclaves within a vast complex of gneisses, migmatites and granites, generally termed the Peninsular Gneiss Complex. The Sargur Group is represented by a series of orthoquartzites, calc-silicate rocks, marbles, metapelites and banded magnetite quartzites, interlayered with basic metavolcanic rocks (Janardhan et al. 1978). This lithological assemblage indicates an environment typical of intra-cratonic basins or stable continental margins. The Sargur supracrustal rocks have been affected by at least two major events of deformation and metamorphism, ca. 3.0 and 2.6 Ga ago (Chadwick et al. 1978; Janardhan et al. 1979). The type area of these supracrustal series is the high-grade amphibolite to granulite facies terrane around Sargur (Swami Nath et al. 1974; Janardhan et al. 1978, Fig. 1), buth equivalent belts have been recognized in the lower grade part of the craton, e.g., the Holenarsipur, Nuggihalli and Nagamangala belts (Swami Nath et al. 1974). In all these Sargur-type schist belts, a typical lithologic component are chromite-bearing ultramafic and mafic igneous complexes.

The Dharwar supracrustal succession, mainly exposed in the northern part of the craton, comprises quartzites, banded iron formations, phyllites and meta-greywackes, interlayered with basic to acid volcanic rocks (Swami Nath et al. 1974). The sequence has been deposited unconformably upon the migmatite-gneiss basement with the enclaves of Sargur supracrustal rocks (Chadwick et al. 1978, 1981). The Dharwar succession thus may have formed in a continental basin environment, perhaps related to intracratonic rifting. The whole unit un-derwent deformation and greenschist to amphibolite facies metamorphism around 2.5. Ga ago.

1.3 Structure of the Sargur Region

Structural investigations in the Sargur region have shown that the supracrustal rocks, together with the ultramafic to mafic complexes, have been affected by three major deformational episodes and two phases of high-grade metamor-phism and migmatization (Chadwick et al. 1978; Janardhan et al. 1979; Srikan-tappa 1979).

During the first deformational episode tight to isoclinal F_1 folds, trending N-S to 20°E and plunging towards the north, were formed. The second deforma-tional episode produced tight to open F_2 folds which are coaxial with the F_1 folds. The regional conformable schistosity which is now parallel to bedding following isoclinal folding, was developed during this episode, including high-grade meta-morphism and migmatization. The third phase of folding is represented by F_3 folds, trending NE to E-W, and has affected the entire Sargur region. Develop-ment of ENE to E-W trending faults subsequently resulted in displacement and rotation of the rock units (Fig. 1). Emplacement of pink porphyritic granites is related to this episode.

The original relationship between the high-grade Sargur supracrustals, the as-sociated ultramafic to mafic complexes and the surrounding Peninsular gneisses is difficult to establish because of polyphase deformation and high-grade metamorphism. Field observations, mainly in the well exposed northern part of the Sargur region, show that ultramafic to mafic bodies are almost always bordered by thin bands of metasediments, e.g. banded magnetite quartzite, marble, biotite-silimanite-graphite bearing mica schist and garnet-biotite-sillimanite gneiss. Locally, some of these metasediments occur as lenses or even as bands within the meta-igneous bodies. Alteration by metasomatism has mainly affected the carbonate rocks, leading to the formation of calc-silicate assemblages. Ultramafic rocks, emplaced into graphite-bearing metapelites, have assimilated graphite (Rama Rao 1926).

The structural evidence indicates that the emplacement and differentiation of the ultramafic and mafic complexes took place before or during the first deformational episode. The bodies now occur along the axial planes of the F_1 folds. The chromitite layers in the ultramafic rocks as well as the layering and gradation observed in the mafic rocks are generally conformable to the bedding and regional schistosity. The igneous bodies have been tilted and rotated during the second and third deformational episode. This structural pattern is well exposed in the area near Motha (Fig. 1). Possible signs of contact metamorphism connected with the emplacement of the igneous complexes, if present, must have been destroyed as a result of the superimposed upper amphibolite to granulite facies metamorphism.

A reconstructed lithostratigraphic "model" succession of the Sargur region is given in Table 1.

Table 1. Reconstructed lithostratigraphic model succession of the Sargur region based on Janardhan et al. (1979) and Srikantappa (1979)

	Dolerite dykes
	Pink granites
	Upper amphibolite to granulite facies metamorphism (formation of charnockites and migmatitic gneisses)
	Late gabbros (two pyroxene granulites)
Stratified igneous complexes	Mafic gabbros/gabbroic anorthosites
	Ultramafics (dunite-harzburgite-pyroxenites)
	Probable metamorphic event?
Sargur supracrustal sequence	Iron stones (cummingtonite/grunerite + garnet + orthopyroxene + magnetite)
	Manganiferous horizons
	Amphibolite and pyribolites
	Carbonates (calcite + dolomite + diopside ± sphene)
	Pelites (kyanite + sillimanite ± corundum + graphite)
	Quartzites (sillimanite ± kyanite ± fuchsite)
	Sialic basement? (quartzofeldspathic gneisses not distinguished from the Peninsular Gneiss)

2 Field Relations and Petrography of the Ultramafic and Mafic Rocks

2.1 Ultramafic Unit

The ultramafic unit is composed of spinel dunite, harzburgite, minor lherzolite, bronzite peridotite and bronzitite. The various rock types exhibit gradational contacts and locally show compositional layering and cumulus textures. Chrome spinel, characteristic in dunites and harzburgites, occurs either concentrated in layers or as disseminated crystals as in the Sinduvalli ultramafic body (Fig. 1 and Srikantappa et al. 1980). The chromite layers vary in thickness from a few mm to a maximum of $3-4$ cm and alternate in a rhythmic fashion with thicker olivine-rich (Fo_{93-95}) layers. Five to ten such individual layers have been recognized within a thickness of 15 m. The occurrence of spinel layers, grading upwards into layers of olivine with interstitial spinel to pure olivine layers, indicates relic igneous layering textures. These chromitite layers exhibit a constant way up stratigraphy with a sharp base and a gradational top, indicating gravitational segregation of chromite from the magma − a feature of many layered ultramafic complexes (Irvine 1967; Thayer 1970). Some of the chromitite layers show gradation in grain size varying from 0.45 mm to 2.8 mm. Nodular chromitites are typically absent. Instead, chromitites with antileopard structure are common.

Apart from the chromite layering, the ultramafic rocks of the Sargur region show other relic igneous features. Near Sinduvalli a thick zone of chromite-bearing harzburgite (~ 100 m), now considerably serpentinized, grades into bronzite peridotite (~ 20 m) and bronzitite ($20-25$ m). In the peridotite the modal amount of olivine decreases with increase in orthopyroxene (En_{78}). This zone is characterized by large crystals of bronzite poikilitically enclosing olivine (Fo_{73}) and minor interstitial plagioclase (An_{60}). The pyroxenites show cumulus orthopyroxene (En_{84}) with fine exsolution lamellae of clinopyroxene and an intercumulus assemblage consisting of untwinned plagioclase (An_{70}) and secondary brownish hornblende. This sort of gradation is observed in many ultramafic bodies in the Sargur region as, for example, in the Doddakanya and Mavinahalli areas.

Although there are occurrences of little deformed and less altered ultramafic bodies in the Sargur region, in general they have undergone considerable alteration during regional high-grade metamorphism. Small enclaves of dunite and pyroxenite in the migmatized gneisses are commonly represented by Ca-amphibole + talc + chlorite ± spinel rocks. Some of the isolated ultramafic bodies, however, contain the assemblage orthopyroxene + olivine + magnesio-hornblende + magnesite + herzynitic spinel (Janardhan and Srikantappa 1977). In their field relations and mineralogy these rocks resemble the sagvandites of the type area in Norway (Schreyer et al. 1972; Moore 1977) and from northern Finland (Papunen et al. 1979). The development of the orthopyroxene + magnesite + olivine assemblages in the Sargur region probably took place at similar P-T conditions but at higher fugacities of CO_2 than that of the Ca-amphibole + talc + chlorite assemblages (cf. Ohnmacht 1974).

2.2 Mafic Unit

Overlying the ultramafic rocks is a series of mafic rocks represented by mafic gabbros, gabbros, gabbroic anorthosites, anorthositic gabbros and minor anorthosites[5]. These rocks occur either as concordant sheets, often showing gradational contacts with the ultramafic rocks, or as lenses and layers in the adjacent migmatitic gneisses. Gradations from gabbro to anorthositic gabbro and gabbroic anorthosite are exposed in the Sinduvalli area. The gabbroic rock types often show ellipsoidal aggregates of plagioclase (An_{80}) surrounded by hornblende, and these are interpreted as original cumulus textures. The gabbroic rocks in this area do not show conspicuous mineral layering. On the other hand, numerous concordant sheets of mafic gabbros, anorthositic gabbros and anorthosites exposed near Hullahalli (Fig. 1) show cumulus plagioclase and crude layering. Chromitite layers, common in other Archaean anorthosites as, for example, in Sittampundi (Subramaniam 1956; Ramadurai et al. 1975) and in the Fiskenaesset complex of West Greenland (Ghisler 1976), were not found in these rocks. In these respects and in their highly deformed nature, the gabbroic members of the Sargur region resemble the gabbroic anorthosites of the Scourie area (Bowes et al. 1964; Sills et al. 1982) and the anorthositic rocks of the Limpopo belt (Robertson 1973; Hor et al. 1975).

High-grade metamorphism in the Sargur region has led to thorough recrystallization of the igneous assemblages in the mafic rocks. Under hydrous conditions of the amphibolite facies, the main alteration was the replacement of the original clinopyroxene by brownish-green hornblende. Under less hydrous conditions garnet developed at the expense of orthopyroxene and plagioclase, leading to garnet + clinopyroxene + orthopyroxene + plagioclase + hornblende assemblages typical of the hornblende granulite facies. Primary plagioclase is often recrystallized to a polygonal mosaic and generally exhibits compositional zoning (An_{75} core to An_{58} rim). The formation of scapolite (Me_{64}) in the assemblage with plagioclase (An_{70}) and brownish-green hornblende observed in some of the gabbros indicates a rather high CO_2 fugacity during metamorphism (Goldsmith and Newton 1977).

2.3 Late Gabbros

There are gabbros ranging in composition from mafic gabbro to norite that cut across the Sargur supracrustal rocks as well as the ultramafic-mafic complexes. They postdate the F_1 folding and have been subjected to high-grade metamorphism during the F_2 episode. They also occur as en-echelon, dyke-like bodies or as patches within the younger migmatic gneisses. These gabbros show well preserved gabbroic texture and original layering now accentuated by metamorphic segregation of minerals. Porphyritic orthopyroxene crystals in norites

---

5 Gabbros are classified according to Windley et al. (1973) into mafic gabbros (65 – 90); gabbros (35 – 65); anorthositic gabbros (20 – 35); gabbroic anorthosite (10 – 20) and anorthosites (0 – 10) based on the modal amounts of ferromagnesian minerals.

show bent and broken lamellae, and with increase in deformation and recrystalli- zation a granoblastic texture develops with garnet growing at the expense of plagioclase and pyroxene. These gabbros are now represented by hornblendites (Hbl + Pl ± Cpx) and two-pyroxene granulites (Opx + Cpx + Gar + Pl + Hbl ± Qtz).

2.4 Metamorphism

The standard methods of geothermometry/geobarometry, developed in recent years by various workers (see review by Essene 1982), were used in this study to examine the P-T conditions of metamorphism. The garnet-two-pyroxene bearing assemblages of the mafic rock types are ideal for estimating the P-T conditions of metamorphism. Temperature and pressure estimates obtained using various models are reported in Table 2.

Table 2. Temperature and pressure estimates for the Sargur mafic rocks

Sample		Temperature (°C)			Pressures (kb)		
		Opx-Cpx Wood and Banno (1973)[a]	Wells (1977)[a]	Powell (1978)	Gt-Cpx Ellis and Green (1979)	Opx (eqv. 7) Perkins and Newton (1981)	Cpx (eqv. 9) Perkins and Newton (1981)
S_3	Core	752	728	689	729	9.10	8.25
	Rim	743	713	678	707	9.03	8.01
S_4	Core	719	711	667	667	9.08	8.11
	Rim	718	677	634	635	8.98	7.90
S_5	Core	746	742	712	688	9.36	8.91
	Rim	741	676	669	666	8.94	7.73
S_{11}	Core	731	718	717			
	Rim	730	712	633			
$\bar{x}$	Core	737	724	696	695	9.18	8.42
	Rim	733	694	653	669	8.98	7.88

[a] temperature values corrected by −60°C

The mean temperature estimates derived from the two-pyroxene thermo- meters of Wood and Banno (1973) and Wells (1977), corrected for −60°C (see Raith et al. 1983), are 743 ± 9 °C and 729 ± 17 °C, respectively, for the cores and 738 ± 11 °C and 700 ± 17 °C, respectively, for the rims. Powell's (1978) two-pyro- xene model gives temperature estimates of 706 ± 12 °C for the core and 660 ± 19 °C for the rim composition, and these are still within the range of error quoted for the individual models. Temperatures obtained for the garnet-clino- pyroxene thermometer of Ellis and Green (1979) give mean values of 708 °C for the core and 686 °C for the rim compositions, and these are in agreement with the results of two-pyroxene thermometry.

The mean pressure estimates derived for the garnet-bearing granulites, fol- lowing equations 7 and 9 of Perkins and Newton (1981), are 9 kb and 8 ± 1 kb, respectively.

Based on the results of geothermobarometry, metamorphic conditions of 700 $\pm 50\,°C$ and 9 ± 1 kb are thus obtained for the Sargur region. These P-T data match fairly well the estimate of $690 \pm 60\,°C$ and 7.6 kb (rim compositions) reported previously by Raith et al. (1983) for the Sargur area.

3 Geochemistry

Workers on rocks of the Sargur Group have recently claimed the occurrence of komatiites in the basal sections of the Holenarsipur and Nuggihalli schist belts (Naqvi 1976; Naqvi and Hussain 1979; Hussain et al. 1982). The recognition of the komatiitic nature of these rocks was based mainly on geochemical arguments. Their chemical data were compared with those of the classical greenstone belts of the Barberton Mountain Land, South Africa. However, the identification of volcanic features in rocks of many greenstone belts that have undergone intense deformation and high-grade metamorphism is difficult, if not impossible. In the absence of primary volcanic features the recognition of the ultramafic-mafic complexes as komatiites must therefore rely entirely on geochemical data, particularly on the less mobile trace elements that may have chondritic abundances (Nesbitt et al. 1979).

It was the aim of the present geochemical study to find arguments for the presence or absence of komatiitic rocks in the ultramafic-mafic complexes of the Sargur region.

3.1 Major Elements

Major element analyses for 21 ultramafic rocks, 8 gabbros and gabbroic anorthosites, 5 late gabbros and 1 olivine dolerite are presented in Table 3. Most of the rocks are olivine normative. One sample from a highly migmatized area has normative quartz, possibly due to metamorphic introduction of SiO_2 (Fig. 2).

The ultramafic rocks have peridotitic to harzburgitic compositions with MgO ranging from 38 to 44%, Al_2O_3 from 1 to 3% and CaO from 0.2 to 2%. The CaO/Al_2O_3 ratio of the ultramafic rocks varies from 0.1 to 1, the average value being 0.8. The pyroxenites are characterized by a CaO/Al_2O_3 ratio of 1.0.

The mafic gabbros have CaO/Al_2O_3 ratios lower than 1. Such low ratios are characteristic of tholeiitic magmas, whereas basaltic komatiites are expected to have CaO/Al_2O_3 ratios higher than 1 (Viljoen and Viljoen 1969). Thus, the data for the mafic gabbros do not favour a komatiitic origin. The relationship to the tholeiitic magmas is also evident from the $MgO-CaO-Al_2O_3$ diagram (Fig. 3), originally used by Viljoen and Viljoen (1969) for the distinction of various types of komatiites.

Nesbitt et al. (1979) suggest that two other major element ratios, CaO/TiO_2 and Al_2O_3/TiO_2, could be used for the recognition of peridotitic komatiites since they should show chondritic abundances. At high degrees of partial melting of

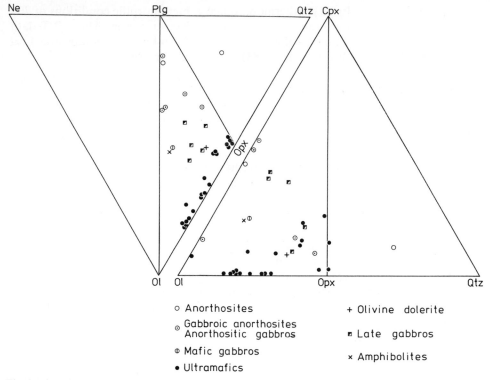

Fig. 2. Plot of normative mineral composition of analyzed samples. Ultramafic and mafic rocks show differentiation trend in both diagrams. Few samples show quartz normative composition indicating addition of quartz during migmatization

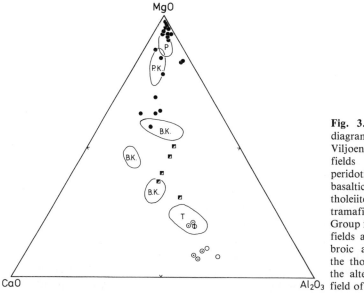

Fig. 3. $MgO - CaO - Al_2O_3$ diagram after Viljoen and Viljoen (1969) indicating fields of peridotite (*P*); peridotitic komatiite (*P.K.*); basaltic komatiite (*B.K.*) and tholeiite (*T*). Note that the ultramafic rocks of the Sargur Group fall into the peridotitic fields and gabbros and gabbroic anorthosites fall into the tholeiitic field. Some of the altered types fall in the field of basaltic komatiite

Table 3. Average major element composition of ultramafic to mafic rocks from Sargur

No. of samples	1[a] 10	2 1	3 1	4 9	5 1	6 2	7 3	8 2	9 5	10 1	11 14
SiO_2	38.56	47.14	53.62	51.80	48.08	47.85	45.38	54.79	49.70	50.30	51.20
TiO_2	0.10	0.62	0.30	0.31	0.22	0.33	0.76	0.54	0.80	0.44	0.86
Al_2O_3	1.67	4.55	2.45	4.45	15.55	20.43	23.29	20.37	10.56	8.76	14.05
Fe_2O_3	4.59	0.87	1.07								1.99
FeO	4.51	12.90	8.76	9.10[b]	8.88[b]	6.43[b]	5.04[b]	5.17[b]	12.03[b]	11.43[b]	9.04
MnO	0.17	0.25	0.22	0.18	0.15	0.09	0.07	0.07	0.17	0.15	0.20
MgO	40.86	23.59	28.17	25.31	13.25	8.60	8.62	3.32	13.88	18.77	7.25
CaO	0.66	6.17	2.76	4.67	10.70	12.55	14.21	9.77	9.03	8.20	10.06
Na_2O	0.11	0.51	0.46	0.61	1.80	2.41	1.38	4.63	2.12	1.55	2.15
K_2O	0.03	0.37	0.04	0.05	0.09	0.40	0.21	0.40	0.23	0.37	0.40
P_2O_5	0.01	0.05	0.22	0.06	0.01	0.05	0.01	0.03	0.10	0.03	0.14
H_2O	5.04	1.57	1.24	3.87	1.03	0.93	1.20	0.80	0.54	–	2.00
CO_2	3.31	1.06	–	–	0.27	0.05	0.10	0.66	0.28	–	–
Total	99.62	99.65	99.31	100.41	100.03	100.12	100.27	100.55	99.44	100.00	99.34

[a] 1 Dunitel harzburgite; 2 bronzite peridotite, Sinduvalli; 3 bronzitite, Sinduvalli; 4 pyroxenites (act/tre + talc with relict opx/cpx); 5 mafic gabbros; 6 anorthositic gabbros; 7 gabbroic anorthosites; 8 anorthosite; 9 late gabbros; 10 olivine dolerite; 11 metatholeiites associated with metasediments

[b] Total iron as FeO

peridotite CaO, TiO_2 and Al_2O_3 completely enter the melt, provided that olivine is the only residual phase. The TiO_2/Al_2O_3 and TiO_2/CaO ratios for the ultramafic rocks of the Sargur area are clearly not chondritic (Fig. 4) and thus differ from those of peridotitic komatiites in the neighbouring Holenarsipur schist belt (Hussain et al. 1982).

However, the main argument against the komatiitic, i.e. volcanic, origin of the ultramafic to mafic rocks from the Sargur region remains the total absence of quench textures which could have been preserved even at amphibolite facies metamorphism (cf. Arndt et al. 1979). In contrast, these rocks are characterized by cumulus and poikilitic textures, indicative of formation under normal plutonic conditions. Objections against a komatiitic nature also arise from the occurrence of thick chromitite layers in the Sinduvalli area. Thick layered chromitite bodies have not yet been reported from any known komatiite sequence. Srikantappa et al. (1980) have shown that the unaltered chromitites from Sinduvalli display relic cryptic layering and are closely related to chromitites of the layered stratiform complexes. The dunite-harzburgite rocks of the Sargur area are chemically similar to dunites and harzburgites of the Bay of Island layered igneous complex (Bowes et al. 1970) and to metaperidotites of the Hunter River belt, Canadia Shield (Collerson et al. 1976). The unaltered bronzitite layers are also similar to the bronzitites of the Bushveld Igneous Complex, South Africa (Bowes et al. 1970).

The mafic gabbros and gabbroic anorthosites have variable CaO and Al_2O_3 contents due to variable amounts of modal plagioclase and clinopyroxene. Their

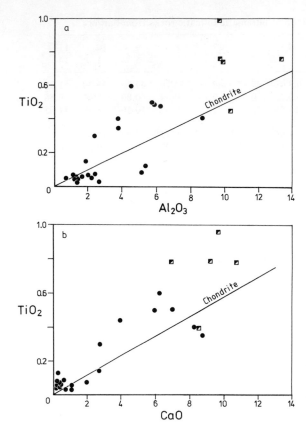

Fig. 4a, b. TiO$_2$ vs Al$_2$O$_3$ (a) and TiO$_2$ vs CaO (b) diagram illustrating non-chondritic nature of ultramafics and late gabbros

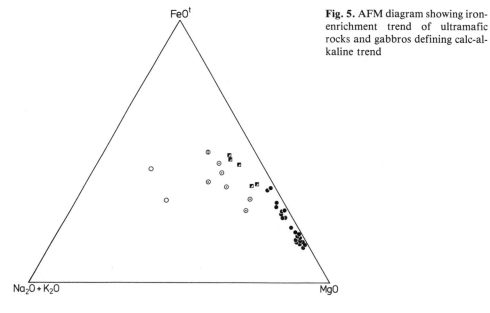

Fig. 5. AFM diagram showing iron-enrichment trend of ultramafic rocks and gabbros defining calc-alkaline trend

K_2O/CaO ratio (~0.04) is lower than that in Proterozoic anorthosites (0.10; Romey 1967). Samples from a sequence of mafic gabbros, anorthositic gabbros and gabbroic anorthosites (e.g. a section near Sinduvalli) display increasing enrichment of sodium and potassium with stratigraphic level, Ni and Cr decreasing in the same direction. This could be taken as evidence for an original igneous stratigraphy. Similar observations have been made by Sills et al. (1982) in the Lewisian of northwestern Scotland. The gabbroic anorthosites have high CaO/Na_2O and Na_2O/K_2O ratios. Except for high Na_2O values the gabbroic anorthosites are comparable to the Archaean anorthosites from the Fiskenaesset area, West Greenland (Windley et al. 1973), and the Sittampundi Complex, southern India (Janardhan and Leake 1975). They show some tendency to develop a calc-alkaline differentiation trend in the AFM diagram of Fig. 5. On the whole it appears that the rocks originated largely from liquids with variable cumulate components.

3.2 Trace Elements

Alkali and alkaline earth cations seem to be mobile during weathering as well as hydrothermal and metamorphic alteration of the rocks (Condie et al. 1977). These elements are therefore largely unsuitable as petrogenetic indicators. In contrast to this, highly charged cations such as Sc, Y, REE, Cr, V, Ti, and Zr are rather immobile under most metamorphic conditions (e.g. Pearce and Cann 1973; Hamilton et al. 1979) and hence can be used as petrogenetic indicators. Average trace element analyses are reported in Table 4.

The Ni and Cr versus MgO correlation of the rocks (Fig. 6a, b) is consistent with the early separation of olivine and chromite from the magma that resulted in low concentrations of these elements in the late differentiates. Ti and V display a strong negative correlation with MgO (Fig. 6c, d). Their contents increase in the melts during the early fractionation of olivine and chromite. Both elements pref-

Table 4. Average trace element composition of ultramafic to mafic rocks from Sargur

| No. of | 1[a] | 2 | 3 | 4 | 5 | 6 | 7 | 8 | 9 | 10 | 11 |
samples	9	1	1	9	1	2	3	2	5	1	1
Cr	5862	2507	3839	1406	531	313	358	62	875	1300	288
Ni	2143	1416	1277	957	124	76	116	55	458	600	124
Co	Nd	Nd	Nd	33	44	23	22	18	60	Nd	Nd
V	89	138	81	106	144	127	61	86	215	188	217
Cu	17	Nd	Nd	44	5	7	21	12	104	79	Nd
Zr	28	57	6	14	80	31	32	75	193	47	40
Y	10	10	3	16	14	10	5	8	15	29	18
Sc	17	22	11	8	37	21	17	13	29	14	31
Rb	2	Nd	Nd	1	Nd	4	4	6	5	Nd	Nd
Sr	39	234	135	69	51	175	156	169	202	Nd	236
Ba	30	397	60	53	34	43	81	68	63	Nd	385

[a] Column numbers as in Table 3

erentially enter the lattices of magnetite and ilmenite during magmatic crystallization (cf. Irvine 1967). As there is neither ilmenite nor another Ti-bearing
mineral associated with the early chromitite layers, it is suggested that Ti and V
were incorporated into magnetite and ilmenite during a late stage of differentiation, i.e. the formation of the mafic gabbros.

Ramakrishnan et al. (1978) found thin bands of titanomagnetite in the gabbroic anorthosites near Hullahalli. Titanium and vanadium-bearing magnetite
bodies showing cumulus textures and layering were seen to grade into gabbroic
anorthosite layers in the Nuggihalli schist belt (Vasudev and Srinivasan 1979).
This indicates that similar processes of late stage crystallization of magnetite
have also been operative in the Nuggihalli region.

Nesbitt et al. (1979) haven shown that peridotitic komatiites from Archaean
greenstone belts characteristically show chondritic ratios of immobile elements

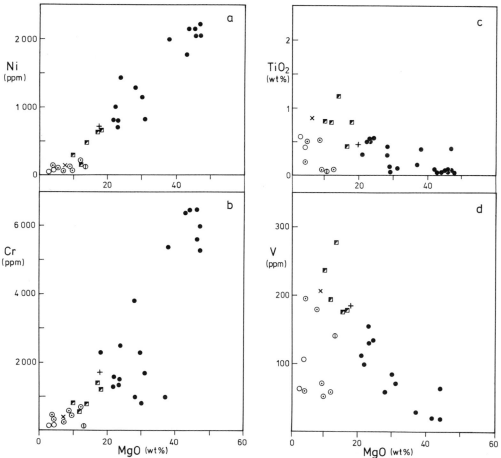

Fig. 6a – d. Ni vs MgO (**a**) and Cr vs MgO (**b**) diagram of ultramafics, gabbros and late gabbros
showing good positive correlation. The V vs MgO (**c**) and TiO$_2$ vs MgO (**d**) diagrams show a negative
correlation

such as Ti/Zr (110), Ti/Y (290), Y/Zr (0.39) and Sc/Zr (1.4). The ultramafic rocks of the Sargur region, however, show Ti/Zr and Y/Zr ratios higher than, and Ti/Y and Sc/Zr ratios lower than, in peridotitic komatiites (Fig. 7).

The present data indicate that considerable fractionation must have taken place during the crystallization process although a definite line of descent cannot be detected. The Sr and Ba contents of the mafic gabbros and gabbroic anorthosites vary from 50 to 200 ppm and 34 to 100 ppm, respectively. Both elements display a fairly good correlation with normative plagioclase. The late gabbros have characteristically low Rb (~5 ppm), Sr (~200 ppm) and Ba (~63 ppm) contents and resemble modern arc tholeiites (Condie and Harrison 1976).

3.3 Distribution of the Rare Earth Elements (REE)

REE abundances in gabbroic anorthosites and late gabbros are represented in Table 5. Chondrite-normalized patterns are plotted in Fig. 8. Gabbroic anortho-

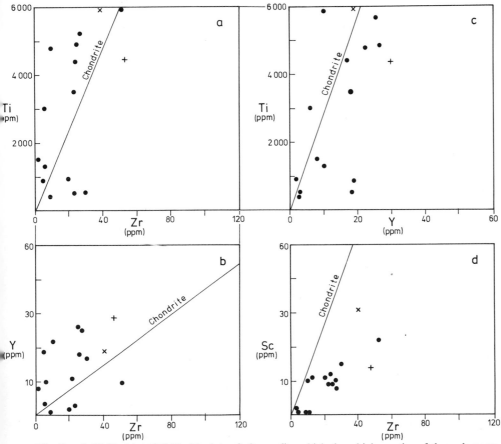

Fig. 7a−d. Ti/Zr (**a**) and Y/Zr (**b**) plots of ultramafics which show higher ratios of these elements when compared to chondrite. The plots of ultramafics on Ti/Y (**c**) and Sc/Zr (**d**) diagrams indicate their low ratios when compared to chondrite. Trace element compositions of chondrites from Nesbitt and Sun (1980)

Table 5. REE analyses and ratios for anorthositic gabbros and late gabbros from Sargur

	1 [a] S_5	2 S_{27}	3 S_6	4 S_{9a}	5 S_9	6 S_3	7 S_4	8 S_{11}	9 S_{39c}	10 S_{39f}
La	1.60	2.31	1.33	4.30	3.00	8.38	7.55	14.56	10.88	5.03
Ce	4.20	5.11	2.93	10.07	5.51	10.19	18.45	44.96	22.20	15.60
Pr	0.68	0.81	0.48	1.10	0.75	2.29	2.21	7.80	3.72	1.26
Nd	4.64	3.95	1.86	4.87	3.65	12.44	10.58	32.23	17.58	6.57
Sm	1.33	1.03	0.23	0.95	1.16	2.91	2.31	8.12	3.81	1.52
Eu	0.39	0.62	0.24	0.30	0.59	1.23	0.96	2.08	1.54	0.675
Gd	6.0	1.72	0.72	1.18	1.40	3.87	2.99	9.91	5.04	2.49
Dy	2.58	1.96	0.81	1.20	1.48	4.01	2.84	11.10	4.56	2.58
Ho	0.58	0.44	0.21	0.25	0.29	0.83	0.57	2.31	0.91	0.56
Er	1.71	1.07	0.48	0.64	0.93	2.27	1.42	6.73	2.25	1.51
Yb	1.49	1.15	0.48	0.68	0.91	2.17	1.27	6.37	2.04	1.50
Lu	0.24	0.22	0.10	0.12	0.15	0.36	0.24	1.12	0.32	0.24
(La/Sm)N	0.74	1.38	3.57	2.79	1.60	1.80	2.01	1.10	1.76	2.04
(Ce/Yb)N	0.71	1.16	3.00	2.49	1.54	1.63	3.69	1.30	3.55	2.02
(Yb/Gd)N	0.89	0.84	0.83	0.72	0.81	0.70	0.53	0.80	0.51	0.76
(Gd/Yb)N	1.12	1.19	1.19	1.38	1.22	1.42	1.90	1.24	1.97	1.33

[a] *1* Mafic gabbro; *2* anorthositic gabbro; *3* und *4* gabbroic anorthosites; *5* anorthosite; *6, 7, 8, 9* and *10* late gabbros

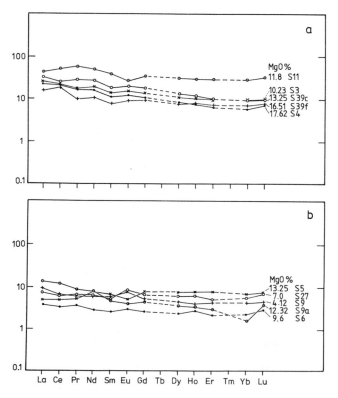

Fig. 8a, b. Chondrite-normalized REE plots for (**a**) late gabbros and (**b**) gabbroic anorthosites. Normalized to C1 chondrite of Nakamura (1974)

sites show a very slight enrichment in LREE relative to HREE and the overall patterns show 5 to 15 times the chondritic abundances. The Eu-anomalies are either slightly negative or positive. Therefore, it is suggested that plagioclase played only a minor role during fractionation or that the oxygen fugacity of the melt was variable.

There is a tendency for the REE abundances to increase with decreasing $Mg/(Mg + Fe)$ ratio of the rocks, indicating olivine, and pyroxene fractionation. This is also consistent with the variation of the major element composition of the rocks (Fig. 2).

The REE distribution patterns of the late gabbros show enrichments of $15 - 40$ times the chondritic abundance, the LREE being slightly enriched relative to the HREE.

The La_N/Sm_N ratios of the late gabbros vary from 1.1 to 3.57, thus covering the total range reported for Archaean basalts (Sun and Nesbitt 1978). The Ce_N/Yb_N ratios of the gabbroic anorthosites have values from 0.7 to 0.3 (average 1.7) which correspond well to those of the Archaean anorthosite complexes of Shawmere, Ontario (Simmons et al. 1980), and Amalrik, West Greenland (O'Nions and Pankhurst 1974), but differ significantly from those of the highly fractionated meta-anorthosites from the Archaean of northwest Norway (Green et al. 1972). The late gabbros have Ce_N/Yb_N ratios slightly higher than the gabbroic anorthosites $(1.3 - 3.55)$.

4 Discussion and Conclusions

The ultramafic to mafic complexes and associated supracrustal rocks of the Sargur area show many similarities with the other schist belts of the Karnataka craton, e.g. Nuggihalli and Nagamangala. The Sargur and Nuggihalli belts are arranged along a narrow arcuate structure and are both characterized by a dominance of ultramafic-mafic rocks. Therefore, the Nuggihalli belt, for which chemical data are available, is also discussed in this context.

Like for the Sargur area, there are different interpretations for the genesis of the ultramafic to mafic complexes in the Nuggihalli region. Pichamuthu (1956) and Varadarajan (1970) regarded these complexes as Alpine-type bodies because of the concordance with the country rocks, the high degree of serpentinization, the absence of contact metamorphic aureoles and the occurrence of podiform-like chromite deposits. An opposite view was taken by Viswanathan (1974), Sreenivas and Srinivasan (1974) and Naqvi and Hussain (1979) who interpreted the ultramafic rocks as a metavolcanic sequence with komatiitic to tholeiitic affinities. They found rocks with apparent pillow-like structures and microspinifex textures. Ramakrishnan (1981) later argued that these structures are the result of a deformation process. Furthermore, the occurrence of chromitite layers in rhythmic fashion with way up stratigraphy and cumulus textures in the same area contradicts the volcanic nature of these rocks. The chemistry of the chromitites from Nuggihalli is typical of the stratiform igneous bodies (Nijagunappa and Naganna 1983). Layering and gradational contacts in the gabbros and gabbroic anorthosites have also been observed in the Nuggihalli area (Vasudev and Srinivasan 1979). Stratiform titanomagnetite deposits are typically associated with the gabbros. The fine-grained anorthositic gabbros from Holenarsipur and Nuggi-halli, formerly interpreted as volcanic flows (Drury et al. 1978; Naqvi and Hussain 1979), rather represent highly deformed and recrystallized cumulate rocks (Janardhan 1979; Bhaskara Rao and Veerabhadrappa 1982).

Thus, comparing the Nuggihalli and Sargur belts, there is a striking similarity of rock types, indicating that the ultramafic to mafic rocks of both areas represent stratiform layered intrusions. This is supported by the petrographic and geochemical data of the present study.

The key to the recognition of the geological environment of the intrusions is the nature of country rocks into which these complexes have been emplaced. The ultramafic to mafic rocks of the Sargur region have been emplaced into the Sargur supracrustal sequence consisting of quartzites, banded iron formations, K-rich metapelites, calc-silicate rocks and marbles interlayered with tholeiitic metabasic rocks. Recent studies have shown the occurrence of significant manganese-bearing quartzites in the Sargur region, similar to the khondalite terrane of the Eastern Ghats (Janardhan et al. 1981). The character of the sedimentary sequence indicates an intra-continental or shelf facies environment rather than an oceanic setting. The source region for the clastic sediments was probably a sialic terrane with minor ultramafic to mafic rocks. This may be inferred from the characteristic heavy mineral association consisting of zircon, rutile, tourmaline, magnetite and chromite in many Sargur-type schist belts (Ramiengar et al. 1978; Ramakrishnan 1981).

The metabasalts associated with the metasediments could give further information about the geotectonic setting. Limited major and trace element data on the amphibolites that form conformable layers within the metasediments show affinities with modern ocean floor basalts (Janardhan et al. 1978). Similar low K-tholeiites from the Holenarsipur schist belt show chemical similarities with basalts from modern arc, rise and marginal basin regions (Ramakrishnan 1981). They have all the characteristics of Archaean metatholeiites (Srikantappa 1979; Venkataramana 1982).

On the basis of the present data it is suggested that the deposition of the Sargur supracrustal sequence took place in a shallow-water intra-crustal environment similar to the models proposed by Kröner (1982) for the Limpopo belt. The intrusion of the ultramafic to mafic complexes into the Sargur supracrustal rocks and their differentiation took place before or during the first deformational episode. The occurrence of the meta-igneous complexes along a specific linear zone suggests that the emplacement of the magmas followed deep-seated N-S trending lineaments that may have penetrated the entire Archaean continental crust of southern India. Subsequently, the entire terrane has been affected by various phases of deformation and by metamorphism of amphibolite to granulite facies. The P-T conditions of the last high-grade metamorphic event ($700 \pm 50\,°C$, 9 ± 1 kb) indicate a minimum crustal thickness of ca. 35 km about 2.6 Ga ago.

Acknowledgements. We are grateful to B. P. Radhakrishna, A. S. Janardhan, C. R. L. Friend, P. Raase, and J. D. Sills for their comments to the manuscript and to D. Ackermand for the help in microprobe analysis.

Financial support of the German Academic Exchange Service (DAAD) through a post-doctoral Research Fellowship to C. Srikantappa is gratefully acknowledged. The first author thanks M. N.

Viswanathia and V. Venkatachalapathy for their encouragement and support during the course of investigation.

The manuscript has benefited considerably from the criticisms and suggestions by A. Kröner, J. Tarney, and B. F. Windley.

References

Arndt NT, Naldrett AJ, Pyke DR (1977) Komatiitic and iron rich tholeiitic lavas of Munro Township, Northeast Ontario. J Petrol 18:319–369

Arndt NT, Francis D, Hynes AJ (1979) The field characteristics and petrology of Archaean and Proterozoic komatiites. Can Mineral 17:147–163

Bhaskara Rao B, Veerabhadrappa SM (1982) Geochemistry and genesis of Archaean meta-anorthositic rocks from the Southern Karnataka, South India. N Jahrb Miner Mh 3:133–144

Bickle MJ, Martin A, Nisbet EG (1975) Basaltic and peridotitic komatiites and stromatolites above a basal unconformity in the Belingwe greenstone belt, Rhodesia. Earth Plan Sci Lett 27:155–162

Blais S, Auvray B, Capdevila R, Jahn BM, Hameurt S, Bertrand TM (1978) The Archaean greenstone belts of Karelia (eastern Finland) and their komatiitic and tholeiitic series. In: Windley BF, Naqvi SN (eds) Developments in Precambrian geology 1. Archaean Geochemistry. Elsevier, Amsterdam, pp 87–108

Bowes DR, Wright AE, Park RG (1964) Layered intrusive rocks in the Lewisian of the northwest Highlands of Scotland. Q J Geol Soc Lond 120:153–191

Bowes DR, Skinner WR, Wright AE (1970) Petrochemical comparison of the Bushveld igneous complex with some other mafic complexes. Geol Soc S Afr Spec Publ 1:425–440

Broekaert JAC, Hörmann PK (1981) Separation of yttrium and rare earth elements from geological materials. An Chem Acta 124:421–425

Chadwick B, Ramakrishnan M, Viswanatha MN, Murthy VS (1978) Structural studies in the Archaean Sargur and Dharwar supracrustal rocks of the Karnataka craton. J Geol Soc India 19:531–549

Chadwick B, Ramakrishnan M, Viswanatha MN (1981) Structural and metamorphic relations between Sargur and Dharwar supracrustal rocks and Peninsular gneiss in central Karnataka. J Geol Soc India 22:557–569

Collerson KD, Jesseau WC, Bridgwater D (1976) Contrasting types of bladed olivine in ultramafic rocks from the Archaean of Labrador. Can J Earth Sci 13: 442–450

Condie KC, Harrison NM (1976) Geochemistry of the Archaen Bulawayan group, midlands greenstone belt, Rhodesia. Precambrian Res 3:253–271

Condie KG, Viljoen MJ, Kable EJD (1977) Effects of alteration on element distributions in Archaean tholeiites from the Barberton greenstone belt, South Africa. Contrib Mineral Petrol 64:75–89

Drury SA, Naqvi SM, Hussain SM (1978) REE distributions in basaltic anorthosites from the Holenarasipur greenstone belt, Karnataka, South India. In: Windley BF, Naqvi SM (eds) Archaean Geochemistry. Elsevier, Amsterdam, pp 363–374

Ellis DJ, Green DH (1979) An experimental study of the effect of Ca upon garnet-clinopyroxene Fe-Mg exchange equilibria. Contrib Mineral Petrol 71:13–22

Essene EJ (1982) Geologic thermometry and barometry. Reviews in mineralogy, Min Soc Am, Vol 10:153–206

Gansser A, Dietrich VJ, Cameron WE (1979) Paleogene komatiites from Gorgona island. Nature 278:545–546

Ghisler M (1976) The geology, mineralogy and geochemistry of the pre-orogenic Archaean stratiform chromite deposits of Fiskenaesset, Western Greenland. Rapp Grønlands Geol Under 12:36 p

Glickson AY (1977) Evidence on the radius of the Precambrian earth. Bur Miner Resour J Aust Geol Geophys 2:229–232

Glickson AY (1982) The early Precambrian crust with reference to the Indian Shield: An essay. J Geol Soc India 23:581–603

Goldsmith JR, Newton RC (1977) Scapolite-plagioclase stability relations at high pressures and temperatures in the system $NaAlSi_3O_8$-$CaAl_2Si_2O_8$-$CaCO_3$-$CaSO_4$. Am Mineralogist 62:1063–1081

Green TH, Brunfelt AO, Heier KS (1972) Rare-earth element distribution and K/Rb ratios in granulites, mangarites and anorthosites, Lofoten-Vesteraalen, Norway. Geochim Cosmochim Acta 36:241 – 257

Hamilton PJ, Evenson NM, O'Nions RK, Tarney J (1979) Sm-Nd systematics of Lewisian gneisses, implications for the origin of granulites. Nature 277:25 – 28

Henderson JB (1981) Archaean basin evolution in the Slave province, Canada. In: Kröner A (ed) Precambrian plate tectonics. Elsevier, Amsterdam, pp 213 – 233

Hörmann PK, Pichler H, Zeil W (1973) New data on the young volcanism in the Puna of NW Argentina. Geol Rundsch 62:397 – 418

Hörmann PK, Eulert H (1980) Ion exchange separation of Li and Rb using Amberlite IR 120. Working methods. Min Inst Univ Kiel (copies available on request).

Hor AK, Hutt DK, Smith JV, Wakefield J, Windley BF (1975) Petrochemistry and mineralogy of early Precambrian anorthositic rocks of the Limpopo belt, southern Africa. Lithos 8:297 – 310

Hussain SM, Naqvi SM, Gnaneshwar Rao T (1982) Geochemistry and significance of mafic-ultramafic rocks from the southern part of the Holenarasipur schist belt, Karnataka. J Geol Soc India 23:19 – 31

Irvine TN (1967) Chromian spinel as a petrogenetic indicator: Part 2, Petrologic applications. Can J Earth Sci 4:71 – 103

Jackson ED, Thayer TP (1972) Some criteria for distinguishing between stratiform, concentric and alpine peridotite – gabbro complexes. Proc 24th I G C Sec 2. Petrology, pp 289 – 296

Jafri SH, Subba Rao DV, Ahmad SM, Saxena R (1982) Geochemistry of spinifex textured peridotitic komatiites from Nuggihalli and Holenarsipur (northern part) schist belts, Dharwar craton, Karnataka, India. Abs. Indo-US workshop on Precambrians of South India. Hyderabad, pp 9 – 10

Janardhan AS, Leake BE (1975) The origin of the meta-anorthositic gabbros and garnetiferous granulites of the Sittampundi complex, Madras, India. J Geol Soc India 16:391 – 408

Janardhan AS, Srikantappa C (1977) Carbonate-orthopyroxene-chrome tremolite rocks (Sagvandite) from Mavinahalli, Mysore district, Karnataka State. J Geol Soc India 18:617 – 622

Janardhan AS, Srikantappa C, Ramachandra HM (1978) The Sargur schist complex – an Archaean high-grade terrain in southern India. In: Windley BF, Naqvi SM (eds) Developments in Precambrian geology I. Archaean geochemistry. Elsevier, Amsterdam, pp 127 – 150

Janardhan AS, Ramachandra HM, Ravindra Kumar GR (1979) Structural history of the Sargur supracrustals and associated gneisses, south-western part of Mysore, Karanataka. J Geol Soc India 20:61 – 72

Janardhan AS, Sadakshara Swamy N, Ravindra Kumar GR (1981) Petrological and structural studies of the manganiferous horizons and recrystallised ultramafics around Gundlupet, Karnataka. J Geol Soc India 22:103 – 111

Kröner A, Puustinen K, Hickman M (1981) Geochronology of an Archaean tonalitic gneiss dome in northern Finland and its relation with an unusual overlying volcanic conglomerate and Komatiitic greenstone. Contrib Mineral Petrol 76:33 – 41

Kröner A (1982) Precambrian plate tectonics. In: Kröner A (ed) Precambrian plate tectonics. Elsevier, Amsterdam, pp 58 – 90

Moore AC (1977) The Petrography and possible regional significance of the Hjelmkona ultramafic body (Sagvandite). Nordmore, Norway. Nor Geol Tidsskr 57:55 – 64

Nakamura N (1974) Determination of REE, Ba, Fe, Mg, Na and K in carbonaceous and ordinary chondrites. Geochim Cosmochim Acta 38:757 – 775

Naqvi SM (1976) Physico-chemical conditions during the Archaean as indicated by Dharwar Geochemistry. In: Windley BF (ed) The early history of the Earth, Wiley, London, pp 289 – 298

Naqvi SM, Hussain SM (1979) Geochemistry of metaanorthosites from a greenstone belt in Karnataka, India. Can J Earth Sci 16:1254 – 1264

Nesbitt RW, Sun SS, Purvis AC (1979) Komatiites: Geochemistry and genesis. Can Mineral 17:165 – 186

Nesbitt RW, Sun SS (1980) Geochemical features of some Archaean and post-Archaean high-magnesian – low-alkali liquids. Phil Trans Roy Soc 297A:365 – 381

Nijagunappa R, Naganna C (1983) Nuggahalli schist belt in the Karnataka craton: in Archaean layered complex as interpreted from chromite distribution. Econ Geol 78:507 – 513

Ohnmacht W (1974) Petrogenesis of carbonate-orthopyroxenites (Sagvandites) and related rocks from Troms, Northern Norway. J Petrol 15:303 – 324

O'Nions RK, Pankurst RJ (1974) Rare earth element distribution in Archaean gneisses and anortho-sites, Godthaab area, West-Greenland. Earth Planet Sci Lett 22:328–338

Papunen H, Häkli TA, Idman H (1979) Geological geochemical and mineralogical features of sulfide-bearing ultramafic rocks in Finland. Can Mineral 17:217–232

Pearce JA, Cann JR (1973) Tectonic setting of basic volcanic rocks determined using trace element analyses. Earth Planet Sci Lett 19:290–300

Perkins D III, Newton RC (1981) Charnockite geobarometers based on coexisting garnet-pyroxene-plagioclase-quartz. Nature 292:144–146

Pichamuthu CS (1956) The problem of the ultrabasic rocks. Proc Mys Geol Assoc, 12 pp

Powell R (1978) The thermodynamics of pyroxene geotherms. Philos Trans R Soc Lond A 288:457–469

Raith M, Raase P, Ackermand D, Lal RK (1983) Regional geothermobarometry in the granulite facies terrane of South India. Trans R Soc Edinburgh, Earth Sci 73:221–244

Rama Rao B (1926) On the correlation of the magnesite-chromite bearing rocks of Mysore district. Rec Mys Geol Dept 23:109–116

Ramadurai S, Sankaran M, Selvan TA, Windley BF (1975) The stratigraphy and structure of the Sittampundi Complex, Tamil Nadu, India. J Geol Soc India 16:409–414

Ramakrishnan M, Viswanatha MN, Chayapathi N, Narayanan Kutty TR (1978) Geology and geochemistry of anorthosites of Karnataka craton and their tectonic significance. J Geol Soc India 19:115–134

Ramakrishnan M (1981) Review of geochronology and geochemistry. In: Swami Nath T, Ramakrishnan M (eds) Early Precambrian supracrustals of southern Karnataka. Mem Geol Surv India 112:249–259

Ramakrishnan M (1981) Nuggihalli Krishnarajpet-belts. in: Swami Nath J, Ramakrishnan M (eds) Early Precambrian supracrustals of southern Karnataka. Mem Geol Surv India 112:61–70

Ramiengar AS, Devudu GR, Viswanatha MN, Chayapathi N, Ramakrishnan M (1978) Banded chromite-fuchsite quartzite in the older supracrustal sequence of Karnataka. J Geol Soc India 19:577–582

Robertson IDM (1973) The geology of the country around Mount Towla, Gwanda district. Rhod Geol Surv Bull 68:168 pp

Romey WD (1967) An evaluation of some differences between anorthosite in massifs and in layered complexes. Lithos 1:230–240

Schreyer W, Ohnmacht W, Mannchen J (1972) Carbonate-orthopyroxenites (Sagvandites) from Troms, Norway. Lithos 5:70–85

Sills JD, Savage D, Watson JV, Windley B (1982) Layered ultramafic-gabbro bodies in the Lewisian of northwest Scotland: Geochemistry and petrogenesis. Earth Planet Sci Lett 58:345–360

Simmons EC, Hanson GN, Lumbers SB (1980) Geochemistry of the Shawmere Anorthosite Complex, Kapuskasing structural zone, Ontario. Precambrian Res 11:43–71

Sreenivas BL, Srinivasan R (1974) Geochemistry of granite-greenstone terrain of South India. J Geol Soc India 15:390–406

Srikantappa C (1979) Petrology and geochemistry of ultramafic rocks around Sinduvalli, Karnataka State. Unpubl Ph D thesis Univ of Mysore, 171 pp

Srikantappa C, Friend CRL, Janardhan AS (1980) Petrochemical studies on chromites from Sinduvalli, Karnataka, India. J Geol Soc India 21:473–483

Srikantappa C, Raith M, Ackermand D (1984) High-grade regional metamorphism of ultramafic and mafic rocks from the Archaean Sargur terrane, Karnataka, South India, submitted to Precambrian Research.

Subramaniam AP (1956) Mineralogy and petrology of the Sittampundi complex, Salem district, Madras State, India. Geol Soc Am Bull 67:317–379

Sun SS, Nesbitt RW (1978) Petrogenesis of Archaean ultrabasic and basic volcanics: Evidence from rare earth elements. Contrib Mineral Petrol 65:301–325

Swami Nath J, Ramakrishnan M, Viswanatha MN (1974) Dharwar stratigraphic model and Karnataka craton evolution. Rec Geol Surv India 107(2):149–175

Thayer TP (1970) Chromite segregations as petrogenetic indicators. Geol Soc S Afr Spec Publ 1:380–390

Varadarajan S (1970) Emplacement of chromite bearing ultramafic rocks of Mysore State, India. Proc Symp U M P 2nd, Hyderabad, 441–454

Vasudev VN, Srinivasan R (1979) Vanadium bearing titaniferous magnetite deposits of Karnataka, India. J Geol Soc India 20:170–178

Venkataramana P (1982) Chemical remnants of the Archaean protocrust in the Sargur schist belt of Karnataka craton, India. Precambrian Res 19:51–74

Viljoen MG, Viljoen RP (1969) A collection of 9 papers on many aspects of the Barberton granite – greenstone belt, South Africa. Geol Soc S Afr Spec Publ 2:295 pp

Viswanathan S (1974) Basaltic komatiite occurrences in the Early Precambrian of India: their geochemistry and geological and economic significance. Abstr. 61st Ind Sci Congr Nagapur, Proc Pt 3:161

Viswanatha MN, Ramakrishnan M (1981) Sargur and allied belts. In: Swami Neth J, Ramakrishnan M (eds) Early Precambrian supracrustals of southern Karnataka. Mem Geol Surv India 112:61–70

Wells PRA (1977) Pyroxene thermometry in simple and complex systems. Contrib Mineral Petrol 62:129–139

Wood BJ, Banno S (1973) Garnet-orthopyroxene and orthopyroxene and clinopyroxene relationships in simple and complex systems. Contrib Mineral Petrol 42:109–124

Windley BF, Herd RK, Bowden AA (1973) The Fiskenaesset complex, West Greenland, Pt. 1: A preliminary study of stratigraphy, petrology and whole rock chemistry from Qeqertarssuatsiaq. Grønlands Geol Under Bull 106:1–80

Windley BF, Smith JV (1976) Archaean high-grade complexes and modern continental margins. Nature 260:671–675

Windley BF, Bishop FC, Smith JV (1981) Metamorphosed layered igneous complexes in Archaean granulite-gneiss belts. Ann Rev Earth Planet Sci 9:175–178

Pressures, Temperatures and Metamorphic Fluids Across an Unbroken Amphibolite Facies to Granulite Facies Transition in Southern Karnataka, India

E. C. Hansen[1], R. C. Newton[1] and A. S. Janardhan[2]

Contents

1 Introduction 162
1.1 Nature of Granulite Terranes 162
1.2 The Transition Regions 163
1.3 The South India High-Grade Terrane 164
1.4 Scope of the Present Work 164
2 Regional Geology 165
2.1 Lithologic and Metamorphic Relations 165
2.2 Structures 166
3 Fluid Inclusions 167
3.1 Description 167
3.2 Microthermometry 170
4 Chemical Analyses 170
4.1 Mineral Analysis 170
4.2 Whole Rock Analysis of K and Rb 170
5 Geobarometry 173
5.1 Mineralogic Geobarometry 173
5.2 CO₂ Inclusion Geobarometry 175
6 Discussion and Conclusions 176
6.1 Continuity of Metamorphic Terrane 176
6.2 Evidence for a Discrete Metamorphic Event 177
6.3 The Role of Carbonic Fluids 177
7 Tectonic Interpretations 178
7.1 Hot Spots 178
7.2 Infracontinental Disruptions 178
7.3 Continental Collision 179
References 180

Abstract

A sampling traverse has been made across the late Archaean regional amphibolite-facies to granulite-facies transition in southern Karnataka, India. The traverse extends from the Peninsular Gneiss-Closepet Granite terrane in the north, through the incipient charnockite localities near Kabbaldurga and southwards into the charnockite massifs of the Biligirirangan and Andhiyur Hills.

Geobarometry based on the temperature-insensitive indicator assemblage garnet-pyroxene-plagioclase-quartz shows that there is a smooth increase of pressure from near 5 kb at Channapatna ($12°40'N$) to 7.5 kb in the highest-grade

---

1 Department of Geophysical Sciences, University of Chicago, Chicago, IL 60637, USA
2 Department of Geology, University of Mysore, Mysore, Karnataka 570006, India

Archaean Geochemistry (ed. by A. Kröner et al.)
© Springer-Verlag Berlin Heidelberg 1984

massif areas near 12°00′N. Geobarometry based on the densities of CO_2 fluid inclusions in quartz, determined from hundreds of homogenization temperature measurements at seven localities in the transition zone, agrees quantitatively with the mineralogic geobarometry. The average palaeo-temperature in the transition zone is 750°C, determined from garnet-pyroxene Fe/Mg distributions. The quality of the geothermometry is not sufficient to establish with certainty a temperature increase southward over the transition zone.

A reconnaissance profiling of the K/Rb whole rock ratios of charnockites and gray gneisses shows normal upper-crustal ratios near 300 for gneisses *and* charnockites in the northern half of the traverse and onset of a few very high ratios, indicating extreme depletion of Rb southward from about 12°20′N where charnockite becomes the dominant country rock. Even in the southernmost high-grade massif areas the depletion is patchy, with many rocks in the relatively low range 300 – 600. No correlation of Rb depletion with Fe/Mg ratio, which is a good indicator of partial melt/magmatic residue relations, was found.

The results establish that the metamorphic gradient south of Kabbal is an unbroken prograde transition to granulite facies. A depth-zone arrangement of amphibolite facies and granulite facies is indicated. Most of the CO_2 inclusions observed were entrapped at peak metamorphic pressures. Therefore CO_2 streaming is implicated as the major agency of metamorphic dehydration (charnockitization). The source of CO_2 was deep-seated, below the deepest level sampled in this traverse. Partial melting, manifest in migmatization in the transition interval, probably helped lower H_2O activity and may have been instrumental in Rb depletion, but this process is, *by itself,* inadequate to account for charnockite formation and Rb depletion patterns. Carbonic fluid streaming was a more fundamental agency.

1 Introduction

1.1 Nature of Granulite Terranes

The highest grade rocks of the continental crust are the granulites. Eskola (1939) first gave granulites the status of a metamorphic facies on the basis of a peculiar mineralogy in which the dense, anhydrous mafic silicates orthopyroxene, clinopyroxene and garnet are prominent in many lithologies, in contrast to the micas, amphiboles and other hydrous minerals characteristic of lower grade metamorphic rocks. Such rocks are now known to occupy large areas of the Precambrian shields of every continent.

A number of petrologic features of granulite terranes are so widespread as to be considered generally definitive. Early large-scale recumbent overfolds and horizontal tectonics have been documented in many well-studied granulite terranes. Examples are the Scourie terrane of Scotland (Khoury 1968) and the Adirondack Highlands of New York (McLelland and Isachsen 1980). Another definitive feature is interlayered rocks of obvious supracrustal origin, which are present in every known granulite terrane. Variable to extreme depletion of the large ion lithophile (LIL) elements, such as Rb, U, and Th, relative to common upper crustal rocks (Lambert and Heier 1968) is characteristic. Recent geothermometry and geobarometry indicate a general temperature range of $700° - 900°C$ and a pressure range of $5 - 10$ kb (Perkins and Newton 1981). The temperatures are higher than those of most younger metamorphic terranes. An additional characteristic shown by many granulites is the presence of low H_2O, dominantly CO_2-rich fluid inclusions in minerals (Touret 1981). The low H_2O activity allowed metamorphism to proceed in a P-T range where crustal rocks with typical H_2O contents would be melted to a considerable extent. Any theory of the origin of granulite terranes needs to take these features into account.

1.2 The Transition Regions

Many or most exposed granulite terranes are bordered by regions of lower metamorphic grade. There is usually a progression through the upper amphibolite facies, including some melting or migmatization, the granulite facies being introduced by the first appearance of orthopyroxene. In some regions orthopyroxene isograds may be mapped, examples being the Bamble terrane of southern Norway (Touret 1974), the granulite terrane of northern Finland (Hörmann et al. 1980), and the Adirondacks of New York (Buddington 1963). Unlike in many younger progressive metamorphic terranes, migmatization does not increase to culmination in granitic batholiths, but rather seems to be arrested with or by the advent of the granulite facies (Touret 1974). The incipient granulites of the transitional zone are commonly not depleted in the LIL elements and may actually be enriched in some LIL (Janardhan et al. 1982). In some terranes identifiable structures such as volcanic-sedimentary belts (greenstone belts) may be traced from the lower grade region into the high-grade region as attenuated relics, as in the Pikwitonei Belt of northern Manitoba (Hubregtse 1980) and the northern margin of the Limpopo Belt (Mason 1973).

Monitoring of chemical and petrologic features across transition zones is of obvious importance to the problem of the origin of granulites. In many or most transitional regions such studies are hampered by the existence of a structural dislocation which separates the very high-grade, LIL-depleted granulite terrane from adjacent lower grade rocks. Such dislocation zones, which may be thrusts or shears, are found in the Bamble area (Touret 1974), the Limpopo Belt (Mason 1973) and the Adirondacks (Romey et al. 1980). Perhaps partly for this reason granulite terranes have sometimes come to be regarded as isolated entities, which are not obviously relatable to surrounding provinces.

An important indicator of conditions during progressive granulite metamorphism is the nature of fluid inclusions, especially composition and density. Indirect deductions of the composition of metamorphic fluids across progressive granulite tracts have been made by Hörmann et al. (1980) and Phillips (1980). In both studies evidence was found for large decrease of the H_2O activity with increasing penetration into the granulite facies terranes. To the present authors' knowledge no detailed study has been made heretofore of fluid inclusions across a continuous progressive granulite interval into a very high-grade area.

1.3 The South India High-Grade Terrane

The late Archaean high-grade terrane of southern India is one of the most extensive granulite provinces known, with granulite-facies rocks occupying thousands
of square kilometres in the hill regions of Karnataka, Tamil Nadu, Kerala, and
Sri Lanka. Pichamuthu (1953, 1965) first recognized that this extensive region of
charnockites is the culmination of a progression of metamorphic grade which
begins with the granite-greenstone terrane of central Karnataka. He attributed
the metamorphic gradient to exposure of progressively deeper levels of the crust
going southward. However, he thought that some granulites, characterized by
granulitic (granoblastic) textures and commonly outcropping in the higher-grade
terranes, were older than other granulites which have granitic (coarse xeno-
blastic) textures with sutured mineral boundaries and are commonly found in the
transitional regions. This implied age discontinuity between the higher and lower
grade rocks has been restated by others (for instance Viswanathan 1969). Some
workers have stated that the southern Karnataka amphibolite facies gneisses were
derived from older charnockites by retrogression (Devaraju and Sadashivaiah
1969; Ray 1972).
     Very recent work has tended to converge on the viewpoint that the meta-
morphic gradation in southern Karnataka and northern Tamil Nadu is the result
of a single prograde metamorphic episode (*see also Condie and Allen, this Vol.,
eds.*). Ramiengar et al. (1978) studied the field relationships between the charnockites and amphibolite facies gneisses and reviewed the available geochronological data. They concluded that a prograde metamorphic event culminating in
the granulite facies occurred in the late Archaean, 2500 – 2700 Ma ago.
Janardhan et al. (1982) produced mineralogic and some chemical evidence that
the progression is prograde in the incipient charnockite areas. Harris et al. (1982)
did reconnaissance geothermometry and geobarometry in the South Indian high-
grade area and found a relatively large range of pressures (5 to 9 kbar) together
with a relatively small range of temperatures ($700° - 800°C$). From this they
deduced an advective geotherm for the late Archaean metamorphism, with small
temperature variation over a large depth range, such as might be produced by a
broad subcrustal source of heat, resulting from a degassing mantle diapir with
heat transported pervasively upward by volatile streaming. Raith et al. (1982)
found a similar distribution of palaeo-temperatures and -pressures in the South
Indian localities of their study.

1.4 Scope of the Present Work

This study presents data on mineral chemistry, fluid inclusions and reconnaissance whole rocks measurements of the K-Rb ratio, a key indicator of LIL depletion, in a traverse across the northern margin of the amphibolite facies to
granulite facies progression in southern Karnataka. The mineral analyses provide
data for geothermometry and geobarometry, and the fluid inclusions provide additional geobarometry. An invariant pressure, or a smooth pressure increase
without interruption, would indicate a continuous metamorphic progression in

the area. If fluid inclusions yield the same general pressures as the minerals, this may be evidence that the fluids represent those associated with the metamorphism. The K/Rb analysis can provide evidence on whether the depletion process is pervasive or selective, and this may provide insight into the postulated mechanisms of depletion and granulite formation.

2 Regional Geology

2.1 Lithologic and Metamorphic Relations

Figure 1 is a map of the area studied. It was constructed from the maps of Rama Rao (1945), Suryanarayana (1960) and Devaraju and Sadashivaiah (1969) and from Landsat photos and observations made during approximately 6 weeks field

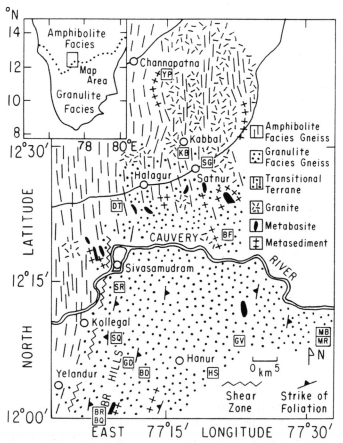

Fig. 1. The amphibolite facies to granulite facies progression in southern Karnataka, Channapatna to BR Hills section. Sources of areal geology given in text. Regional granulite-in line in southern India after Harris et al. (1982)

work for this study. The major lithologic units are tonalitic to trondhjemitic amphibolite facies gneiss (Peninsular Gneiss), granite, and granulite facies (orthopyroxene-bearing) quartzofeldspathic rock (charnockite). The Peninsular Gneiss in this area is locally invaded by leucogranite seams and patches, coalescing in places to form broad outcrops and discrete stocks, especially at the southern end of an elongate N-S granite-rich terrane (Closepet Granite) which can be traced northward for hundreds of kilometres.

The stone quarries near Yelachipalaiyam and Kabbal (YP and KB, Fig. 1), with excellent fresh exposures, lie near the margin of this granite terrane. In these localities charnockite makes its appearance as dark stringers and patches overprinting both Peninsular Gneiss and granite veins (Pichamuthu 1960). The field relationships indicate that anatexis producing the leucogranitic veins is generally somewhat earlier than the dehydration that produced the charnockite. However, there is also no doubt of the near synchroneity of the two processes: both are closely related in origin by the same thermal/volatile episode (Friend 1981). Adjacent samples of the amphibolite facies gneiss and charnockite from Kabbal were found to be nearly identical in major and minor element chemistry by Janardhan et al. (1982), indicating that the formation of charnockite did not directly involve the extraction of a partial melt. They invoked the action of a low-P_{H_2O} fluid phase, probably dominantly CO_2, in the development of the charnockite. The postulated H_2O-poor fluids closely followed H_2O-richer fluids which promoted anatexis. Access of the fluids was facilitated by zones of shear and/or foliation in the gneisses.

South of Kabbal the metamorphic grade increases somewhat irregularly, with amphibolite facies gneisses and charnockites both occupying extensive outcrop areas near Satnur. This is the kind of terrane marked as transitional in Fig. 1. Pink granite of Closepet affinity is still present in this area (Suryanarayana 1960). Still farther south the South Indian hill terrane is entered in the Basavanbetta State Forest and charnockite is dominant there, although occasional pink granites are found. The Karnataka high-grade charnockitic massifs, including the Andhiyur Hills, the Madeswara Malai (MM) Hills, and the Biligirirangan (BR) Hills, follow directly southward.

Amphibolites and basic granulites occur throughout the area both as thin lenses and large bodies up to several km^2 in area. Metasediments have been found sparsely throughout the area, including quartzites, marbles, calc-silicates and metapelites, and are probably much more common in the charnockite area than indicated in Fig. 1, because outcrops and quarries where reliable observations may be made are very few in this heavily forested hill country.

2.2 Structures

The trend of the foliation and strike of metabasic enclaves is roughly north-south in both the amphibolite and granulite facies portions of the area. The charnockite terrane is discernible on Landsat images as dark rugged areas traversed by bold N-S ridges which reflect the underlying structures. In several places the ridges curve around, forming the noses of broad folds. The Earth Resources and

Technology Satellite image #E-2719-04141-701 shows no E-W linear features which could correspond to a boundary fault between the massif and transitional terranes, nor was any evidence of such a break seen during our field work. On the scale of an outcrop the structural style is characterized by dominantly steeply dipping foliation planes and tight isoclinal folds. This same structural style is seen in the high-grade, transitional and amphibolite facies areas.

Evidence for a system of N-S shears along the west side of the BR Hills, extending at least as far north as the bend of the Cauvery River near Sivasamudram, was found in the form of mylonites, pseudotachylite, and gouge zones. This could correspond to a proposed regional N-S shear system (Rollinson et al. 1981). To the south of the area of Fig. 1 the BR Hills massif is bounded by a pronounced E-W megashear, the Moyar-Bhavani lineament (Drury and Holt 1980). Preliminary regional geobarometry (Harris et al. 1982) indicates that south of this megashear the high-grade terrane may consist of fault-bounded large regions or blocks of different degrees of uplift.

3 Fluid Inclusions

3.1 Description

Relatively large and abundant fluid inclusions in quartz grains were practically confined to the coarser-grained and less deformed rocks. In the northern most localities of Fig. 1, fluid inclusions were studied in charnockite stringers from within the gneisses (YP, KB), granite pegmatites (YP, SG), and heavily migmatized gneisses (KB and SG). The fluid inclusion suites at all of these localities were basically the same. The fluid inclusions from the four southern localities were studied in coarser-grained charnockites (BF, GD) and in quartz feldspar veins and pegmatites (BF, GV, HS) within the granulites. Nearly all the fluid inclusions in quartz occur in planar arrays (trails) which may extend up to, but usually do not cross, grain boundaries. Thus, the fluid inclusions have formed by the trapping of fluids in fractures in crystals and are, strictly speaking, secondary. However, many of these fractures where healed continued crystal growth, which indicates the possibility that some of them contain fluids trapped near peak metamorphic conditions.

Liquid CO_2 fluid inclusions are dominant in every sample. The most common types in quartz grains are roughly equant to slightly elongated like those pictured at the bottom of Fig. 2A. Large sinuous tubular inclusions like those pictured in Fig. 2E are abundant in some samples from the southern localities. Smaller, more equant inclusions appear to have formed from some of these larger inclusions by a process of necking down and pinching off. In some samples fluid inclusions are found in the feldspars as well as in quartz. These inclusions generally have rectilinear forms like those pictured in Fig. 2B and are almost always associated with small carbonate and opaque inclusions. Although they are spatially associated, the fluid and solid inclusions are rarely in direct contact.

Trails of *small aqueous fluid inclusions,* generally less than 3 microns in diameter, occur in all but one of the samples. The intersections of these trails with

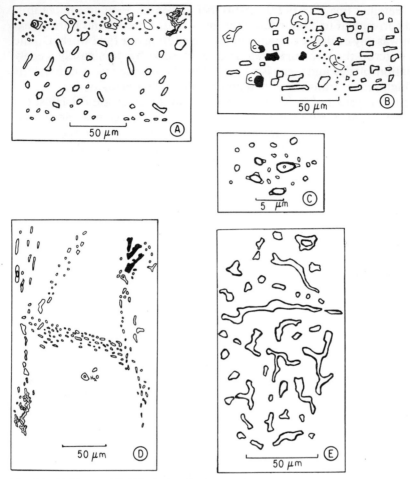

Fig. 2A – E. Types of fluid inclusions in charnockites of the traverse. **A** Typical array of liquid CO_2 inclusions in a quartz grain, intersected by a trail of smaller aqueous inclusions. Note hybrid CO_2-H_2O inclusions and exploded CO_2 inclusion (*upper right*). Kabbal (*KB*); Fig. 1. **B** Array of liquid CO_2 inclusions in a plagioclase crystal, associated with carbonate (*C*) and opaque inclusions, and a trail of small, probably aqueous inclusions. Garnet Valley (*GV*); Fig. 1. **C** Mixed CO_2-H_2O inclusions, probably primary rather than hybrid. Small capillaries on irregular walls of larger inclusions contain adsorbed H_2O. Yelachipalaiyam (*YP*); Fig. 1. **D** Trails of CO_2 inclusions in quartz with a single isolated H_2O inclusion. Kabbal (*KB*); Fig. 1. **E** Long tubular CO_2 inclusions in quartz, probably in the process of fragmenting. Garnet Valley (*GV*); Fig. 1

trails of larger liquid CO_2 inclusions are frequently marked by *mixed H_2O-CO_2, low density liquid + vapor CO_2*, or decrepitated inclusions (Fig. 2A). These features all suggest that the trails of aqueous inclusions are later than the CO_2 inclusions. Although most of the mixed H_2O-CO_2 inclusions are limited to such intersection zones, a few trails in the samples from the northern lower grade localities are made up almost entirely of *H_2O-CO_2 inclusions*. A few such inclusions are shown in Fig. 2C: the water component preferentially wets the walls and is most conspicuous when it collects in irregularities in the walls of larger inclusions. The water is estimated to be less than 10 vol% and can be virtually undetectable in many inclusions dominated by CO_2. Thus, some of the CO_2-rich inclusions from our northern localities may have roughly 30 mol% H_2O (Burruss 1981).

A small number of *relatively large aqueous inclusions,* occurring alone or in small clusters, were seen in some samples from the northern localities (Fig. 2D). Such inclusions are very rare and their origin is problematical; however, they may possibly be representative of the early aqueous phase postulated by Friend (1981) and Janardhan et al. (1982) to have preceded the CO_2-rich phase during the metamorphism.

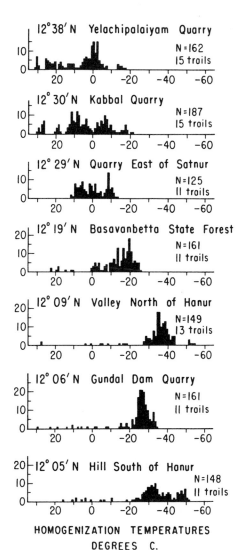

Fig. 3. Histograms of homogenization temperatures of CO_2 inclusions. All samples are charnockites or charnockitically altered gneisses and granites except for Satnur specimen, which is a gray gneiss. The southward increase of average homogenization temperature, and hence density and pressure of entrapment, is evident

Table 2. Pyroxene compositions

	DT	DT	SR	MB	MB	SQ	GV-1	GV-2	GV-2	GV-3	MR	GD-1	GD-2	BD	BR	BQ
	opx	cpx	opx	opx	cpx	opx	opx	opx	cpx	opx	opx	opx	opx	opx	opx	opx
Wt %																
SiO_2	49.41	51.01	48.75	49.51	50.79	48.87	50.58	50.17	50.60	49.35	51.67	48.95	49.67	48.45	49.36	48.44
Al_2O_3	1.67	2.29	1.74	1.83	2.71	2.26	2.36	1.84	2.83	2.13	4.49	2.64	1.42	3.29	2.40	2.43
FeO^a	31.70	12.67	32.79	32.18	13.84	30.88	27.31	16.56	14.01	31.21	21.38	32.40	31.26	30.96	33.86	32.47
MnO	0.71	N.D.	0.63	0.48	N.D.	0.79	N.D.	0.60	N.D.	0.24	N.D.	0.43	0.06	0.46	0.61	0.88
MgO	15.86	11.58	15.09	15.26	10.83	15.87	13.62	16.56	11.19	16.11	23.07	15.19	16.24	16.22	14.44	14.74
CaO	0.45	22.01	0.42	0.52	21.04	0.19	0.55	0.47	20.94	0.30	N.D.	N.D.	0.43	N.D.	N.D.	N.D.
Total	99.80	99.56	99.42	99.78	99.39	98.86	99.43	100.80	99.60	99.34	100.61	99.61	99.08	99.53	100.67	98.96
Atoms/6(O)																
Si	1.94	1.94	1.93	1.94	1.94	1.93	1.93	1.94	1.93	1.93	1.90	1.92	1.94	1.90	1.93	1.92
Al	0.08	0.10	0.08	0.08	0.12	0.11	0.11	0.08	0.13	0.10	0.19	0.12	0.07	0.15	0.11	0.11
Fe	1.04	0.40	1.09	1.06	0.44	1.02	0.89	1.01	0.45	1.02	0.66	1.06	1.03	1.01	1.11	1.08
Mn	0.02	–	0.02	0.02	–	0.03	–	0.02	–	0.01	–	0.01	0.00	0.02	0.02	0.03
Mg	0.92	0.66	0.89	0.89	0.62	0.93	1.06	0.95	0.63	0.94	1.26	0.89	0.95	0.95	0.84	0.87
Ca	0.02	0.90	0.02	0.02	0.86	0.01	0.02	0.02	0.85	0.01	–	–	0.02	–	–	–
No. of anal.	7	7	6	6	8	7	13	5	8	8	8	8	4	7	8	10

^{a} Total iron as FeO; N.D. = Not detected

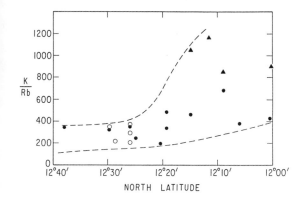

Fig. 4. Whole rock K/Rb ratio of charnockites (*filled symbols*) and amphibolite grade gneisses (*open circles*) as a function of penetration southward into granulite terrane. *Triangles* indicate that K/Rb ratio is a minimum value. The diagram shows that the onset of Rb-depleted rocks occurs at about the latitude where charnockite becomes dominant (see Fig. 1) and that the depletion is irregular rather than pervasive

higher than amphibolite grade and transitional rocks farther to the north. Janardhan et al. (1982) also found variable, rather than pervasive, Rb depletion in a reconnaissance survey of the BR Hills and Nilgiri Hills massif areas (*see also Condie and Allen, this Vol., eds.*).

5 Geobarometry

5.1 Mineralogic Geobarometry

The nature of the pressure, and hence, the depth variation across the amphibolite facies to granulite facies transition is a crucial consideration. To minimize the effects of metamorphic temperature uncertainty, geobarometers should be chosen that have the smallest possible dependence on temperature. Two suitable geobarometers are based on the reactions:

enstatite + anorthite = pyrope + grossular + quartz
diopside + anorthite = pyrope + grossular + quartz

These geobarometers have been calibrated from published thermodynamic data (Perkins and Newton 1981). Although they depend to a considerable extent on assumed values of the free energy of solid solution of garnets, quantities which are still controversial, they have been shown in a number of field studies to give consistent results when applied to closely related rocks having a wide range of mineral chemistry (Harris et al. 1982; Raith et al. 1982), and they have temperature dependences of only about 400 bars per 100°C.

Although the barometers chosen are relatively independent of temperature, an approximate temperature estimate is needed. Several studies of the metamorphic conditions in this general area have arrived at temperatures that fall mostly in the range 700° – 800°C (Rollinson et al. 1981; Harris and Jayaram 1982; Janardhan et al. 1982; Raith et al. 1982; Harris et al. 1982). Temperatures based on the Fe/Mg distribution between garnet and orthopyroxene (Harley 1981) and garnet and clinopyroxene (Ellis and Green 1979) were calculated with the data from Tables 1 and 2 and are presented in Table 3. These temperatures

Table 3. Temperatures and pressures of metamorphism

Sample	DT	SR	MB	SQ	GV-1	GV-2	GV-3	MR	GD-1	GD-2	BD	BR	BQ
N.Lat. (12°)	23'	15'	9'	9'	9'	9'	9'	8'	6'	6'	5'	1'	0'
T_{EG}	680		720			730							
T_{HG}	670	710	690	700	770	660	720	840	750	730	810	790	780
P_{opx}	5.6	6.7	6.2	7.2	7.1	5.9	6.9	6.9	7.1	7.2	6.6	7.8	7.1
P_{cpx}	6.2		6.7			6.9							

T_{EG}: Temp. based on Fe/Mg K_D of garnet/opx (Ellis and Green 1979). $\left.\right\}$ Temp. calculated at 6 kb
T_{HG}: Temp. based on Fe/Mg K_D of garnet/opx (Harley 1981).

P_{opx}: Press. based on gt-plag-opx-qtz (Perkins and Newton 1981). $\left.\right\}$ Press. calculated at 750°C
P_{cpx}: Press. based on gt-plag-cpx-qtz (Perkins and Newton 1981).

also mostly fall in the range 700 °C to 800 °C. It is not certain if there is a definite tendency for the temperatures to increase southward with increasing grade. A reference temperature of 750 °C was chosen for all the present geobarometric calculations. Table 3 gives the garnet-pyroxene-plagioclase-quartz pressures calculated from equations 9 of Perkins and Newton (1981).

5.2 CO_2 Inclusion Geobarometry

The densities of CO_2 fluid inclusions can be inferred from their homogenization temperatures using measured and theoretical PVT data (Touret and Bottinga 1979). If the temperatures of entrapment are known, these densities can be combined with the PVT data to calculate pressures of entrapment. This method depends on the assumption that the fluid inclusions have neither changed in volume nor lost or gained material since their entrapment. There is no absolute way to estimate the temperature of entrapment, but it may be assumed that the fluid inclusions were trapped at temperatures close to those at the peak of metamorphism. If the pressures calculated in this way are consistent with independent estimates of the peak metamorphic pressures, this is strong evidence that the fluid inclusions represent metamorphic fluids of the charnockite-forming event.

The largest low temperature peaks on the histograms of Fig. 3 were used in the pressure calculations, following Coolen (1981). The results are given in Table 4.

The pressures obtained from both the garnet-plagioclase-pyroxene-quartz barometers and the fluid inclusions are plotted as a function of latitude in Fig. 5. No charnockites or metabasites with garnet were found in the northernmost areas which, in itself, may be an indication of lower pressures. We supplement our pressures in these areas with pressures obtained by Harris and Jayaram (1982), based on garnet-plagioclase-sillimanite-quartz and cordierite-garnet-sillimanite-quartz. The fluid inclusion barometer yields pressures that are broadly consistent with the mineralogical barometers. Both types of barometers indicate a southward increase in pressure from about 5 to about 7.5 kb. No large discontinuities in pressure are indicated. The trend is probably not linear but may

Table 4. Homogenization temperatures and calculated entrapment pressures of CO_2 inclusions

Sample local	North latitude	Homogenization T °C		Pressure kbar	
		Range	Peak	Range	Peak
YP	12°38'	−7 to +7	0	4.4−5.6	5.3
KB	12°30'	−1 to −12	−6	5.3−6.1	5.6
SG	12°29'	−5 to −13	−8	5.5−6.1	5.8
BF	12°19'	−9 to −25	−19	5.8−7.0	6.6
GV	12°09'	−27 to −43	−35	7.2−8.4	7.8
GD	12°06'	−21 to −34	−26	6.7−7.8	7.1
HS	12°05'	−21 to −43	−32	6.7−8.4	7.6

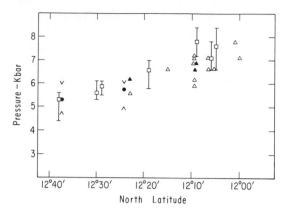

Fig. 5. Palaeo-pressures of late Archaean metamorphism in the granulite facies transition region, calculated from the assemblages garnet-opx-plagioclase-quartz (*open triangles*), garnet-cpx-plagioclase-quartz (*filled triangles*), garnet-sillimanite-plagioclase-quartz (*filled circles*, Harris and Jayaram 1982), and garnet-sillimanite-cordierite-quartz (*Downward-pointing arrowheads* for $P_{H_2O} = P_{total}$ and *upward-pointing arrowheads* for $P_{H_2O} = 0$. Cordierite compositions from Harris and Jayaram 1982). Shown also are palaeo-pressures calculated from measured densities of CO_2 inclusions (see text). The plot shows that the transition region is an apparently unbroken depth profile

show the most rapid rate of rise at about 12°20′ in the Basavanbetta forest region.

6 Discussion and Conclusions

6.1 Continuity of Metamorphic Terrane

The most important finding of the present work is the identification of a progressive metamorphic tract culminating in depleted high-grade granulites over which there is a continuum of palaeo-pressures. This finding is in harmony with evidence from satellite imagery and reconnaissance field work that no major structural breaks occur in the transition interval. There remains little doubt, therefore, that the South Indian high-grade terrane has an intimate connection with the rest of the South Indian Shield, being derived from it by high-grade metamorphic conversion. The idea of a distinctly older age for the granulites of Karnataka than for the adjacent amphibole-bearing gneisses, and the idea that the amphibole-bearing gneisses were derived from the granulites by regional retrogression, are not in accord with the smooth increase of pressure, the appearance of garnet and disappearance of primary amphibole and, to some extent primary biotite, in acid rocks, and the onset of profound Rb depletion in many rocks. The major relations are much more in accord with prograde increase of pressure, low P_{H_2O} volatile activity and, probably, temperature. Minor retrogresse features, which include partial replacement of some orthopyroxene by amphibole, as observed by Devaraju and Sadashivaiah (1969) in the Satnur area charnockites are commonly present also but should not be confused with the dominant prograde features.

6.2 Evidence for a Discrete Metamorphic Event

Although the palaeo-pressures increase with increasing penetration into the high-grade terrane, there is much evidence that this zonation does not merely represent the ambient geothermal condition at some time in the past. The specific action of low-P_{H_2O} fluids seen arrested at Kabbal argues for a discrete chemical and thermal event dated at 2670 Ma ago by whole rock Rb-Sr methods (Venkatasubramanian 1975). Some Peninsular Gneiss in areas to the north of Kabbal has been shown to be as old as 3400 Ma (Beckinsale et al. 1980). It should not be assumed that all of the granite-greenstone terrane is underlain by granulite facies equivalents of the same local age. It is more probable that a profound late Archaean high-grade metamorphism affected much of southern India, converting former lower grade gneisses and supracrustals to granulites.

6.3 The Role of Carbonic Fluids

The close correspondence of the CO_2 inclusion barometry with the mineralogic barometry suggests that CO_2 was captured at or near the pressure-temperature conditions of major mineral growth, i.e., that CO_2 was the dominating ambient fluid species. A similar conclusion was reached by Coolen (1981) for CO_2 inclusions in the high-pressure granulites of the Furua Complex, Tanzania, and by Hörmann et al. (1980) on the basis of CO_2-bearing cordierites of the Finnish Lappland granulites.

The increasing pressure of metamorphism with grade requires that the source of the volatiles be deep-seated: as the depth of the exposed section increases, the former source of the volatiles, as indicated by intensity of action, is approached. Whatever the nature of this source, it probably also corresponded closely to a source of heat to supply the energy of recrystallization and latent heat of dehydration of a large body of rocks. Thermal calculations (Schuiling and Kreulen 1979) show that large-scale CO_2 streaming upward in the crust can transport large quantities of heat. This principle was invoked by Harris et al. (1982) to account for their nearly isothermal South Indian late Archaean geotherm reconstructed from large-scale regional geothermometry/geobarometry.

The observed Rb depletions bear on the nature of the source of CO_2 and heat in ways that are not fully decipherable at present. The non-uniform nature of the depletions suggests that, if volatile action was the main depleting agency, that action was not as pervasive as might be imagined in terms of the model of a large decarbonating mantle upwell affecting subcontinental masses (Harris et al. 1982; Newton et al. 1980). This point was also made by Valley (1982) in connection with the strongly variable $^{18}O/^{16}O$ ratios of Adirondack granulites. He argued for local stratigraphic sources and control of volatiles. On the other hand, it could be argued that the very depleted rocks are those through which depleting volatiles have had the most extensive circulation because of propitious permeability factors, or that these rocks have undergone more extensive anatexis than neighbouring rocks, perhaps by virtue of greater initial hydration. However, we found no evidence for correlation of K/Rb with decrease of SiO_2 or increase in

Mg/Fe ratio, Ni, or Cr, which components have been used to distinguish magmatic residues. The same situation was found by Weaver (1980) for the Madras charnockites and by Rollinson and Windley (1980) for the Scourie granulites, leading these authors to conclude that vapor transport processes had affected the LIL elements, with the possible exception of the rare earths, to at least the same extent as removal of a partial melt. At present knowledge the depletion process appears to be complex and not characterizable by any single process such as partial melting.

7 Tectonic Interpretations

The south Indian high-grade terrane has not been intensively mapped, and the major structural features remain unknown. However, some discussion based on the present findings may be made relative to the major current models of granulite terranes:

7.1 Hot Spots

Harris et al. (1982), in their reconnaissance geothermometry and geobarometry of southern India, deduced a late Archaean advective geotherm in which the deeper and middle portions of the crust were nearly isothermal at $700° - 800°C$. The major heat transporting agency was thought to be pervasive CO_2, liberated from a decarbonating mantle diapir of broad extent. A variant of this hypothesis might be that of Wells (1979) who envisioned magmatic overplating or mid-crustal injection of massive magmatic units, perhaps tonalite sheets, to produce a thickened crust, in accounting for the high pressures of some of the metamorphic rocks of southwest Greenland, and this type of overplating could also have contributed to a flat dT/dP gradient. Wells (1979) invoked tonalitic magmas, degassing as they froze, as a source of CO_2 for granulite metamorphism.

The correlation of increasing CO_2 activity and increasing pressure found in the present study indicates that the source of the CO_2 was deep-seated rather than superimposed above the metamorphic rocks, and this could favour the Harris et al. (1982) concept.

7.2 Intracontinental Disruptions

Kröner (1980) envisioned a disruption within the southern African cratonal group to produce the Archaean high-grade Limpopo Belt, without much relative separation of the Kaapvaal and Rhodesian fragments. The disruption might have been generated by intracontinental subduction or A-subduction. This mechanism could give rise to a depth-zone relation of metamorphic intensity and, although Kröner (1980) did not discuss a possible role of CO_2 action, CO_2 could conceivably be generated by destruction of subducted basinal carbonate deposits accumulated in the disruption zone. Another type of disruption might be large-scale transcurrent faulting in the southern Indian craton, driven by plate tectonics

(Katz 1976). The ways in which the transcurrent mechanism could give a depth zone relation of metamorphic intensity or increased CO_2 action in the higher-grade products are not obvious.

7.3 Continental Collision

The major cause of ancient granulite metamorphism was suggested to be collisions of ancient continents or continental fragments by McLelland and Isachsen (1980) for the Adirondacks and Falkum and Petersen (1980) for the Rogaland area of southwestern Norway. This model has the advantages of explaining the ubiquitous continental shelf sedimentary suite (quartzites and marbles) as basin deposits caught up between approaching continental masses, and a thickened crust to produce high-pressure metamorphism without invoking magmatic over-plating. If the collision resulted in flat underthrusting similar to that pictured for the modern collision of India with Asia (Powell and Conaghan 1975; Barazangi and Ni 1982), the thickening could be due essentially to continental doubling. The pressures generated on continental shelf sediments caught in the continent-continent interface would tend to range about 8 kb, or equivalent to burial under a continental thickness (Newton and Perkins 1981). The near horizontal primary foliation and nappe structures which are frequently observed in granulite ter-ranes would be plausibly explained. Shelf carbonates and evaporites could pro-vide a certain source of CO_2 for granulite metamorphism, though probably not sufficient to charnockitize the whole of the lower crust. A considerable difficulty with this model applied to southern India lies in the generation of temperatures as high as 800 °C over very large areas. According to the thermal modelling of Thompson (1981), temperatures of metamorphism at a depth of 30 km at the megathrust interface would probably not be high enough to produce granulite metamorphism, even with somewhat higher geotherms in the Archaean than in later geologic time. Thus, some additional source of heat, such as CO_2 transport or magmatic underplating, would seem to be required to augment the tempera-tures of metamorphism.

Attempts to apply any of these models to the granulite terrane of South India and, in particular, to the transitional region of Karnataka, would be premature considering the present lack of structural and geophysical data. The contribution of the present work is further documentation that a massive source of CO_2 must be regarded as a boundary condition to be satisfied by any tectonic model which proposes to explain the Archaean granulite facies terrane in southern Karnataka.

Acknowledgements. The authors express gratitude to D. Gopalkrishna for his great help in the field operations. Professor V. Venkatachalapathy kindly placed the resources of the Department of Geology, Mysore University, at our disposal. Much advice, encouragement and logistic support was given by J. J. W. Rogers to whom the authors are very grateful. J. Touret provided us with a fluid inclusion calibration specimen and helpful advice on microthermometry.

Reviews of the original manuscript were provided by W. L. Griffin, M. Raith and W. Schreyer, which resulted in substantial changes.

The field work of this research was supported by a National Science Foundation grant, INT 78-17128 (J. J. W. Rogers and S. M. Naqvi, principal investigators). Laboratory operations were sup-ported by grants EAR 81-07110 (RCN) and EAR 82-03628 (RCN), and EAR 79-05723 (Rogers and Naqvi).

References

Barazangi M, Ni J (1982) Velocities and propagation characteristics of P_n and S_n beneath the Himalayan arc and Tibetan plateau: Possible evidence for underthrusting of Indian continental lithosphere beneath Tibet. Geology (Boulder) 10:179 – 185

Beckinsale RD, Drury SA, Holt RW (1980) 3,360-Myr old gneisses from the South Indian Craton. Nature 283:5746, 469 – 470

Buddington AF (1963) Isograds and the role of H_2O in metamorphic facies of orthogneisses of the northwest Adirondack area. N Y Geol Soc Am Bull 74:1155 – 1182

Burruss RC (1981) Analysis of phase equilibrium in C-O-H-S fluid inclusions. In: Hollister LS, Crawford ML (eds) Short Course in Fluid Inclusions: Applications to Petrology. Min Assoc Can: 39 – 74

Condie KC, Allen P (1984) Origin of Archaean charnockites from southern India, 182 – 203, this Vol.

Coolen JJMMM (1981) Carbonic fluid inclusions in granulites from Tanzania – a comparison of geobarometric methods based on fluid density and mineral chemistry. Chem Geol 37:59 – 77

Devaraju TC, Sadashivaiah MS (1969) The charnockites of Satnur-Halaguru area, Mysore State. Ind Mineral 10:67 – 88

Drury SA, Holt RW (1980) The tectonic framework of the South Indian Craton: A reconnaissance involving LANDSAT imagery. Tectonophys 65:T1 – T15

Ellis DJ, Green DH (1979) An experimental study of the effect of Ca upon garnet-clinopyroxene Fe-Mg exchange equilibria. Contrib Mineral Petrol 71:13 – 22

Eskola P (1939) Die metamorphen Gesteine. In: Barth TFW, Correns CW, Eskola P (eds) Die Entstehung der Gesteine. Springer, Berlin Heidelberg New York, pp 263 – 407

Falkum T, Petersen JS (1980) The Sveconorwegian Orogenic Belt, a case of Late-Proterozoic plate-collision. Geol Rundsch 69:622 – 647

Friend CRL (1981) The timing of charnockite and granite formation in relation to influx of CO_2 at Kabbaldurga, Karnataka, South India. Nature 294:550 – 552

Harley SL (1981) Garnet-orthopyroxene assemblages as pressure-temperature indicators. Ph D thesis, Univ of Tasmania

Harris NBW, Jayaram S (1982) Metamorphism of cordierite gneisses from the Bangalore region of the Indian Archean. Lithos 15:89 – 97

Harris NBW, Holt RW, Drury SA (1982) Geobarometry, geothermometry, and late Archean geotherms from the granulite facies terrain of South India. J Geol 90:509 – 527

Hörmann PK, Raith M, Raase P, Ackermand D, Seifert F (1980) The granulite complex of Finnish Lappland: Petrology and metamorphic conditions in the Ivalojoki-Inarijärvi area. Geol Surv Finl Bull 308:1 – 95

Hubregtse JJMW (1980) The Archean Pikwitonei granulite domain and its position at the margin of the northwestern Superior Province (Central Manitoba). Manit Geol Surv Pap GP80-3, pp 1 – 16

Janardhan AS, Newton RC, Hansen EC (1982) The transformation of amphibolite facies gneiss to charnockite in southern Karnataka and northern Tamil Nadu, India. Contrib Mineral Petrol 79:130 – 149

Katz MB (1976) Early Precambrian granulites-greenstones, transform mobile belts and ridge-rifts on early crust? In: Windley BF (ed) The Early History of the Earth. Wiley, pp 147 – 158

Khoury SG (1968) Structural analysis of complex fold belts in the Lewisian north of Kylesku, Sutherland, Scotland. Scot J Geol 4:109 – 120

Kröner A (1980) New aspects of craton-mobile belt relationships in the Archaean and Early Proterozoic: examples from southern Africa and Finland. In: Closs H, Gehlen K v, Illies H, Kuntz E, Neumann J, Seibold E (eds) Mobile Earth. Boldt Verlag, pp 225 – 234

Lambert IB, Heider KS (1968) Chemical investigations of deep-seated rocks in the Australian Shield. Lithos 1:30 – 53

Mason R (1973) The Limpopo mobile belt – southern Africa. Philos Trans R Soc Lond A273: 463 – 485

McLelland J, Isachsen Y (1980) Structural analysis of the southern and central Adirondacks: A model for the Adirondacks as a whole and plate-tectonics interpretations. Geol Soc Am Bull 91: (Pt. 1)68 – 72

Newton RC, Perkins D III (1981) Ancient granulite terrains — "eight kbar metamorphism". EOS 62:420

Newton RC, Smith JV, Windley BF (1980) Carbonic metamorphism, granulites, and crustal growth. Nature 288:45—50

Perkins D, Newton RC (1981) Charnockite geobarometers based on coexisting garnet-pyroxene-plagioclase-quartz. Nature 292:144—146

Phillips GN (1980) Water activity changes across an amphibolite-granulite facies, Broken Hill, Australia. Contrib Mineral Petrol 75:377—386

Pichamuthu CS (1953) The charnockite problem. Mys Geol Assoc (Bangalore), 178 pp

Pichamuthu CS (1960) Charnockite in the making. Nature 188:135—136

Pichamuthu CS (1965) Regional metamorphism and charnockitization in Mysore State, India. Ind Mineral 6:119—126

Powell CM, Conaghan PF (1975) Tectonic models for the Tibetan Plateau. Geology (Boulder) 3:727—731

Raith M, Raase P, Ackermand D, Lal RK (1982) Regional geothermobarometry in the granulite facies terrain of South India. Trans R Soc Edinb 73 (Earth Sci):221—244

Rama Rao B (1945) The charnockite rocks of Mysore (southern India). Mys Geol Dept Bull 18:1—199

Ramiengar AS, Ramakrishnan M, Viswanatha MN (1978) Charnockite-gneiss-complex relationship in southern Karnataka. J Geol Soc India 19:411—419

Ray S (1972) "Charnockite" of Kabbal, Mysore — a brief study. Q J Geol Mineral Met Soc India 44:163—166

Reed SJB, Ware NG (1975) Quantitative electron microprobe analysis of silicates using energy-dispersive X-ray spectrometry. J Petrol 16:449—519

Rollinson HR, Windley BF (1980) Selective elemental depletion during metamorphism of Archaean granulites, Scourie, NW Scotland. Contrib Mineral Petrol 72:257—263

Rollinson HR, Windley BF, Ramakrishnan M (1981) Contrasting high and intermediate pressures of metamorphism in the Archaean Sargur Schists of southern India. Contrib Mineral Petrol 76:420—429

Romey WD, Elberty WT, Jacoby RS, Christofferson R, Shrier T, Tietbohl D (1980) A structural model for the northwestern Adirondacks based on leucogranite gneisses near Canton and Pyrites, New York. Geol Soc Am Bull Pt I, 90:97—100

Schuiling RD, Kreulen R (1979) Are thermal domes heated by CO_2-rich fluids from the mantle? Earth Planet Sci Lett 43:298—302

Suryanarayana KV (1960) The Closepet granite and associated rocks. Ind Mineral 1:86—100

Thompson AB (1981) The pressure-temperature (P,T) plane viewed by geophysicists and petrologists. Terra Cognita 1:11—20

Touret J (1974) Le faciès granulite en Norvège méridionale I. Les associations minéralogiques. Lithos 4:239—249

Touret J (1974) Faciès granulite et fluides carboniques. An Soc Geol Belgique, P Michot volume, pp 267—287

Touret J (1981) Fluid inclusions in high grade metamorphic rocks. In: Hollister LS, Crawford ML (eds) Short course in fluid inclusions: Application to petrology. Min Assoc Can, pp 182—208

Touret J, Bottinga Y (1979) Équation d'état pour le CO_2; application aux inclusions carboniques. Bull Minéral 102:577—583

Valley JW (1982) Fluid heterogeneity in the lower crust. EOS 63:448

Venkatasubramanian VS (1975) Studies in the geochronology of the Mysore Craton. Geophys Res Bull N G R India 13:239—246

Viswanathan TV (1969) The granulitic rocks of the Indian Precambrian Shield. Geol Surv Ind Mem 100:1

Weaver BL (1980) Rare-earth element geochemistry of Madras granulites. Contrib Mineral Petrol 71:271—279

Wells PRA (1979) Chemical and thermal evolution of Archaean sialic crust, southern West Greenland. J Petrol 20:187—226

Werre RW, Bodnar RJ, Bethke PM, Barton PB (1979) A novel gas-flow fluid inclusion heating/freezing stage. Geol Soc Am Abstr w/Prog 11:539

Origin of Archaean Charnockites from Southern India

K. C. CONDIE and P. ALLEN[1]

Contents

1 Introduction 183
2 Petrographic Features 185
3 Analytical Methods 185
4 Major Element Compositions 188
5 K, Rb, Cs, Th, U, and Pb 189
6 Ba and Sr 193
7 Rare Earth Elements 193
8 Other Trace Elements 193
9 Origin of the Charnockite Protoliths 197
10 Discussion 198
References 201

Abstract

Charnockites in southern India range from low (5 – 6 kb) to high-pressure (7.5 – 8 kb) types. Intermediate chemical compositions are infrequent, although a continuous series exists from mafic to tonalite-trondhjemite end members. With the exception of K_2O, major element distributions reflect the original igneous parentage of the rocks. Rb, Cs, Th, U, and Pb are depleted in charnockites although the degree of depletion does not correlate well with inferred P-T conditions. These depletions appear to have developed during fluid-phase metamorphism. The lack of K depletion and of extreme Rb depletion may reflect the presence of a fluid phase with relatively high CO_2/H_2O ratios. Other trace elements including REE do not appear to have been redistributed by the fluid and reflect igneous fractionation trends of the protoliths. Although the distribution of some trace elements shows a geographic coherence, trace element distributions do not generally correlate with metamorphic grade.

Geochemical modelling indicates that hornblende and/or garnet fractionation, and in some cases plagioclase fractionation, were important in the formation of the charnockite protoliths. The sparsity of mafic cumulates and the relatively high content of transition metals in the charnockites do not favour fractional crystallization models. Progressive batch melting of a mafic source may account for the diversity of compositions in the charnockites. The source, however, must be undepleted in incompatible elements relative to both TH1 and N-MORB compositions.

1 Department of Geoscience, New Mexico Institute of Mining and Technology, Socorro, NM 87801, USA

Archaean Geochemistry (ed. by A. Kröner et al.)
© Springer-Verlag Berlin Heidelberg 1984

Although neither the average low or high-pressure charnockites from southern India can represent residues left after extraction of granitic magma, some charnockites from the transition zone between granulite and amphibolite facies terranes may represent such residues. Extensive migmatization in the transition zone is also consistent with granites being produced at intermediate rather than deep crustal levels.

1 Introduction

The origin of Archaean charnockites in southern India has been the subject of numerous investigations (as summarized in Pichamuthu 1953), but only recently has the subject been approached using modern geochemical and experimental methods (Weaver and Tarney 1980; Janardhan et al. 1982; Condie et al. 1982). A transition zone exists between a granite-greenstone terrane to the north and a charnockite (granulite) terrane to the south (Fig. 1). This transition zone, across which rocks change from amphibolite to granulite facies, is 20 to 35 km in outcrop width and has been the subject of several detailed studies (Janardhan et al. 1979; Friend 1981; Condie et al. 1982; *see also Hansen et al., this Vol., eds.*). Field and geochemical data strongly suggest that charnockites in the transition

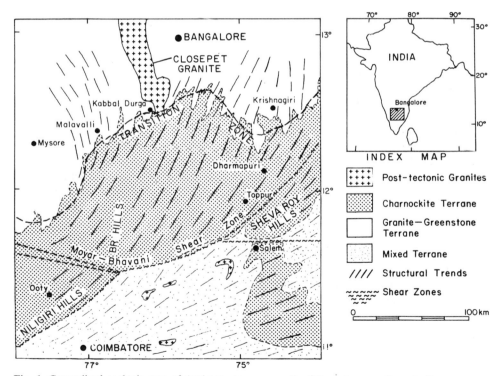

Fig. 1. Generalized geologic map of Archaean terranes south of Bangalore, southern India

zone developed from tonalitic precursors similar in composition to tonalites and trondhjemites in the granite-greenstone terrane to the north. To more fully evaluate geochemical differences between the low-pressure charnockites in the transition zone and high-pressure charnockites from several massifs south of the transition zone, we have collected and analyzed a large suite of samples ranging from felsic to mafic in composition. Although it is generally agreed upon by most investigators that granulites are produced in the lower crust in the presence of a fluid phase relatively rich in CO_2 and perhaps in halogens (Collerson and Fryer 1978; Newton et al. 1980), it is not as yet clear which elements are remobilized under such conditions. Refinements in the calibration of geobarometers and geothermometers in the past few years make it possible to estimate P-T conditions in granulite-facies terranes (Newton and Perkins 1982; Harris et al. 1982). Some studies have focused on charnockites in southern India, and it is possible with existing data to broadly characterize the P-T conditions of some terranes in this area (*see Hansen et al., this Vol., eds.*).

For our investigation we have collected samples in areas where such P-T estimates are available, concentrating on the transition zone southeast and southwest of Bangalore and the high-pressure terranes farther to the south. The low-pressure charnockites in the transition zone are mixed with amphibolite-facies rocks and reflect pressures of about 5 – 6 kb and temperatures of 600 – 700 °C (Harris and Jayaram 1982; Harris et al. 1982; Raith et al. 1983; *Hansen et al., this Vol., eds.*). Field relations are described in previous publications (Pitchamuthu 1953; Janardhan et al. 1979, 1982; Friend 1981; Condie et al. 1982). The most extensive sampling of the transition zone, in conjunction with detailed mapping, was conducted in the Krishnagiri-Dharmapuri area (Fig. 1). Evidence for both prograde and retrograde metamorphic processes are present in charnockites from the transition zone; the geochemical relationships of these processes are the subject of a separate investigation (Allen et al. 1984). Most charnockites in the transition zone are interspersed in tonalitic gneiss from which they have developed. Mafic charnockites, which comprise a minor part of the charnockite terrane, will be considered in a separate study.

Charnockites from the medium and high-pressure terranes were collected from the charnockite massifs along and to the south of the Moyar-Bhavani shear zone (Fig. 1). Charnockites from the south and central Nilgiri Hills, where medium P-T conditions appear to have existed (6 – 7 kb, 735 – 750 °C; Newton and Perkins 1982; Harris et al. 1982; Raith et al. 1983), are massive and commonly contain garnet. Higher pressure terranes, representing P-T conditions of 7.5 – 8 kb and 680 – 800 °C, were sampled in the Biligirinrangan (BR) Hills, the Shevaroy Hills, and in the Toppur area (Fig. 1). The highest pressure terrane reported in south India, the north slopes of the Nilgiri Hills (~9 kb, 750 – 900 °C; Janardhan et al. 1982; Raith et al. 1983) is not represented by samples in this study.

2 Petrographic Features

In terms of modal minerals in the charnockites there is no simple relationship to metamorphic grade (Table 1). Textures are granoblastic and range from massive to foliated. Plagioclase ($An_{20} - An_{50}$) is the most abundant mineral with smaller amounts of quartz, biotite, hornblende, and ortho and clinopyroxene. Accessory phases include apatite, zircon, rutile, and iron oxides. Plagioclase is commonly antiperthitic and contains dusty granular inclusions. Twins are variably developed and often bent and fractured, especially in the medium and high-pressure samples. K-feldspar (microcline) is relatively uncommon in most samples ($\leq 5\%$) and occurs as late prophyroblasts which engulf other minerals. Biotite is yellow to brown and laths are typically bent. Pyroxene is dominantly orthopyroxene ($1 - 10\%$) and occurs as pink to pale green, partially altered crystals. Clinopyroxene is of minor importance in most felsic charnockites. Although textural evidence is often ambiguous, most data from charnockites in the transition zone indicate that pyroxenes developed at the expense of biotite and hornblende. In medium and high-pressure samples, however, most of the biotite and some of the hornblende appear to have developed late, during retrogrde metamorphism; bent biotite laths typically occur along mylonitic micro-shears, and biotite and secondary hornblende replace orthopyroxene and garnet along cleavage planes and grain margins. Garnet occurs in many medium and high-pressure charnockites that are relatively high in Mg and Fe, and it is particularly common in samples from the south and central Nilgiri Hills and the Shevaroy Hills where it comprises up to 35% of some samples but has a rather inhomogeneous distribution, occurring chiefly in local lenses and bands. Minor secondary minerals, chiefly of retrograde origin, include chlorite, cummingtonite, sericite, iron oxides and carbonate. Carbonate occurs as small veinlets and patches in plagioclase and along micro-shears and as an alteration product, together with chlorite and iron oxides, developed in pyroxenes and garnet.

Some coherent geographic variations occur in mineral assemblages. For instance, samples from the BR Hills do not contain garnet and are consistently high in plagioclase. The samples from the central and south Nilgiri Hills, on the other hand, contain significant amounts of garnet and orthopyroxene and unusually large amounts of K-feldspar ($5 - 10\%$).

In terms of petrography high and medium-pressure charnockites differ from transition-zone charnockites as follows: (1) they commonly contain garnet; (2) evidence for retrograde reactions is widespread; (3) samples commonly exhibit mylonitic and flaser textures and thus reflect greater ductile deformation than low-P samples; and (4) plagioclase has poorly developed twins and ubiquitous antiperthitic textures.

3 Analytical Methods

Model analyses were made with a Swift automatic point counter, and identification and reproducibility were checked by two or more investigators. Element

Table 1. Representative chemical and modal analyses of Archaean charnockites from southern India (major element oxides in wt %, trace elements in ppm and modes in vol %

	Low grade Transition zone					Medium-grade Nilgiri Hills (south and central)			
	120-1	122-1	NC1	66	68	478	480	567	569
SiO_2	68.61	63.44	68.68	64.78	57.98	72.61	67.96	59.39	70.24
TiO_2	0.44	0.73	0.36	0.54	0.82	0.37	0.54	0.68	0.49
Al_2O_3	16.33	17.42	16.55	16.30	17.02	11.85	14.72	16.84	12.03
Fe_2O_3T	3.40	5.72	2.93	4.62	7.16	7.15	5.11	10.70	7.64
MnO	0.03	0.08	0.04	0.07	0.10	0.09	0.05	0.13	0.10
MgO	1.34	1.65	1.07	2.67	3.26	2.28	2.74	4.01	3.14
CaO	3.88	4.83	4.51	4.71	6.13	3.53	2.94	3.80	2.67
Na_2O	4.60	5.07	5.06	4.20	4.90	2.02	3.57	2.62	1.97
K_2O	1.23	0.87	0.52	1.20	1.54	0.49	2.30	1.02	0.97
P_2O_5	0.16	0.26	0.05	0.22	0.33	0.07	0.15	0.14	0.09
LOI	0.48	0.04	0.41	0.41	0.64	0.79	0.51	0.27	0.24
Total	100.50	100.11	100.18	99.72	99.88	101.25	100.59	99.61	99.57
Rb	22	41	3.6	12	36	7	71	24	18
Cs	0.6	0.4	0.47	0.63	1.0	0.2	0.1	0.1	0.2
Sr	449	487	418	572	728	167	363	227	227
Ba	654	217	353	873	388	90	792	305	309
Sc	7	11	3.8	13	19	12	8	25	15
Y	10	9.3	<2	12	24	16	13	30	23
Zr	158	161	136	156	174	95	82	68	127
Hf	3.4	6.9	4.0	6.0	5.3	2.1	2.6	1.9	3
Nb	4.6	6.4	2.3	5.9	13	6.2	6.6	7.9	6.8
Ta			3.8	2.5	1.9	0.4	0.3	0.3	0.3
Cr	30	25	4	38	29	108	65	268	207
Ni	14	<43		13	21	19	25	136	111
Co			32	21	25	17	16	36	21
Pb	<2	<2	5.0	<2	2	11	13	<2	<2
U	<0.3	2	0.13[a]	0.14[a]	0.63[a]	0.2	0.2	0.46[a]	0.43[a]
Th	1	3	0.3	0.5	2.2	1	4	1.0	2.3
La	27	22	13	35	50	19	37	22	18
Ce	51	47	16	52	85		64	39	32
Sm	3.0	5.2	0.70	5.1	9.0	2.8	4.1	4.1	2.9
Eu	1.3	1.9	1.5	1.8	2.6	0.54	1.3	1.1	0.93
Tb	0.36	0.65	0.16	0.80	1.5	0.51	0.55	0.73	0.62
Yb	0.62	2.4	0.37	1.3	2.2	1.7	1.6	2.8	1.9
Lu	0.10	0.36	0.05	0.15	0.31	0.25	0.24	0.40	0.28
Qtz	15.4	16.1	31.3	18.4	5.4	40.5	20.6	0.5	34.5
Plg	71.3	72.9	55.9	64.2	62.1	29.8	53.1	47.5	39.6
KF	tr	–	1.3	0.2	0.9	0.5	11.2	9.6	7.7
Bi	6.4	1.3	5.4	7.8	15.8	2.6	6.9	15.3	0.5
Hb	tr	–	1.9	0.6	0.8	0.1	0.1	–	–
Opx	5.6	7.8	1.6	5.2	2.4	6.5	5.5	1.4	8.2
Cpx	–	–	0.2	0.8	9.3	–	–	–	–
Opq	1.2	2.4	2.1	2.2	3.1	0.5	0.4	1.8	0.6
Ap	0.1	tr	0.2	0.4	0.1	0.2	tr	tr	tr
Zir	tr	tr	0.1	tr	tr	tr	tr	tr	tr
Gar	–	–	–	–	–	18.4	2.2	23.9	8.2
Other[b]	tr	tr	tr	tr	tr	0.9	tr	tr	0.4

[a] Determined by delayed neutron activation analysis
[b] Chiefly carbonate, epidote, sericite, cholorite, cummingtonite and sphene

Table 1 (continued)

High-Grade

	Toppur area		Shevaroy Hills			BR Hills (south)			
	485-2	486-1	526-1	528	535-2	553	554	556	558
SiO_2	60.7	56.47	71.26	63.73	66.53	69.43	69.51	69.85	65.57
TiO_2	0.72	0.88	0.30	0.67	0.60	0.31	0.41	0.40	0.63
Al_2O_3	15.77	15.89	15.52	16.28	17.30	15.89	15.98	16.16	15.60
Fe_2O_3T	6.71	8.51	2.83	6.78	3.85	2.85	2.88	2.20	4.46
MnO	0.09	0.12	0.03	0.09	0.06	0.03	0.03	0.02	0.05
MgO	3.80	5.00	0.92	3.10	1.41	0.91	1.00	0.89	2.21
CaO	6.15	7.45	3.62	4.02	4.45	3.00	3.51	3.60	3.73
Na_2O	4.17	4.20	4.84	4.05	3.93	5.12	5.04	5.03	5.69
K_2O	1.52	0.82	1.31	1.35	1.27	1.35	1.12	1.14	1.08
P_2O_5	0.31	0.28	0.10	0.16	0.19	0.11	0.11	0.10	0.23
LOI	0.45	0.67	0.31	0.29	0.28	0.66	0.22	0.33	0.30
Total	100.39	100.31	101.04	100.53	99.85	99.65	99.82	99.72	99.56
Rb	10	2	8	29	8	23	7	5	7
Cs	0.3	0.2	0.1	0.5	<0.03	0.04	<0.03	<0.03	0.05
Sr	812	784	555	537	879	590	661	981	916
Ba	909	418	547	467	600	405	417	615	362
Sc	18	20	1.9	16	5.3	2.9	2.8	2.7	4.7
Y	24	26	2.8	18	5.8	4.3	2.7	1.5	4.7
Zr	124	193	154	124	188	138	161	134	147
Hf	3.8	5.3	3.3	3	4.5	3.1	3.7	3.5	3.6
Nb	6.3	8.7	3.2	7.8	4.4	3.0	3.3	3.0	6.1
Ta	0.1	0.3	0.06	0.2	0.04	0.05	0.06	0.04	0.1
Cr	105	133	9	127	19	9	10	8	93
Ni	29	61	8.7	54	40	3	16	15	49
Co	21	30	7.3	24	11	6	6	6	17
Pb	<2	<2	<2	<2	<2	<2	<2	<2	<2
U	0.1	0.3	0.12[a]	0.16[a]	0.13[a]	0.13[a]	0.09[a]	0.12[a]	0.17[a]
Th	<0.3	0.2	0.1	0.3	4.7	3.8	2.5	0.2	0.6
La	40	36	25	32	49	32	31	26	33
Ce	84	69	38	58	88	55	53	41	62
Sm	9.1	8.4	1.7	4.4	5.1	3.0	1.9	1.4	3.3
Eu	1.9	1.9	1.3	1.5	1.7	0.97	0.92	1.5	1.5
Tb	1.1	0.99	0.16	0.53	0.35	0.21	0.13	0.08	0.24
Yb	1.5	2.0	0.19	1.8	0.46	0.32	0.19	0.19	0.49
Lu	0.20	0.28	0.02	0.30	0.06	0.04	0.02	0.02	0.09
Qtz	10.2	5.9	25.8	19.2	26.8	20.8	21.6	20.7	12.4
Plg	58.5	62.5	69.3	64.1	64.0	74.1	70.8	74.5	60.7
KF	3.8	–	0.2	–	–	0.4	–	1.1	–
Bi	0.3	0.1	0.3	1.2	0.8	2.2	1.0	0.5	7.5
Hb	5.7	0.4	0.4	–	0.1	0.2	0.3	–	2.4
Opx	11.7	14.9	1.4	7.2	1.0	1.2	4.9	1.6	12.5
Cpx	6.6	11.5	–	–	–	0.2	–	–	0.1
Opq	2.9	2.2	1.8	1.3	2.7	0.3	1.1	1.6	2.3
Apt	0.3	0.6	0.4	0.1	tr	0.1	0.1	tr	–
Zir	tr	0.2	tr	–	tr	tr	0.1	tr	–
Gar	tr	tr	tr	6.9	4.6	tr	tr	tr	0.3

concentrations were determined with an automated Rigaku X-ray spectrometer (major elements, Mn, P, Rb, Sr, Zr, Y, Nb) and with a Nuclear Data 6600 gamma-ray spectrometer by neutron activation using an intrinsic Ge detector (7 REE, Na, Co, Ni, Cr, Rb, U, Th, Cs, Ba, Zr, Hf, Ta, Sc). U, at levels below 1 ppm, was determined by the delayed neutron method at Los Alamos National Laboratory. Reproducibility and estimated analytical errors are less than 5% for major elements and less than 10 – 15% for most trace elements. Details of analytical procedures and assessment of errors are described in Norrish and Hutton (1969); Condie and Lo (1971) and Gordon et al. (1968).

4 Major Element Compositions

Major element distributions in the charnockites analyzed in this study define continuous trends from mafic to felsic end members. Compositional continuity between mafic and felsic end members, which may be an artifact of deformation, is also reported in Archaean Lewisian granulites from Scotland (Rollinson and Windley 1980a). Other granulite facies terranes, such as the Madras granulites (Weaver 1980), appear to be bimodal with intermediate compositions very rare. If our sampling in the region south of Bangalore is representative, rocks with intermediate compositions (55 – 65% SiO_2) also seem to be less abundant than mafic and felsic end members. On chemical variation diagrams most major elements define igneous-like fractionation trends. For instance, with the exception of some samples from the Nilgiris, Fe_2O_3T, CaO, MgO, and TiO_2 decrease in a linear fashion with increasing SiO_2, exhibiting calc-alkaline-like fractionation trends (Fig. 2). Although such trends may result from tectonic mixing of mafic and felsic end members (Tarney 1976), in the transition zone south of Bangalore clearly defined mappable units exist and mixing appears to be limited to local shear zones and migmatites. An exception to the igneous rock trends is K_2O, which for most rocks, regardless of silica content, ranges between 0.5 and 1.5% (Fig. 2). As previously described (Janardhan et al. 1982; Condie et al. 1982), and as reported in other granulite terranes (Sheraton et al. 1973; Tarney 1976), such low and relatively constant K_2O values appear to reflect K-depletion by a fluid phase during granulite facies metamorphism. The K-enrichment trend observed in Lewisian granites and granodiorites (Tarney and Windley 1977) is not found in the Indian charnockites except locally in granitic gneisses from the transition zone (Condie et al. 1982). That other major elements have not been remobilized by fluids is attested to by the lack of a correlation between metamorphic grade and element concentration. Low, medium and high-pressure charnockites are indistinguishable in terms of major element distributions as shown on silica variation diagrams. It is noteworthy, however, that there are some geographic groupings of composition, as illustrated for example by the samples from the BR Hills (Fig. 2), which probably represent original compositional variations of the protolith.

In terms of the system Ab-Or-Q-An-H_2O the Indian charnockites fall on the tonalite-trondhjemite trend, characteristic also of Archaean tonalites and tonalit-

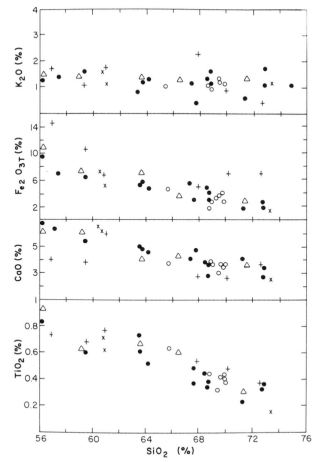

Fig. 2. Silica variation diagrams for K_2O, CaO, Fe_2O_{3T}, and TiO_2 in Archaean charnockites from southern India. *Symbols:* ● transition zone; + Nilgiri Hills; ○ BR Hills; × Toppur area; △ Shevaroy Hills

ic granulites from the Lewisian in Scotland (Rollinson and Windley 1980a). Such a distribution is clearly consistent with field relations and other geochemical data that suggest tonalite-trondhjemite precursors for most of the granulites in this part of India.

5 K, Rb, Cs, Th, U, and Pb

As with many granulite terranes (Lambert and Heier 1968; Sighinolfi 1971; Tarney and Windley 1977; Newton et al. 1980), significant depletion in Rb, Cs, Th, and U occurs in the Archaean charnockites from southern India. These depletions are illustrated in Fig. 3 by comparison with Th, Rb, Cs, and U abundances in tonalitic precursors. Rb, Cs, and U decrease with decreasing Th, with high-pressure charnockites often exhibiting the greatest degree of depletion. The Th-Rb and Th-Cs values for the transition charnockites predominantly fall between the tonalitic gneisses from north of the transition zone and the medium and high-

1981; Condie et al. 1982; Hansen et al., Chap. 8, this Vol.). The relative abundance of CO_2 in fluid inclusions from granulite terranes is generally thought to reflect the importance of CO_2 in the fluid phase. Investigations of fluid inclusions in the charnockites from southern India also imply a fluid with high CO_2/H_2O ratios (*Hansen et al., this Vol., eds.*; R. C. Newton, pers. commun.); however, these fluid inclusions are not totally devoid of water, even in high-pressure charnockites (Allen et al. 1984). The absence of significant K-depletion and of extreme Rb-depletion and the poor correlation of Rb, Cs, U, and Th depletion with metamorphic grade in the Indian charnockites requires some comment in light of the fluid-flux model. Since there is no evidence to suggest that the Indian charnockites had protoliths that were more K-rich than Lewisian protoliths, the explanation may be related to the volume of fluid phase that passed through the rock. Rollinson and Windley (1980b) suggest that up to 2 equivalent rock volumes of aqueous fluid were necessary to produce the extreme Rb depletion in the Lewisian granulites. Employing their model, only 0.3 of a rock volume of fluid is necessary to produce the greatest Rb depletion observed in the Indian charnockites. Perhaps a relatively small volume of fluid or, alternately, a fluid with a high CO_2/H_2O ratio, is responsible for the lack of K depletion and of extreme Rb depletion in the Indian charnockites. As U readily forms soluble carbonate complexes (Namouv 1959), a fluid with a relatively high CO_2/H_2O ratio may also explain the rather abrupt depletion in U in some of the charnockites, particularly if significant differences in CO_2/H_2O ratio existed locally in the transition zone. The poor correlation of metamorphic grade with degree of element depletion may also reflect variable CO_2/H_2O ratios in the fluid

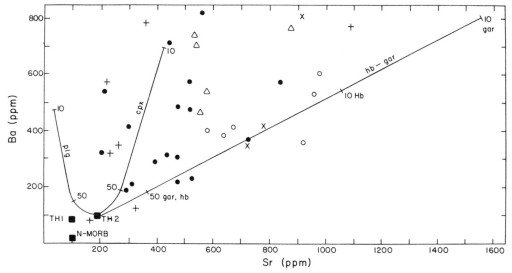

Fig. 5. Distribution of Indian charnockites on a Ba-Sr plot. Symbols given in Fig. 2. Also shown are liquid trends for batch melting of plagioclase (*plg*), clinopyroxene (*cpx*), hornblende (*hb*) and garnet (*gar*). Numbers indicate percent melting. *TH1* and *TH2* are average compositions of two types of Archaean tholeiite (Condie 1981a) and *N-MORB*, the average composition of normal ocean-ridge basalt. Distribution coefficients from Arth (1976)

phase. This would imply that the composition of the fluid phase was not related to burial depth or temperature in any simple manner.

6 Ba and Sr

Both Ba and Sr generally have concentrations of >300 ppm in the Indian charnockites (Fig. 5). Ba may range as high as 800 ppm and Sr 1000 ppm, and thus the rocks fall within the broad range of other Archaean granulites (Tarney 1976). No simple correlation exists between the abundances of the two elements, even in samples from the same locality. Ba/Sr ratios range from about 0.5, which characterizes samples from the BR Hills, to about 2, which characterizes some of the Nilgiris samples. Not surprisingly, samples with the highest Ba/Sr ratios also generally have the greatest K-feldspar or biotite contents. Ba-Sr distributions are similar to those in tonalite protoliths north of the transition zone, suggesting that these elements were not greatly disturbed during fluid-phase metamorphism.

7 Rare Earth Elements

The REE distributions in the Indian charnockites described here are similar to those reported in other granulites (Weaver 1980; Weaver and Tarney 1980; Pride and Muecke 1980; Condie et al. 1982). Felsic and intermediate charnockites have light REE-enriched patterns, and the overall concentrations of REE exhibit an approximately inverse correlation with silica content (Fig. 6). Eu-anomalies range from slightly negative to strongly positive and also tend to correlate with REE concentrations as shown by a plot of Eu/Eu* versus Sm_N (Fig. 7). Those samples with negative Eu-anomalies are also generally low in plagioclase. La_N/Sm_N ratios commonly range from 2 to 6 with most samples from the BR Hills having values as high as 10. Heavy REE concentrations vary more than light REE with La_N/Yb_N ratios ranging from about 10 to 175. Although there are some geographic groups in REE distributions, such as samples from the BR Hills, the Nilgiri Hills and the Toppur area (Fig. 6), there is no clear relationship between relative or absolute REE concentrations and metamorphic grade. Similar REE patterns and correlations are found in the tonalite protoliths from within and to the north of the transition zone; this indicates that REE were not significantly remobilized during fluid-phase metamorphism. Even in samples from areas of high-grade metamorphism (7.5 – 8, 680 – 800 °C) the effect of fluids rich in CO_2 does not appear to have disturbed the REE.

8 Other Trace Elements

The high field strength (HFS) elements (Ti, Nb, Y, Zr, Hf, Ta) occur in the same concentration ranges in the charnockites as in tonalite protoliths to the north

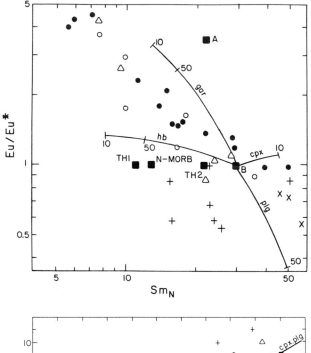

Fig. 7. Distribution of Indian charnockites on a Eu/Eu* vs Sm_N diagram. Symbols and references given in Figs. 2 and 5. Distribution coefficients from Arth (1976) and Condie and Hunter (1976). Hypothetical mafic sources *A* and *B* discussed in the text

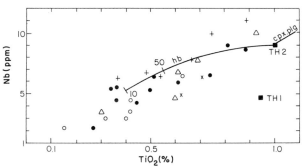

Fig. 8. Distribution of Indian charnockites on a Nb vs TiO_2 plot. Symbols and references given in Figs. 2 and 5. N-MORB falls off the figure at about 3–4 ppm Nb and 1.5% TiO_2. Distribution coefficients from Pearce and Norry (1979)

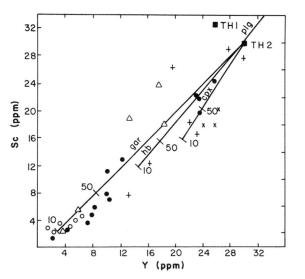

Fig. 9. Distribution of Indian charnockites on a Sc vs Y plot. Symbols and references given in Figs. 2 and 5. N-MORB falls off the diagram at 40–45 ppm Sc and 30–35 ppm Y. Distribution coefficients from Pearce and Norry (1979), Irving (1978), and Irving and Frey (1978)

tends to increase with Ta. Also, Y exhibits a rough positive correlation with Sc (Fig. 9) and with Yb (and other heavy REE). Samples with the highest Y, Sc, and heavy REE contents tend to be garnet-bearing and relatively rich in Fe and Mg.

Transition metals (Ni, Co, Cr) exhibit relatively high, although variable, concentrations in the charnockites, a feature which again appears to be inherited from their tonalite protoliths. In some cases geographic groupings are evident, as shown by the relatively low transition metal contents in samples from the BR Hills and the high contents of the Nilgiris samples.

9 Origin of the Charnockite Protoliths

Evidence at hand clearly indicates that, with the exception of Rb, Cs, U, Pb, Th, and perhaps K, major and trace element distributions in intermediate to felsic Indian charnockites reflect their igneous parentage. This same conclusion has been advocated for the charnockites from the transition zone (Condie et al. 1982) and (with exception of Pb) for granulites from other areas both in India and elsewhere (Weaver 1980; Field et al. 1980; Pride and Muecke 1980; Weaver and Tarney 1981). This is not to imply that supracrustal protoliths are not present, but that they are minor in abundance and were not included in these investigations. Intermediate and felsic charnockites appear to have formed from tonalite or trondhjemite precursors by the action of a fluid phase with variable CO_2/H_2O ratios (Janardhan et al. 1982; Condie et al. 1982). The charnockite protoliths appear to have been produced either by partial melting of a mafic source rich in hornblende and/or garnet (Weaver and Tarney 1980; Condie et al. 1982) or by fractional crystallization of wet basaltic magma (Barker and Arth 1976; Arth et al. 1978). Some investigators have suggested that felsic charnockites represent the residuum remaining after removal of granitic magma (Pride and Muecke 1980) or are produced by deep crustal fractional crystallization of dacitic magma (Field et al. 1980). However, when all geochemical and field data are considered, neither of these processes appear to have been important in the production of the south Indian charnockites for the reasons discussed in Weaver and Tarney (1981), Condie et al. (1982), and later in this paper.

The effect of hornblende, clinopyroxene, garnet and plagioclase fractionation on liquid composition during partial melting of a mafic source is illustrated for various trace elements in Figs. 5, 7, 8, and 9. Shown in these figures are batch melting curves for the various minerals. Fractional crystallization vectors exhibit similar trends. In terms of Eu/Eu* versus Sm_N (Fig. 7) it is clear that most of the charnockite protoliths can be produced by a combination of hornblende and garnet fractionation with variable contributions of clinopyroxene and plagioclase, beginning with a mafic source represented by point B. A greater contribution of plagioclase in the residue results in liquids with negative Eu-anomalies as represented by samples from the Nilgiri Hills and Toppur area. Although enriched Archaean tholeiite (Th2, Condie 1981a) may also be a satisfactory starting material, sources with compositions similar to depleted Archaean tholeiite (Th1) or normal ocean ridge basalt (N-MORB) are clearly unacceptable. An alternate and less satisfactory source (point A), as suggested by Condie et al. (1982), can explain the array of samples with Eu/Eu* ≥ 1 by varying the proportions of hornblende, plagioclase and clinopyroxene in the residue. Both sources A and B are undepleted in incompatible elements relative to N-MORB, and source A represents a plagioclase-rich cumulate. One disadvantage of source A is that derivative liquids have significantly lower Sr contents than observed in the charnockites. For either source plagioclase must have played an important role in production of the Nilgiris samples that are low in Sr and exhibit negative Eu-anomalies. As previously discussed (Weaver and Tarney 1980) heavy REE distributions in felsic and intermediate charnockites also reflect the role of hornblende and/or garnet fractionation in the production of the protoliths.

The importance of hornblende and/or garnet in the origin of the charnockite protoliths is also attested to by the distribution of samples on Nb-Ti and Sc-Y plots (Figs. 8 and 9). The mafic source in each case must be chemically similar to Th2 and not to Th1 or N-MORB sources which are depleted in Nb relative to Ti. In terms of Ba and Sr distributions (Fig. 5) most samples can be explained by hornblende-garnet-clinopyroxene fractionation. Small contributions of plagioclase or a greater contribution of clinopyroxene to the melt can, in turn, account for the high-Ba, variable-Sr samples.

10 Discussion

Results of geochemical model studies indicate the importance of hornblende and/or garnet fractionation in the production of the charnockite protoliths. The Eu and Sr contents of some protoliths also demand that plagioclase is important in some cases. Whether fractional crystallization or partial melting was most important in protolith production relies chiefly on non-geochemical and circumstantial evidence. Most data seem to favour a partial melting origin as discussed by Weaver and Tarney (1980, p 290) and Condie et al. (1982) (*see also Jahn and Zhang, this Vol., eds.*). Of particular importance in not favouring fractional crystallization are (1) the sparsity or absence of mafic cumulates which must have been produced in large volumes at depths ≤ 45 km (i.e. within the crust) for plagioclase to be stable, and (2) the relatively high contents of transition metals in the charnockites which demand unreasonably low K_d values to prevent complete exhaustion of these elements in the liquid before reaching the tonalite fractionation stage.

Geochemical results imply that the source of the charnockite protoliths, whether liquid or solid, is undepleted or perhaps enriched in incompatible elements relative to N-MORB and Th1. A source similar in composition to Th2 is acceptable for most elements (Condie et al. 1982). Considering the large volume of tonalite-trondhjemite that was produced in the Archaean (perhaps 50% of the volume of the present continents) this undepleted source must have been relatively abundant.

Granulites in the lower crust have been considered by some investigators as a residue remaining after extraction of granitic magmas (Pride and Muecke 1980;

Condie 1981b). It has been suggested by Taylor and McLennan (1982) that the negative Eu-anomalies in Phanerozoic sediments, which are probably inherited from the erosion of granites, are balanced by positive Eu-anomalies in the lower crust. As summarized by Weaver and Tarney (1981) this is clearly not the case for Lewisian granulites. These same arguments, for the most part, also apply to the southern India charnockites. If the lower crust is not the source of Archaean granites, a mantle source must be advocated; however, a satisfactory model incorporating such a source is faced with apparently insurmountable experimental and geochemical difficulties (Condie and Hunter 1976; Wyllie 1977a). Although one may also produce granites by fractional crystallization of basalt at crustal depths, the absence of large volumes of mafic cumulates in the Archaean crust, at any erosion level, does not favour such an origin.

Let us examine in more detail the parent-daughter-residue relationships in reference to the Archaean charnockites in southern India. First of all, the low contents of Rb, Cs, Th, Pb, and U in the charnockites render them unlikely sources for granite magma without calling upon unrealistically small amounts of partial melting. Most data suggest that tonalite is a major source for Archaean granites (Condie 1981a, b). Approximately 20% melting of average Archaean tonalite will produce granite which is similar in trace element distributions to average Archaean granite. This is illustrated in Fig. 10 by plotting relatively incompatible elements in order of decreasing incompatibility for a felsic granulite source as a function of their tonalite-normalized concentrations. The only major deviations are in Zr and Ti, which are lower in average Archaean granite than in calculated granite, a difference that can readily be eliminated if minor amounts of zircon and Ti-rich magnetite occur as residual phases in the source. If we compare average low- and high-pressure charnockites from southern India to the calculated residue, major differences are apparent (Fig. 10B). The residue is more depleted in incompatible elements than either charnockite average, indicating that neither can be considered as residue after granite extraction. It is noteworthy, however, that element distributions in some samples of charnockite from the transition zone match the distribution in the calculated residue quite well as illustrated, for instance, by sample NC1 (Fig. 10B). This suggests that although the charnockite terrane as a whole cannot be considered as residual, portions of the terrane, at least in the transition zone, may represent the residues of partial melting. Favouring this possibility is the fact that granites and interlayered granite-tonalite migmatites are common in the transition zone.

Although field and geochemical data clearly indicate that some of these granites are of metasomatic origin (Condie et al. 1982), others may be partial melts and still others involve both processes (Friend 1981). Field relationships along the transition zone strongly suggest that a fluid phase relatively rich in CO_2 purged H_2O from the system and concentrated it in relatively narrow regions where partial melting produced migmatites with granite leucosomes (Friend 1981; Janardhan et al. 1982). In some areas along the transition zone major plutons have formed and one major batholith, the Closepet granite, may also have been produced at this crustal level.

Thus, it may be that granite formation in the Archaean crust is closely related to charnockitization and is localized, for the most part, at intermediate crustal

Gordon GE, Randle K, Goles G, Corliss J, Beeson M, Oxley S (1968) Instrumental neutron activation analysis of standard rocks with high resolution gamma-ray detectors. Geochim Cosmochim Acta 32:369–396

Hansen EC, Newton RC, Janardhan AS (1984) Pressures, temperatures and metamorphic fluids across an unbroken amphibolite-facies to granulite-facies transition in southern Karnataka, India. This Vol., 161–181

Harris NBW, Jayaram S (1982) Metamorphism of cordierite gneisses from the Bangalore region of the Indian Archaean. Lithos 15:89–97

Harris NBW, Holt RW, Drury SA (1982) Geobarometry, geothermometry, and late Archaean geotherms from the granulite facies terrain of south India. J Geol 90:509–527

Hunter DR (1974) Crustal development in the Kaapvaal craton, II. The Proterozoic. Precambrian Res 1:295–326

Irving AJ (1978) A review of experimental studies of crystal/liquid trace element partitioning. Geochim Cosmochim Acta 42:743–770

Irving AJ, Fey FA (1978) Distribution of trace elements between garnet megacrysts and host volcanic liquids of kimberlitic to rhyolitic composition. Geochim Cosmochim Acta 42:771–788

Jahn BM, Zhang ZQ (1984) Radiometric ages (Rb-Sr, Sm-Nd, U-Pb) and REE geochemistry of Archaean granulite gneisses from eastern Hebei Province, China. This Vol., 204–234

Janardhan AS, Newton RC, Smith JV (1979) Ancient crustal metamorphism at low pH$_2$O: Charnockite formation at Kabbaldurga, South India. Nature 278:511–514

Janardhan AS, Newton RC, Hansen EC (1982) The transformation of amphibolite facies gneiss to charnockite in southern Karnataka and northern Tamil Nadu, India. Contrib Mineral Petrol 79:130–149

Lambert JB, Heier KS (1968) Geochemical investigations of deep-seated rocks in the Australian shield. Lithos 1:30–53

Moorbath S, Welke H, Gale NH (1969) The significance of lead isotope studies in ancient high-grade metamorphic basement complexes, as exemplified by the Lewisian rocks of northwest Scotland. Earth Planet Sci Lett 6:245–256

Namouv GB (1959) Transportation of uranium in hydrothermal solutions as a carbonate. Geochemistry 4:5–20

Newton RC, Perkins D (1982) Thermodynamic calibration of geobarometers based on the assemblages garnet-plagioclase-orthopyroxene (clinopyroxene)-quartz. Am Mineral 67:203–222

Newton RC, Smith JV, Windley BF (1980) Carbonic metamorphism, granulites and crustal growth. Nature 288:45–49

Norrish K, Hutton JT (1969) An accurate X-ray spectrographic method for the analysis of a wide range of geological samples. Geochim Cosmochim Acta 33:431–453

Pearce JA, Norry MJ (1979) Petrogenetic implications of Ti, Zr, Y, and Nb variations in volcanic rocks. Contrib Mineral Petrol 69:33–47

Pichamuthu CS (1953) The Charnockite Problem. Mys Geol Assoc Publ, 170 pp

Pride C, Muecke GK (1980) Rare earth element geochemistry of the Scourian complex NW Scotland – evidence for the granite-granulite link. Contrib Mineral Petrol 73:403–412

Raith M, Raase P, Ackermand D, Lal RK (1983) Regional geothermobarometry in the granulite facies terrain of South India. Trans R Soc Edinb 73:221–244

Rollinson HR, Windley BF (1980a) An Archaean granulite-grade tonalite-trondhjemite-granite suite from Scouria, NW Scotland: geochemistry and origin. Contrib Mineral Petrol 72:265–281

Rollinson HR, Windley BF (1980b) Selective elemental depletion during metamorphism of Archaean granulites, Scourie, NW Scotland. Contrib Mineral Petrol 72:257–263

Sheraton JW, Skinner AC, Tarney J (1973) The geochemistry of the Scourian gneisses of the Assynt district. In: Park RG, Tarney J (eds) The Early Precambrian rocks of Scotland related rocks of Greenland. Univ Keele, pp 13–30

Sighinolfi GP (1971) Investigations into deep crustal levels: fractionating effects and geochemical trends related to high-grade metamorphism. Geochim Cosmochim Acta 35:1005–1021

Tarney J (1976) Geochemistry of Archaean high-grade gneisses, with implications as to the origin and evolution of the Precambrian crust. In: Windley BF (ed) The Early History of the Earth, Wiley, New York, pp 405–418

Tarney J, Windley BF (1977) Chemistry, thermal gradients and evolution of the lower crust. J Geol Soc Lond 134:153–172

Taylor SR, McLennan SM (1981) The composition and evolution of the continental crust: rare earth element evidence from sedimentary rocks. Philos Trans R Soc Lond A 301:381 – 399

Weaver BL, Tarney J (1980) Rare earth geochemistry of Lewisian granulite-facies gneisses, northwest Scotland: implications for the petrogenesis of the Archaean lower continental crust. Earth Planet Sci Lett 51:279 – 296

Weaver BL, Tarney J (1981) Lewisian gneiss geochemistry and Archaean crustal development models. Earth Planet Sci Lett 55:171 – 180

Wyllie PJ (1977a) From crucibles through subduction to batholiths. In: Saxena SK, Bhattacharji S (eds) Energetics of Geological Processes. Springer, Berlin Heidelberg New York, pp 389 – 433

Wyllie PJ (1977b) Crustal anatexis: an experimental review. Tectonophys 43:41 – 71

shared by some Chinese geologists. However, our new results of Rb-Sr and Sm-Nd isotopic analyses have indicated that the granulites of the Qianxi Group, especially from the region of Guojiago-Taipingzhai, are not of early Archaean age ($\geqslant 3.5$ Ga). Instead, they are relatively young (≈ 2.5 Ga).

Because the granulites of the Qianxi Group have had a unique importance in previous models of continental evolution in China, a more detailed study of isotopic and geochemical characteristics of these rocks was imperative. This paper outlines the new geochronological results, presents major and trace element (particularly the rare earth elements, REE) data, discusses the petrogenesis of the granulites and associated rocks and finally provides some tectonic implications. The original data of major and trace element abundances can be found in Jahn and Zhang (1984), and the detailed isotopic work will be published elsewhere (Zhang and Jahn in prep.).

2 Geological Setting

According to recent Chinese work (Cheng et al. 1982; Sun and Lu 1983; Sun et al. 1983), the lower Precambrian rocks in the Yenshan region of eastern Hebei Province can be subdivided into four groups (Fig. 1):

Qinglonghe Group
~~~~~ unconformity
Shuanshanzi Group   } Lower Proterozoic
~~~~~ unconformity

Badaohe Group
– – – – –(?)   } Archaean
Qianxi Group

The lowermost Qianxi Group is composed mainly of granulites of basic to acid composition, amphibolites (including those retrograded from granulites) and pyroxene-bearing banded iron formations (BIF). Many of these rocks have been further migmatized. Previous unconfirmed Rb-Sr whole rock isochron ages of about 3.5 Ga (CAGS 1975; ASIG 1978) have been frequently used to suggest that the Qianxi Group is the oldest part of the Chinese Archaean.

The Badaohe Group is composed mainly of amphibolites and biotite-bearing leptinite (or biotite granulitite by Cheng et al. 1982). Banded iron formations also occur in the upper part. The protoliths are believed to be the basic volcanic rocks and semipelitic sediments and have undergone amphibolite facies metamorphism and migmatization (Cheng et al. 1982). In terms of protolith nature the Group shows an upward gradation from basic to intermediate-acid volcanic rocks, whereas the upper Badaohe Group displays an upward transition from basic volcanics to intermediate-acid tuffaceous sediments (Cheng et al. 1982).

The Shuanshanzi Group is separated from the Archaean rocks by an unconformity. It is composed of amphibolite still preserving pillow structure, two-mica leptinite and other metasediments including BIF and mica schist.

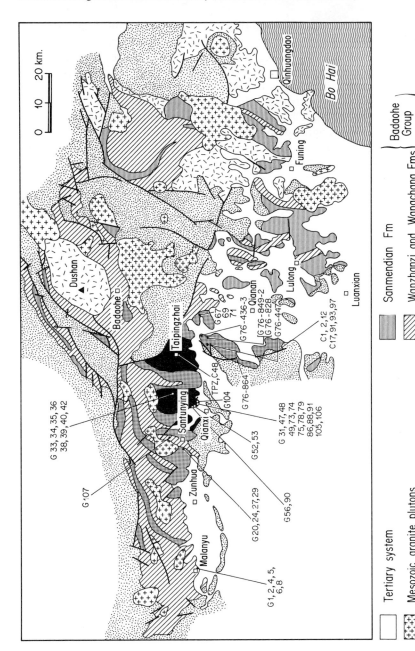

Fig. 1. Simplified geological map after Sun et al. (1983) and sampling localities

The Qinglonghe Group unconformably overlies the Shuangshanzi Group. This Group is essentially composed of metasediments. The base is made up of metaconglomerates and garnet-bearing mica schist. Other metasediments include two-mica leptinite and garnet-two-mica-quartz schist with intercalated BIF. The protoliths were evidently of pelitic composition.

It should be noted that the stratigraphic scheme used here is the one recently adopted by the school of the Chinese Academy of Geological Sciences. Other Chinese geologists may not agree with this scheme and may have their own terminology and classification. In our view, especially after the acquisition of a large amount of isotopic and trace element data in relation to rock ages and petrogeneses, the implicit layered cake stratigraphy as outlined above cannot be supported. Nevertheless, we will use the above scheme as a guideline for the following discussion.

3 Petrographic Description

The majority (nearly 90%) of samples were collected from the Qianxi Group, of which 60% are from the Shangchuang Formation and 40% from the Santunyin Formation (Fig. 1). According to the mineral paragenesis the analyzed rocks can be grouped into four categories as follows.

3.1. Granulites with Orthopyroxene

In this category the principal mineral assemblage are opx + cpx + plag + qz + gt + hb + bio + opaques, and the accessory minerals include apatite and zircon. Generally the rocks have equigranular granoblastic textures. Rarely the granoblastic textures have been modified to show foliation as a result of post-granulitization deformation. The mineral composition suggests that the protoliths of these granulites are basic to intermediate volcanic rocks. Samples belonging to this category include G05, G29, G31, G34, G38, G49, G61, G71, G88, G101, C44, C97.

3.2 Granulites Without Orthopyroxene

Two classes of samples may be distinguished: (1) basic granulites, including C1, C2, C12, C48, and C91, and (2) acid granulites, including G74, G78, G91, and G102.

In basic granulites the texture varies from polygonal granoblastic to gneissic. The development of foliation appears to be resulted from retromorphism to amphibolite facies. The principal mineral assemblages are: hb + plag + qz + cpx + bio + gt + opaques, and the accessory minerals include sphene, apatite, calcite, and zircon. Note that these rocks do not contain the characteristic orthopyroxene of the granulite facies. The only evidence of having been in granulite facies conditions is their polygonal granoblastic texture.

Regarding the acid granulites, foliated gneissic texture dominates, and their principal mineral assemblages are composed mainly of non-coloured minerals (>90%) such as plag + K-feldspar + qz. The minor coloured minerals include hb + diop + bio + opaques + gt + opx, and the accessory minerals are apatite, zircon, and exceptionally, allanite (G 78) and rutile (G 102).

The original characteristic minerals of the granulite facies are mostly destroyed by retrograde metamorphism. However, the presence of diopside, garnet, and orthopyroxene relicts attests to the stage of granulite facies for these rocks.

3.3 Amphibolites

These rocks, represented by C 17 and C 93, were collected near the village of Tsaozhuan. They are included in the Santunyin Formation as shown in Fig. 1. However, these rocks and other amphibolites in the vicinity, together with some iron formations, have been assigned by other geologists to the Tsaozhuan Group and are believed to be equivalent to the Shangchuang Formation of the Qianxi Group (Bai et al. 1980).

The amphibolites in the Tsaozhuan area occur as enclaves in tonalite and biotite leptinite or as dikes in tonalitic bodies. Petrographically, hornblende and plagioclase constitute more than 95% of the bulk, and the accessory minerals are opaques, sphene, quartz, apatite, and zircon.

3.4 Ultrabasic Rocks

Small ultrabasic bodies, including dunite, harzburgite, and rocks of possible komatiitic composition, are intercalated at the base of the Group. By contrast, metamorphosed iron formations are well developed in the upper part of the Qianxi Group. In the Louzishan area a rare rock type, eulysite, has been found and described (Zhang et al. 1981). It is composed essentially of eulite (Fe-rich opx), almandine and quartz. Locally it also contains ferrifayalite [$Fe_{4-x}(SiO_4)_2$], magnetite, ferroaugite, biotite, graphite, and apatite. Relict plagioclase is surrounded by a reaction rim of fine-grained garnet (Zhang et al. 1981). This rock is believed to be formed from Fe-rich sediment (probably BIF) under granulite facies metamorphism. Associated with eulysite in the Louzishan area are biotite-garnet gneiss, kyanite-garnet gneiss and two-pyroxene granulite.

The P-T conditions of the granulite metamorphism for the Qianxi Group has been obtained using mineral chemistry data (Zhang and Cong 1982; Auvray unpubl. data). The two-pyroxene equilibrium temperatures for the granulites (e.g. C97) are $850 \pm 50°C$ and for the eulysite (e.g. G71) $740 \pm 50°C$. These temperatures estimates are also in agreement with those obtained from the cpx-ga pair (Zhang and Cong 1982). It is quite often observed that metasediments recorded lower temperatures than meta-igneous rocks. The lower temperatures are believed to reflect the temperature conditions of retromorphism (Barbey 1982). Pressure estimates cannot be made really independently. Using the garnet-opx geobarometer, the Qianxi basic granulites were estimated to have formed at about 11 – 13 kb (Zhang and Cong 1982). It must be emphasized here that the very high-pressure condition recorded in the Qianxi granulites is truly exceptional. Another documented example of high-pressure granulites (≥10 kb) is found in the Furua granulite complex of Tanzania (Coolen 1980).

4 Geochronological Data

The age data presented below were determined on rocks almost exclusively from the Shangchuang Formations (black regions of Fig. 1), long believed to be the oldest of the Qianxi Group.

4.1 Rb-Sr Whole Rock and Mineral Data

The Rb-Sr whole rock isotopic data are displayed in Fig. 2. Except for five acid granulites collected from the Guojiago area, the data for the granulites of the Qianxi Group give an isochron age of 2480 ± 70 Ma with an initial $^{87}Sr/^{86}Sr$ ratio (designated hereafter I_{Sr}) of 0.70174 ± 6 (2 σ). This age is identical to:

1. The zircon U-Pb age of 2480 ± 20 Ma obtained by Pidgeon (1980) for a two-pyroxene tonalitic gneiss and a two-pyroxene basic granulite;
2. The new zircon U-Pb age of 2510 ± 14 Ma for a granulite from Taipingzhai (Liu et al. 1984) (Fig. 4);
3. The Rb-Sr whole rock isochron age of 2517 ± 94 Ma by Chung et al. (1979); and
4. The new Sm-Nd isochron age to be presented below.

However, it is entirely different from the two Rb-Sr ages of about 3.5 Ga published earlier (CAGS 1975; ASIG 1978). We suspect that the Rb-Sr system

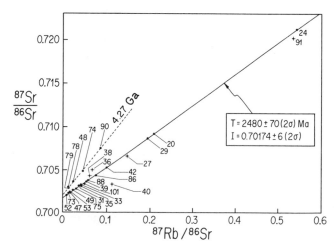

Fig. 2. Whole rock Rb-Sr isochron for granulites from the Qianxi Group (exclusively from the Guo-jiago-Taipingzhai-Xiachuang area). The alignment of some acid rocks on the 4.2 Ga reference line has no chronological significance (see text for explanation). $\lambda(^{87}Rb) = 0.0142 \, Ga^{-1}$

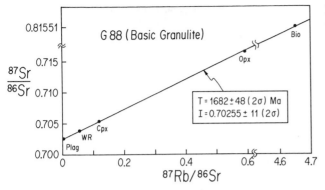

Fig. 3. Rb-Sr mineral isochron for a basic two-pyroxene granulite

was not completely closed during the granulite metamorphism. Isotopic perturbation is clearly seen in Fig. 2, especially for the data points with ^{87}Rb/^{86}Sr ratios less than 0.1. The analyses of ASIG (1978), though reasonably precise, may reflect a problem of sampling bias, especially because their data points are limited to a very narrow range of ^{87}Rb/^{86}Sr ratios between 0.02 and 0.05.

Concerning the light-coloured acid granulites (G48, 74, 78, 79, 90) it is surprising to see the well aligned data array which corresponds to an apparent age of 4.27 Ga (Fig. 2). These rocks contain 60 to 75% SiO_2 (except G90 which is a meta-quartzite), but their Rb concentrations are very low (2 – 6 ppm) and hence their Rb/Sr ratios. It is likely that these rocks have experienced post-granulite loss of alkalis, thus causing significant shift in Rb/Sr ratios. The observed alignment is believed to be purely accidental with no chronological meaning.

A two-pyroxene granulite (G88) was analyzed for its mineral constituents. The results yielded an isochron (Fig. 3) with $T = 1682 \pm 48$ Ma and $I_{Sr} = 0.70255 \pm 11$ (2σ). Chung et al. (1979) reported a mineral isochron age of 1740 ± 27 Ma for another two-pyroxene granulite. It appears that an important post-granulite thermal event has been recorded in the granulite samples. The age of about 1.7 Ga corresponds very closely to the boundary of the Sinian and the pre-Sinian systems in China and to the important orogenic Luliang or Chungtiao Movement (Huang 1978; Ma and Wu 1981).

4.2 U-Pb Zircon Data

A granulite sample collected from a quarry near the village of Taipingzhai was analyzed by Dun-yi Liu of the Chinese Academy of Geological Sciences (Liu et al. 1984). The U-Pb isotopic results of five zircon fractions yielded an upper intercept age of 2510 ± 14 Ma (Fig. 4), in good agreement with that determined earlier by Pidgeon (1980).

4.3 Sm-Nd Whole Rock Data

The Sm-Nd isotopic data for the granulites collected from the Xiachuang and the Guojiago areas define an errorchron with $T = 2480 \pm 125$ Ma and initial

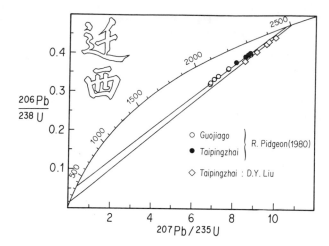

Fig. 4. U-Pb zircon concordia diagram for granulites from the Taipingzhai area

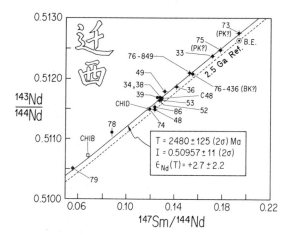

Fig. 5. Whole rock Sm-Nd "isochron" or errorchron for the Qianxi granulites (exclusively from the Guojiago-Taipingzhai-Xiachuang area). *B.E.* = Bulk Earth or the chondritic mantle value. Note that all rocks have very low Rb (average 4.5 ppm, see Fig. 9). *CHIB* and *CHID* are samples taken from the same area and reported in Table 1 of Allègre and Ben Othman (1980). $\lambda(^{147}Sm) = 0.00654$ Ga^{-1}

$^{143}Nd/^{144}Nd$ ratio (I_{Nd}) of 0.50957 ± 11 (2 σ), corresponding to $\varepsilon_{Nd}(T) = +2.7 \pm 2.2$ (Fig. 5). Note that all the analyzed rocks have low Rb concentrations ranging from 0.8 to 16 ppm, with a mean of 4.5 ppm. The cause for the scatter of data points is not clear. Archaean granulites from Lofoten, Norway, and from Enderby Land, Antarctica, also show scattering in their Sm-Nd isotopic data (Jacobsen and Wasserburg 1978; DePaolo et al. 1982). Although REE patterns and hence Sm/Nd ratios are generally regarded as immobile during metamorphism, a slight modification of LREE in some rocks during granulite metamorphism is possible. However, in the Qianxi case the scatter is small enough to suggest the general immobility of REE in granulite facies metamorphism. Of course, apart from analytical errors, the scatter could well be due to the heterogeneity of I_{Nd} values.

4.4 Summary

A summary of available geochronological results for the granulites of the Qianxi Group is given in Table 1. We interpret the coherent age of about 2.5 Ga determined by the Rb-Sr, U-Pb and Sm-Nd methods as the time of granulite facies metamorphism as well as the time of primary protolith emplacement. This means that the primary emplacement was followed shortly, perhaps within less than 100 Ma, by the granulite facies metamorphism. These two events were closely spaced in time and cannot be distinguished by the present results and techniques.

Our conclusion is not entirely in agreement with some Chinese geologists who maintain that the age of 2.5 Ga reflects only the important granulite facies metamorphism and that the time of protolith emplacement must be much older, most likely greater than 3 Ga (Cheng et al. 1982; Ma and Wu 1981). As far as reliable isotopic data are concerned, we believe that the protolith emplacement (or primary magmatism) could not be much older than 2.5 Ga for the following reasons:

1. Assuming the primary age was 3.5 Ga, and the primary isochron had been rotated during granulite facies metamorphism 2.5 Ga ago, what would have been the I_{Sr} value at 3.5 Ga? Since the granulitic gneisses of the Qianxi Group have an average composition of intermediate rocks, it is not unreasonable to assume a Rb/Sr ratio of 0.1 as the mean value for the protoliths. Using the boundary value of $I_{Sr} = 0.7017$ at 2.5 Ga ago (Fig. 2), the calculated I_{Sr} value for $T > 3.0$ Ga would be less than 0.700. Furthermore, the I_{Sr} value of 0.7017 ($T = 2.5$ Ga) is very close to the Main Path value of mantle evolution (Jahn and Nyquist 1976). On the other hand, if $T = 3.67$ Ga and $I_{Sr} = 0.7010$ (ASIG 1978) were correctly interpreted as the age of primary emplacement, the calculated I_{Sr} at $T = 2.5$ Ga (again assuming Rb/Sr = 0.1 for the protoliths) would be 0.7058. Evidently this value is much higher than that obtained in this study and by Chung et al. (1979).

2. Similarly, if the Sm-Nd isochron shown in Fig. 5 really resulted from rotation of the 3.5 Ga isochron, what would be the expected I_{Nd} or ε_{Nd} value for the reset or rotated 2.5 Ga isochron? Since all the granulitic gneisses of basic to acid compositions show LREE-enriched patterns (see a later section), a value of 0.13 can be estimated for the average $^{147}Sm/^{144}Nd$ ratio (Fig. 5) of the granulites. REE distribution patterns are customarily assumed to remain unaffected even under granulite facies metamorphism (Hamilton et al. 1979a; Tarney and Windley 1977; Drury 1978; Weaver and Tarney 1980, 1981; *see also Condie and Allen, this Vol., eds.*). A simple calculation using the equation $\varepsilon_{Nd} = Q_{Nd} \cdot f_{Sm/Nd} \cdot T$ (DePaolo and Wasserburg 1976) indicates that the ε_{Nd} value at $T = 2.5$ Ga would be -7 instead of the near chondritic or slightly positive value as shown in Fig. 4. The fact that the isochron defines $\varepsilon_{Nd}(T) = +2.7 \pm 2.2$ already suggests that the protoliths of these granulites could not be much older than 2.5 Ga.

3. Many of the analyzed rocks have unusually low Rb contents and hence very high K/Rb ratios. This is generally not considered a primary igneous feature but is regarded as a result of preferential Rb loss during granulite facies metamorphism (*see also Condie and Allen, this Vol., eds.*). Any Rb-Sr isochron age determined on rocks which have undergone significant Rb loss (and coupled with Sr isotope homogenization) should reflect the time when Rb ceased to be lost, prob-

Table 1. Summary of geochronological results of granulitic gneisses from the Qianxi region, eastern Hebei Province, China

Method	T (Ma)	I_{Sr} or I_{Nd}	References	Comments
Rb-Sr (WR)	3480 ± 240	0.707 ± 1	CAGS = Chinese Academy of Geol. Sci. (1975)	5 points; T = age of primary emplacement
Rb-Sr (WR)	3670 ± 230	0.70102 ± 12	ASIG = Academia Sinica, Inst. of Geol. (1978)	7 points; T = age of primary emplacement
U-Pb (Zircon)	2480 ± 20		R. T. Pidgeon (1980)	T = age of granulite facies metamorphism
U-Pb (Zircon)	2502 ± 14		D. Y. Liu (pers. comm.)	
Rb-Sr (WR)	2476 ± 67 (2σ)	0.70174 ± 6 (2σ)	This work	17 points; T = age of granulite metamorphism ~ age of protolith emplacement
Sm-Nd (WR)	2480 ± 125 (2σ)	0.50957 ± 11 (2σ) $\varepsilon_{Nd}(T) = +2.7 ± 2.2$ [a]	This work	18 points; T = interpretation same as above
Rb-Sr (Mineral)	1682 ± 48 (2σ)	0.70255 ± 11 (2σ)	This work	5 points; T = age of post-granulite metamorphic event

[a] Calculated with $\varepsilon_{Nd}(0) = 0.51264$ and $^{147}Sm/^{144}Nd(CHUR) = 0.1967$.

ably corresponding to the waning stage of granulite facies metamorphism. It cannot in any circumstance truly represent the time of protolith emplacement. Therefore, the previous interpretation by ASIG (1978) is again in error.

Consequently, we conclude that in the Qianxi region two closely related events – primary magmatism (protolith emplacement) and granulite facies metamorphism – took place in the late Archaean about 2.5 Ga ago. It is unlikely that the primary magmatism producing the rocks discussed above occurred prior to 3.0 Ga ago according to the present data. However, our most recent Sm-Nd results on nine samples of amphibolites enclaves from the Tsaozhuang area, near Qianan (Fig. 1), yielded an isochron age of 3.52 ± 0.12 (2 σ) Ga, with $\varepsilon_{Nd}(T)$ $= 0 \pm 2.5$ (2 σ). This proves that early Archaean rocks do occur in North China. The new age data will be reported elsewhere.

Incidentally, Shen et al. (1981) recently reported a Rb-Sr isochron age of 2.53 ± 0.14 Ga with $I_{Sr} = 0.7014 \pm 48$ for five biotite-bearing leuco-granulites (leptinite ?) collected from the adjacent Sijiaying area near Luanxian (Fig. 1). Chemically, these rocks have high Rb contents (146 – 364 ppm) with $^{87}Rb/^{86}Sr$ ratios varying from 0.98 to 3.11. We think that they are probably metasediments and are hence quite different from the rocks analyzed in the present study (see Fig. 2). Stratigraphically they are intercalated in the ferrosiliceous rock series believed to be equivalent to the upper part of the Badaohe Group. Shen et al. (1981) interpreted the age of 2.5 Ga as the time of retrograde amphibolite facies metamorphism.

5 Geochemistry

5.1 Major Element Geochemistry

Except for the iron formations and quartzites, the granulite gneisses of the Qianxi Group can be conveniently separated into three compositional groups. In addition, a fourth group of ultrabasic composition will also be discussed here in an attempt to address the issue of komatiite occurrence in this region. The division of groups was made using the criteria of normative compositions as follows:

1. Basic granulites: containing normative olivine $<41\%$; mean $SiO_2 = 50.4 \pm 3.2$ (σ).
2. Intermediate granulites: hypersthene normative, with $Q < 10\%$; mean $SiO_2 = 56.5 \pm 1.8$ (σ).
3. Acid granulites: hypersthene normative, with $Q \geqslant 10\%$; mean $SiO_2 = 65.4 \pm 5.3$ (σ).
4. Ultrabasic rocks: normative olivine $>48\%$.

It is interesting to determine whether the protoliths of the granulitic gneisses are of igneous or sedimentary origin. We have used the discriminant function (DF) derived by Shaw (1972):

$$DF = 10.44 - 0.21 \ SiO_2 - 0.32 \ Fe_2O_3 \ (total \ Fe) - 0.98 \ MgO + 0.55 \ CaO + 1.46 \ Na_2O + 0.54 \ K_2O.$$

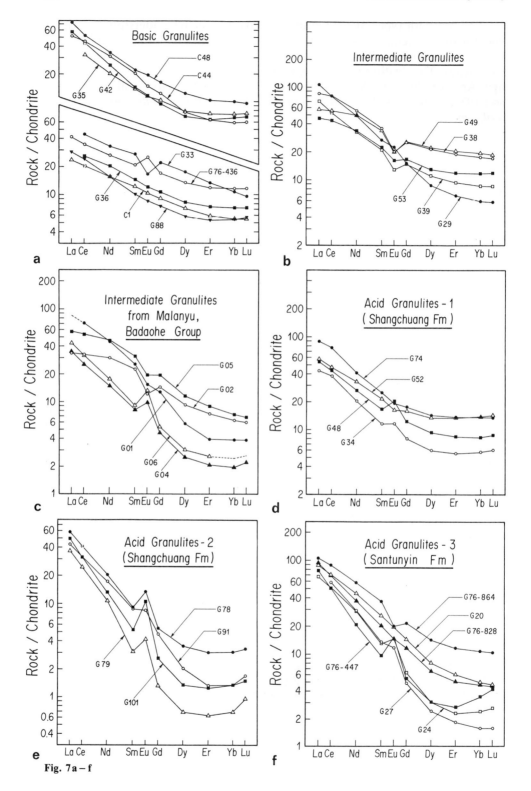

Fig. 7a – f

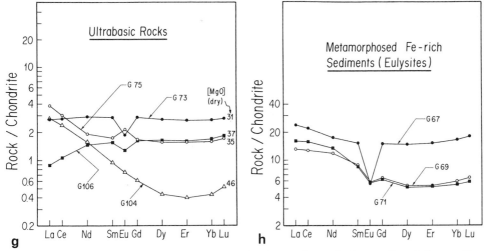

Fig. 7a–h. REE distribution patterns of the Qianxi granulites. **a** Basic granulites; **b** intermediate granulites; **c** intermediate granulites of the Badaohe Group from the Malanyu area; **d, e, f** acid granulites; **g** ultrabasic rocks; **h** metamorphosed Fe-rich sediments (eulysites) from the Louzishan area

McLennan 1981a, b), but the earlier discrimination function analysis suggests that no intermediate granulitic gneiss is likely to have a sedimentary origin.

5.2.3 Acid Granulites (Fig. 7d, e, f)

The acid granulites show moderately (Fig. 7d) to highly fractionated REE patterns (Fig. 7e, f). The moderately fractionated REE patterns are quite similar in shape and concentration level to those in basic and intermediate granulites (Fig. 7a, b). However, most highly fractionated patterns ($(La/Yb)_N > 20$) exhibit positive Eu-anomalies and HREE depletions with a concave minimum or a valley at the Er-Yb position. This resembles the TTG rocks (tonalite-trondhjemite-granodiorite) occurring in Archaean greenstone-granite terranes (Arth and Hanson 1975; Condie and Hunter 1976; Glikson 1976, 1979; Jahn et al. 1981; Martin et al. 1983) as well as in high-grade gneiss terranes (Green et al. 1972; O'Nions and Pankhurst 1974; Compton 1978; Hunter et al. 1978; Weaver 1980; Weaver and Tarney 1980, 1981).

5.2.4 Ultrabasic Rocks (Fig. 7g)

Ultrabasic rocks occur sporadically in the lower part of the Qianxi Group. The existence of komatiitic rocks in this region is not yet certain, although some published chemical data resemble those of basaltic komatiite (Zhang et al. 1980; Sun and Wu 1981; Wang 1982) or peridotitic komatiite (Wang 1982). The major element compositions for G 73, G 75 and Tai-9 are similar to that of peridotitic ko-

5.3 K/Rb Ratios and Inference for the REE Mobility Problem

The K/Rb ratios in granulitic rocks are known to vary widely (e.g. Green et al. 1972; Tarney and Windley 1977). Not uncommonly, they are significantly higher than those in lower grade rocks of corresponding compositions. Figure 9 shows a wide variation of K/Rb ratios for the Qianxi granulites, ranging from about 200 to 4700 (G 52). 40% of the rocks have ratios greater than 1000, and nearly all data points define ratios higher than the main trend (MT) proposed by Shaw (1968) for continental igneous rocks. Oceanic tholeiites (MORB) have high K/Rb ratios (often greater than 1000), but their trend (OT) is clearly distinguished from the trend of Precambrian granulites as shown in Fig. 9. In Archaean greenstone belts basaltic rocks have an average K of 2000 ppm and K/Rb = 350, whereas andesitic and felsic rocks have $K \approx 1\%$ and $K/Rb \approx 300$ (Jahn and Sun 1979). The high ratios and the wide variation in the granulites cannot be primary igneous features. It is generally agreed that this is related to granulite facies metamorphism (Heier 1973; Tarney and Windley 1977; Condie and Allen, this Vol.).

Alkalis, particularly K, Rb, and Cs, are mobile under certain conditions and are often lost during granulite metamorphism. The higher than normal K/Rb ratios in granulites suggest that Rb is preferentially lost relative to K (Heier 1973). The loss is thought to be accomplished not only by the reconstitution of mineral phases less favourable for retention or substitution of K and Rb for major cation sites, but also by the fluid action that removes these elements (Tarney and Windley 1977). Note that lower K/Rb ratios may or may not be re-established during retrograde metamorphism (Moorlock et al. 1972; Drury 1973, 1974). The degree of alkali loss can be estimated roughly by the concentration of Rb and the K/Rb ratio. From Fig. 10 the degree of alkali loss appears to be highly variable. Furthermore, it is interesting to see if the REE patterns have been modified along with the alkalis during granulite metamorphism. We have observed that similar REE patterns for the same group of rocks may be associated with very different Rb concentrations and K/Rb ratios. This implies that the REE pat-

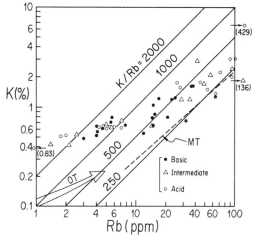

Fig. 9. K-Rb relationship of the Qianxi granulites. Note that many samples have K/Rb ratios greater than 1000, but the trend is sufficiently different from the *OT* (oceanic trend) of Shaw (1968); *MT* main trend of continental igneous rocks (Shaw 1968)

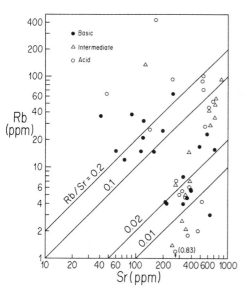

Fig. 10. Rb-Sr relationship of the Qianxi granulites. Many samples have Rb/Sr ratios lower than 0.02, a feature comparable with Lewiasian granulites (Tarney and Windley 1977)

terns have not changed according to the alkali behaviour. It also strongly suggests that the REE patterns are characteristic of the granulite protoliths; this conclusion is essentially in agreement with that reached by many authors (Green et al. 1972; Drury 1978; Weaver and Tarney 1980, 1981; Tarney and Windley 1977; *see also Condie and Allen, this Vol., eds.*) but is at variance with the hypothesis of Collerson and Fryer (1978) who suggested that REE patterns could be modified by preferential solubility of HREE in fluid phases (CO_2, halogens and H_2O) and subsequent removal of the fluid. Furthermore, available data on many eclogites, believed to have formed by basalt transformation at high P-T conditions, show that their REE patterns are similar to the predicted basalt patterns, thus suggesting insignificant mobility of REE during high-pressure metamorphism (J. Bernard-Griffiths, unpubl. data).

5.4 Rb/Sr Ratios

Sr is known to be less mobile than Rb during alteration and metamorphism. In the case where Rb is severely depleted (a few ppm), the Rb/Sr ratios are very low (<0.02) and not unusually less than 0.01 (Fig. 10). Because many of the low Rb/Sr ratios are not primary igneous features, it follows that the whole rock Rb-Sr isochron age for the Qianxi Group granulites (Fig. 2) must represent the time when the new Rb-Sr ratios were re-established and Sr isotopes were re-equilibrated during granulite metamorphism. From consideration of the I_{Sr} value and the Sm/Nd isochron age, together with its I_{Nd} value, it is concluded that the granulite facies metamorphism took place shortly after protolith emplacement.

6.1 Basic Granulites = Basalts or Gabbros

The MgO contents in basic granulites vary from 17% (G 01) to 3.4% (G 35), with the majority ranging between 9 and 5% (Jahn and Zhang 1984). This suggests that most rocks have undergone various extents of crystal fractionation. Except for two cases the basic rocks do not have significant Eu-anomalies, implying that plagioclase fractionation may not be important in their genetic histories, or the plagioclase effect has been offset by concomitant crystallization of clinopyroxene and/or amphibole. Since the MgO contents do not vary systematically with the REE abundance levels, and there is no compelling evidence for a cogenetic relationship for all the analyzed rocks, a quantitative modelling does not seem necessary.

Although these basic rocks do not represent primary liquids, their parental liquids have undoubtedly been derived from the upper mantle. Because all REE patterns (Fig. 7a) are LREE-enriched and significantly fractionated, we conclude that the parental liquids must also have been enriched in LREE for the following reasons.

Olivine, spinel and orthopyroxene have very low K_D values (REE distribution coefficients), hence they cannot induce important REE fractionation during magmatic differentiation. Garnet is not considered as a likely liquidus phase, and plagioclase apparently has not played an important role in the process as evidenced by the general lack of an Eu anomaly. This leaves clinopyroxene as the sole possible phase to fractionate the REE patterns. The variation of Cr abundances suggests that the maximum amount of cpx that could be separated is about 50%. A 50% cpx separation will increase the $(La/Yb)_N$ ratio by less than 20%. Even an impossible 80% cpx separation will only increase the $(La/Yb)_N$ ratio by about 45%. In reality, multiphase differentiation involving olivine will certainly have less effect on the La/Yb fractionation. The significantly fractionated patterns of Fig. 7a are thus inherited from their parental liquids.

The LREE-enriched parental liquids could be produced by two different mechanisms: (1) smaller degrees of melting of mantle peridotites leaving some garnet in the residue, or (2) larger degrees of melting of metasomatized mantle in which LREE and other LIL elements were enriched by a fluid agency prior to the melting events (e.g. Bailey 1982; Chauvel and Jahn 1983). The relative importance of these two processes cannot be determined. However, a high Mg basalt (G 01) from the Badaohe Group (Fig. 7a) has the highest $(La/Yb)_N$ ratio (≈ 22) among the basic rocks. We believe that it is likely derived by partial melting of a metasomatized mantle source. Furthermore, the constraint imposed by the $\varepsilon_{ND}(T)$ value (Fig. 4), that the mantle sources had time-integrated REE patterns which were chondritic or LREE depleted, suggests that mantle metasomatism would have had to occur shortly before the melting event.

6.2 Intermediate Granulites = Andesites or Diorites

Significant negative Eu-anomalies are observed in the intermediate rocks from the Shangchuang Formation (Fig. 7b). These rocks could be derived by fractional crystallization involving significant plagioclase separation from basaltic liquids. A similar process may be applied to two samples (G 2 and G 5) from the Badaohe Group (Fig. 7c). As for G 29 (Fig. 7b), G 04, and G 06 (Fig. 7c), their REE patterns show significant fractionation with $(La/Yb)_N = 17-18$ as well as positive Eu-anomalies. G 04 and G 06 have relative HREE depletion as in some acid rocks. This implies that garnet or amphibole played a role in the REE fractionation process. Thus, these rocks could be derived by partial melting of eclogitic and amphibolitic sources or by amphibole separation from crystallizing liquids.

6.3 Acid Granulites = TTG Rocks or Their Volcanic Equivalents

Petrogenetic discussion of the TTG rocks can be greatly facilitated by using the $(La/Yb)_N$ vs $(Yb)_N$ diagram (Fig. 8) as introduced by Jahn et al. (1981) and Martin et al. (1983). The $(La/Yb)_N$ ratios measure the degree of REE fractionation and the $(Yb)_N$ values refer to the HREE abundances. Hence, each individual data point in this diagram roughly represents a REE pattern.

Superimposed on the diagram (Fig. 8) are the reference fields for the upper mantle (UM), continental flood basalts (CFB), the upper continental crust with the estimated average of Taylor and McLennan (1981a) and the Archaean TTG rocks as compiled by Jahn et al. (1981). Also shown in the diagram are calculated melting trends assuming various sources and different initial $(La/Yb)_N$ ratios.

It is seen that almost all acid granulite data points for the Hebei samples are confined within the field of Archaean TTG rocks (Fig. 8). From the melting trends we conclude that the acid rocks and some intermediate rocks could be derived by partial melting of quartz eclogite, garnet amphibolite or amphibolite with initial $(La/Yb)_N \cong 4$, that is, LREE enriched sources. Melting of an amphibolite source with flat REE pattern will only produce an maximum enrichment in the $(La/Yb)_N$ ratio of about 10 and thus is considered unlikely to generate the acid liquids. We have also observed that there is no problem for the derivation of some acid liquids by remelting of the basic granulites. However, the isotopic constraints require the remelting process to take place shortly before or simultaneously with the granulite facies metamorphism.

7 Crustal Processes and Tectonic Implications

7.1 Time Relationship

The radiometric ages determined by various methods for the Qianxi granulites (Table 1) suggest that the time span from initial crust formation to granulite facies metamorphism was relatively short (probably less than 100 Ma). The geologic processes that could have taken place during this short time span include (not necessarily in exact chronological order):

– melting of mantle peridotite, leading to the production of basaltic rocks;
– fractional crystallization of basaltic liquids;

- possible transformation of basalt to eclogite or amphibolite;
- remelting of eclogitic or amphibolitic material, leading to the production of acid rocks;
- granulite facies metamorphism;
- retrograde metamorphism.

The close temporal relationship for such a series of geologic events is apparently not unique to the Qianxi region. The Lewisian complex has also recorded a short time span (<200 Ma) for initial crustal separation from the mantle through the granulite facies metamorphism (Hamilton et al. 1979a). The ca. 3.7 Ga old Amitsôq gneisses were partly metamorphosed to granulitic rocks, and the granulite facies metamorphism was dated at about 3.6 Ga by the Rb-Sr and common Pb isochron methods (Griffin et al. 1980). Likewise, the granulite complex of Lapland, Finland (T ≈ 2 Ga), reflects a very similar chronological order (Bernard-Griffiths et al. 1983), which is further supported by geochemical and petrographic evidence (Barbey 1982). To our knowledge there exist only two well-documented cases of granulite facies metamorphism taking place long after crustal formation: (1) in Lofoten-Vesteralen, Norway (Griffin et al. 1978; Jacobsen and Wasserburg 1978), and (2) in Enderby Land, Antartica (DePaolo et al. 1982).

Why are the above geologic events in the Qianxi region so closely spaced in time? Numerous hypotheses exist regarding processes of granulite formation and their tectionic implications. The model of crustal underplating proposed by Holland and Lambert (1975) may explain the closely spaced crustal generation and subsequent granulite facies metamorphism for the formation of the Scourie assemblage. They envisaged that basic and intermediate magmas first rose as primary differentiates from the upper mantle and underplated the crust. These magmas then crystallized directly to granulite facies rocks under conditions of high temperature and pressure. This model has been modified later (Tarney and Windley 1977; Weaver and Tarney 1980, 1981) to interpret the origin of the Lewisian complex. Newton et al. (1980) presented two equally plausible models for the origin of granulites using modern tectonic concepts. In both models, named the hot-spot and the plate tectonic, crustal underplating is again emphasized and formation of granulites is achieved by carbonic metamorphism (*see also Hansen et al., this Vol., eds.*).

The most severe drawback of the above models is the negligence of the often close association of granulites of igneous parentage with iron formations or other rock types of clear sedimentary origin. The juxtaposition of widely different lithologies certainly cannot be explained by magmatic crustal underplating alone. A time span of ≈ 100 Ma is short compared to the duration of the Archaean, but numerous important geological events have been recorded within 100 Ma in Phanerozoic terranes. Thus, there was sufficient time for burial and prograde metamorphism en route to granulite formation. However, prograde formation and final uplift or emplacement of granulites at the surface and the process of retrograde metamorphism would be greatly facilitated if intracrustal thrusting is involved, such as observed at the present time in the Southern Appalachians and many other crustal sections (Oliver 1978, 1982; Cook et al. 1979, 1980; Fountain and Salisbury 1981). A dynamic model which involves burial and large-scale thrusting has been proposed to explain the Proterozoic granulite complex of Lapland, Finland (Barbey 1982). A model of tectonic thickening through crustal interstacking has been proposed for the formation of Archaean granulite complexes (Bridgwater et al. 1974; Kröner 1982).

7.2 Intracrustal Melting?

Heat flow and heat balance calculations show that the lower crust is generally depleted in radioactive or heat-producing elements with respect to the upper crust (Heier 1973). Lambert (1976) further showed that unless the radioactive elements were strongly concentrated in the uppermost part of the crust within 50 – 100 Ma after an accretion episode, widespread crustal melting by intrinsic heat production would be inevitable in the Archaean.

The crustal evolution model (island arc or andesite model) of Taylor and coworkers (Taylor 1967, 1977, 1979; Taylor and McLennan 1981a, 1981b), suggests that the upper and the lower continental crust represent the two components, i.e. liquid and residue, respectively, produced by intracrustal melting of island arc or andesite materials, which are considered to represent the bulk composition of the total continental crust. The average REE distribution in the upper crust (granodiorite) is estimated from the element abundances in the sediments, and that of the lower crust could then be calculated by mass balance equations assuming the volume ratio of upper crust to lower crust equal to 1 to 2 (Taylor and McLennan 1981a, b; *see also McLennan and Taylor, this Vol., eds.*). This simplistic model is very elegant but may encounter difficulties when dealing with lower crustal compositions.

It is generally agreed that the lower crust is made up of granulitic rocks, but its bulk composition is still a matter of controversy. In a regional geochemical survey of the Canadian Shield, Eade and Fahrig (1971) have determined that the average for all granulite facies rocks corresponds to the composition of granodiorite, which, surprisingly, is also very close to the average composition of the Canadian Shield and the upper crustal average of Taylor (1967). Fountain and Salisbury (1981) concluded that the large-scale layering in the continental crust is not compositional but metamorphic and that the lower crust is composed of granulites, ranging from mafic to silicic gneisses, the middle crust is made up of migmatites with abundant granite intrusions, and the upper crust consists essentially of little metamorphosed supracrustal rocks plus some tectonically raised lower-middle crustal rocks. In fact, as shown by the present study, granulite terrane often contain a wide variety of bulk compositions, matching the metamorphosed igneous and sedimentary components of lower-grade terrane.

From the REE geochemical studies of clastic sediments Taylor and McLennan (1981b) have emphasized that the general lack of significant negative Eu-anomalies in Archaean sedimentary rocks signifies only minor, shallow intracrustal melting in the differentiation of the Archaean crust. The present geochemical study of granulite gneisses provides an independent check of this conclusion. The geochemical consequence of the andesite model (Taylor and McLennan 1981a) is to have the residual lower crust characterized by (1) a more basic composition with $SiO_2 = 54\%$; (2) a high Al_2O_3 content (19%); (3) the presence of positive Eu-anomalies in the REE patterns; and (4) a near chondritic Sm/Nd ratio (≈ 0.30).

If granulite facies rocks are representative samples of the lower crust, then the andesite model cannot be supported by the available geochemical and geological data of granulite facies rocks from the Qianxi region or elsewhere. Using the presently available data sets it is difficult to obtain a truly weighted average for SiO_2 in granulites, but the Al_2O_3 content is certainly much less than 19% (data from Jahn and Zhang 1984). If only rocks of possible residual nature are considered, namely basic and intermediate granulites, the majority show either no or only negative Eu-anomalies (Fig. 7a, b, c). Moreover, all REE patterns are significantly fractionated, with LREE enrichment and the Sm/Nd ratios much less than the chondritic value (Fig. 5). The persistent positive Eu-anomalies are generally only found in acid granulites which should not be mistaken as residue in intracrustal melting.

A survey of available REE distribution patterns for granulite rocks shows that they are quite similar to upper crustal rocks of corresponding composition. This may suggest that the bulk composition of the lower crust is not so different from that of the upper crust, except for heat-producing elements (K, U, and Th) and Rb (Heier 1973). The occurrence of undepleted granulites (Gray 1977; Sighinolfi et al. 1981) is also not consistent with the hypothesis of intracrustal melting. In conclusion, the formation of granulites, and hence the lower crust, is probably not a result of intracrustal melting, but is more consistent with a model in which intracrustal thrusting and stacking have tectonically mixed rocks of various origin that were eventually subject to carbonic metamorphism (Touret 1971a, b; Newton et al. 1980). The exceptionally high-pressure conditions recorded in the Qianxi basic granulites implies that a thick continental crust (>40 km) must have existed already in the Qianxi area when the granulites were formed. The envisaged tectonic processes involving thrusting and stacking up of crustal slices (e.g. Oliver 1978, 1982) might have facilitated the thickening of crust, thus achieving very high pressures ($\geqslant 10$ kb).

Acknowledgements. We are indebted to the following persons: B. Auvray for this able assistance in petrographic examinations; J. Cornichet for the instructions on chemical separation techniques given to Z. Q. Zhang; J. Macé for the maintenance of mass spectrometers; F. Vidal and M. Lemoin for the chemical analyses by XRF; P. Barbey, G. N. Hanson (New York), and J. Touret (Amsterdam) for constructive comments. The trips of B. M. Jahn to China (1982, 1983) were made possible by a travel grant (ATP Géodynamique II) from the INAG of France. Finally we thank D. Y. Liu of the Chinese Academy of Geological Sciences, Beijing, for permission to publish his U-Pb results in Fig. 4.

References

Allègre CJ, Ben Othman D (1980) Nd-Sr isotopic relationship in granitoid rocks and continental crust development: a chemical approach to orogenesis. Nature 286:335 – 342

Arth JG, Hanson GN (1975) Geochemistry and origin of the early Precambrian crust of northeastern Minnesota. Geochim Cosmochim Acta 39:325 – 362

ASIG (Academy of Science, Institute of Geology, 1978) 3600 Ma old rocks from eastern Hebei: preliminary results of Rb-Sr dating. Kexue Tongbao 23:429 – 431 (in Chinese)

Auvray B, Blais S, Jahn BM, Piquet D (1982) Komatiites and the Komatiitic series of the Finnish greenstone belts. In: Komatiites, Arndt NT, Nisbet EG (eds) Allen and Unwin, London, pp 131 – 146

Bai YL, Li ZZ, Ku TL (1980) Paleo-folds as seen from the metamorphic rocks of the Qianan region, E Hebei Geol Res 3:68 – 90

Bailey DK (1982) Mantle metasomatism – continuing chemical change within the earth, Nature 296:525 – 530

Barbey P (1982) La ceinture des granulites de Laponie (Fennoscandie): une suture de collision continentale d'âge Protérozoïque inférieur (2.3 – 1.9 Ga), reconstitution géochimique et pétrologique. Thèse, Doctorat d'Etat. Univ. de Nancy I et Univ. de Rennes I:346 p

Barbey P, Cuney M (1984) K, Rb, Sr, Ba, U, and Th geochemistry of the Lapland granulites (Fennoscandia): LILE controlling factors. Contrib Mineral Petrol 81:304 – 316

Basaltic Volcanism (1981) Basaltic volcanism on the terrestrial planets. Pergamon, New York, p 1286

Bernard-Griffiths J, Peucat JJ, Postaire B, Vidal P, Convert J, Moreau B (1983) Isotopic data (U-Pb, Rb-Sr, Pb-Pb, and Sm-Nd) on mafic granulites from Finnish Lapland. Precambrian Res 23:325 – 348

Blais S, Auvray B, Capdevila R, Jahn BM, Bertrand JM, Hameurt J (1978) The Archaean greenstone belts of Karelia (eastern Finland) and their komatiitic and tholeiitic series. In: Windley BF, Naqvi SM (eds) Archaean Geochemistry, Elsevier, Amsterdam, pp 87 – 107

Bridgwater D, McGregor VR, Myers JS (1974) A horizontal tectonic regime in the Archaean of Greenland and its implications for early crustal thickening. Precambrian Res 1:179 – 197

CAGS (Chinese Academy of Geological Sciences, 1975) Geochronological study of old metamorphic rocks from the Qianxi-Zunhua region, Hebei Province. Kexue Tongbao 20:29 – 34 (in Chinese)

Chauvel C, Jahn BM (1983) Nd-Sr isotope and REE geochemistry of alkali basalts from the Massif Central, France, Geochim Cosmochim Acta 48:93 – 110

Cheng YQ, Bai J, Sun DZ (1982) The lower Precambrian of China. In: An outline of the stratigraphy in China, Beijing Geol Publ House, 29 pp

Chung FT, Compston W, Foster J, Bai J, Sun DZ (1979) Age of the Qianxi Group, North China. Annu Rep Annu Res School Earth Sci: 147 – 151

Collerson KD, Fryer BJ (1978) The role of fluids in the formation and subsequent development of early continental crust. Contrib Mineral Petrol 67:151 – 167

Compton P (1978) Rare earth evidence for the origin of the Nuk gneisses, Buksefjorden region, southern West Greenland. Contrib Mineral Petrol 66:283 – 294

Condie KC (1976) Trace element geochemistry of Archaean greenstone belts. Earth Sci Rev 12:393 – 417

Condie KC, Hunter DR (1976) Trace element geochemistry of Archaean granitic rocks from the Barberton region, South Africa. Earth Planet Sci Lett 29:389 – 400

Condie KC, Allen P (1984) Origin of Archaean charnockites from southern India. This Vol., 182 – 203

Cook FA, Albaugh DS, Brown LD, Oliver JE, Hatcher RD (1979) Thin-skinned tectonics in the crystalline South Appalachians: COCORP seismic reflection profiling of the Blue Ridge and Piedmont. Geology 7:563 – 567

Cook FA, Brown LD, Oliver JE (1980) The southern Appalachians and the growth of continents. Sci Am 243:124 – 138

Coolen JM (1980) Chemical petrology of the Furua granulite complex, southern Tanzania. GUA Papers of Geology, series 1, No 13 GUA-Amsterdam, pp 258

DePaolo DJ, Wasserburg GJ (1976) Inferences about magma sources and mantle structure from variations of 143Nd/144Nd. Geophys Res Lett 3:743 – 746

DePaolo DJ, Manton WI, Grew ES, Halpern M (1982) Sm-Nd, Rb-Sr, and U-Th-Pb systematics of granulite facies rocks from Fyfe Hills, Enderby Land, Antarctica, Nature 298:614 – 618

Drury SA (1973) The geochemistry of Precambrian granulite facies rocks from the Lewisian complex of Tiree, Inner Hebrides, Scotland. Chem Geol 11:167 – 188

Drury SA (1974) Chemical changes during retrogressive metamorphism of Lewisian granulite facies rocks from Coll and Tiree. Scott J Geol 10:237 – 256

Drury SA (1978) REE distributions in a high-grade Archaean gneiss complex in Scotland: Implications for the genesis of ancient sialic crust. Precambrian Res 7:237 – 257

Eade KE, Fahrig WF (1971) Geochemcial evolutionary trends of continental plates – a preliminary study of the Canadian Shield. Geol Surv Can Bull 179:51 p

Fountain DM, Salisbury MH (1981) Exposed cross-sections through the continental crust: implications for crustal structure, petrology, and evolution. Earth Planet Sci Lett 56:263 – 277

Fryer BJ (1977) Rare earth evidence in iron-formations for changing Precambrian oxidation states. Geochim Cosmochim Acta 41:361 – 367

Gray CM (1977) The geochemistry of central Australian granulites in relation to the chemical and isotopic effects of granulite facies metamorphism. Contrib Mineral Petrol 65:79 – 89

Green TH, Brunfelt AO, Heier KS (1972) Rare-earth element distribution and K/Rb ratios in granulites, mangerites and anorthosites, Lofoten-Vesteraalen, Norway. Geochim Cosmochim Acta 36:241 – 257

Glikson AY (1976) Trace element geochemistry and origin of early Precambrian acid igneous series, Berberton Mountain Land, Transvaal. Geochim Cosmochim Acta 40:1261 – 1280

Glikson AY (1979) Early Precambrian tonalite-trondhjemite sialic nuclei. Earth Sci Rev 15:1 – 73

Griffin WL, Taylor PN, Hakkinen JW, Heier KS, Iden IK, Krogh EJ, Malm O, Olsen KI, Ormaasen DE, Tveten E (1978) Archaean and Proteozoic crustal evolution in Lofoten-Vesteralen, N Norway. J Geol Soc Lond 135:629 – 647

Griffin WL, McGregor VR, Nutman A, Taylor PN, Bridgwater D (1980) Early Archaean granulite-facies metamorphism south of Ameralik, West Greenland. Earth Planet Sci Lett 50:59 – 74

Hamilton PJ, Evensen NM, O'Nions RK, Tarney J (1979) Sm-Nd systematics of Lewisian gneisses: implications for the origin of granulites. Nature 277:25 – 28

Hansen EC, Newton RC, Janardhan AS (1984) Pressures, temperatures and metamorphic fluids across an unbroken amphibolite-facies transition in southern Karnatak, India. This Vol., 161 – 181

Hanson GN (1978) The application of trace elements to the petrogenesis of igneous rocks of granitic composition. Earth Planet Sci Lett 38:26 – 43

Hawkesworth CJ, O'Nions RK (1977) The petrogenesis of some Archaean volcanic rocks from southern Africa. J Petrol 18:487 – 520

Heier KS (1973) Geochemistry of granulite facies rocks and problems of their origin. Philos Trans R Soc Lond A 273:429 – 442

Holland JG, Lambert RSJ (1975) The chemistry and origin of the Lewisian gneisses of the Scottish Mainland: The Scourie and Inver assemblages and sub-crustal accretion. Precambrian Res 2:161 – 188

Huang CC (1978) An outline of the tectonic characteristics of China. Eclogae Geol Helv 71: 611 – 635

Hunter DR, Barker F, Millard HT (1978) The geochemical nature of the Archaean ancient gneiss complex and granodiorite suite, Swaziland: A preliminary study. Precambrian Res 7:105 – 127

Jacobsen SB, Wasserburg CJ (1978) Interpretation of Nd, Sr and Pb isotope data from Archaean migmatites in Lofoten-Vesteralen, Norway. Earth Planet Scie Lett 41:245 – 253

Jahn BM, Nyquist LE (1976) Crustal evolution in the early earth-moon system: constraints from Rb-Sr studies. In: Windley BF (ed) The early history of the earth, Wiley, London, pp 55 – 76

Jahn BM, Sun SS (1979) Trace element distribution and isotopic composition of Archaean greenstones. In: Ahrens LH (ed) Origin and distribution of the elements, 2nd Symposium. Pergamon, Oxford, pp 597 – 618

Jahn BM, Auvray B, Blais S, Capdevila R, Cornichet J, Vidal F, Hameurt J (1980a) Trace element geochemistry and petrogenesis of Finnish greenstone belts. J Petrol 21:201 – 244

Jahn BM, Glikson AY, Peucat JJ, Hickman AH (1981) REE geochemistry and isotopic data of Archaean silicic volcanics and granitoids from the Pilbara Block, Western Australia: implications for the early crustal evolution. Geochim Cosmochim Acta 45:1633 – 1652

Jahn BM, Gruau G, Glikson AY (1982) Komatiites of the Onverwacht Group, S. Africa: REE geochemistry, Sm/Nd age and mantle evolution. Contrib Mineral Petrol 80:25 – 40

Jahn BM, Zhang ZQ (1984) Archaean granulite gneisses from eastern Hebei Province, China: rare earth geochemistry and tectonic implications. Contrib Mineral Petrol 85:224 – 243

James HL (1966) Chemistry of the iron-rich sedimentary rocks. In: Data of geochemistry, 6th edition. US Geol Surv Prof Pap 440-W: 61 p

Kröner A (1982) Archaean to early Proterozoic tectonics and crustal evolution: a review. Rev Bras Geociên 12:15 – 31

Lambert RSJ (1976) Archaean thermal regimes, crustal and upper mantle temperatures, and a progressive evolutionary model for the Earth. In: Windley BF (ed) The early history of the earth. Wiley, London, pp 363 – 373

Liu D, Page RW, Compston W, Wu Z (1984) U-Pb zircon geochronology of late Archaean metamorphic rocks in the Taihangshan-Wutaishan area, North China Prec Research (in press)

Ma XY, Wu ZW (1981) Early tectonic evolution of China. Precambrian Res 14:185 – 202

Martin H, Chauvel C, Jahn BM (1983) Major and trace element geochemistry and crustal evolution of Archaean granodioritic rocks from eastern Finland. Precambrian Res 21:159 – 180

Masuda A, Nakamura N, Tanaka T (1973) Fine structures of mutually normalized rare-earth patterns of chondrites. Geochim Cosmochim Acta 37:239 – 248

McLennan SM, Taylor SR (1984) Archaean sedimentary rocks and their relation to the composition of the Archaean continental crust. This Vol., 47 – 72

Moorlock BSP, Tarney J, Wright AE (1972) K-Rb ratios of intrusive anorthosite veins from Angmagssalik, East Greeland. Earth Planet Sci Lett 14:39 – 46

Nance WB, Taylor SR (1976) Rare earth element patterns and crustal evolution. I Australian post-Archaean sedimentary rocks. Geochim Cosmochim Acta 40:1539 – 1551

Nance WB, Taylor SR (1977) Rare earth element patterns and crustal evolution. II Archaean sedimentary rocks from Kalgoorlie, Australia. Geochim Cosmochim Acta 41:225 – 231

Nesbitt RW, Sun SS (1976) Geochemistry of Archaean spinifex textured peridotites and magnesian and low magnesian tholeiites. Earth Planet Sci Lett 31:433 – 453

Newton RC, Smith JV, Windley BF (1980) Carbonic metamorphism, granulites, and crustal growth. Nature 288:45 – 50

O'Connor JT (1965) A classification for quartz-rich igneous rocks based on feldspar ratio. US Geol Surv Prof Pap 525B:79 – 84

Oliver JE (1978) Exploration of the continental basement by seismic reflection profiling. Nature 257:485 – 488

Oliver J (1982) Probing the structure of the deep continental crust, Science 216:689 – 695

O'Nions RK, Pankhurst RJ (1974) Rare earth element distribution in Archaean gneisses and anorthosites. Godthab area, West Greenland. Earth Planet Sci Lett 22:328 – 338

Pidgeon RT (1980) 2480 Ma old zircons from granulite facies rocks from east Hebei Province, North China. Geol Rev 26:198 – 207

Shaw DM (1968) A review of K-Rb fractionation trends by covariance analysis. Geochim Cosmochim Acta 32:573 – 602

Shaw DM (1972) The origin of the Apsley gneiss, Ontario. Can J Earth Sci 9:18 – 35

Shen QH, Zhang ZQ, Xia MX, Wang XY, Lu JY (1981) Rb-Sr age determination on the late Archaean ferrosiliceous rock series in Sijiaying, Luanxian, Hebei Geol Rev 27:207 – 212 (in Chinese with English abstract)

Sighinolfi GP, Figueredo MCH, Fyfe WS, Kronberg BI, Tanner Oliveira MAF (1981) Geochemistry and petrology of the Jeguie granulitic complex (Brazil): an Archaean basement complex. Contrib Mineral Petrol 78:263 – 271

Sun DZ, Wu CH (1981) The principal geological and geochemical characteristics of the Archaean greenstone-gneiss sequences in North China, Spec Publ Geol Soc Aust 7:121 – 132

Sun DZ, Bai J, Jin WS, Gao F, Wang WY, Wang JL, Gao YD, Yang CL (1983) The geology of early Precambrian of the eastern Hebei, North China. Tianjin Science and Technology Press (in press)

Sun DZ, Lu SN (1983) A subdivision of the Precambrian of China. Precambrian Res (in press, 1984)

Sun SS, Nesbitt RW (1978) Petrogenesis of Archaean ultrabasic and basic volcanics: evidence from rare earth elements. Contrib Mineral Petrol 65:301 – 325

Tarney J, Windley BF (1977) Chemistry, thermal gradients and evolution of the lower continental crust. J Geol Soc Lond 134:153 – 172

Taylor SR (1967) The origin and growth of continents. Tectonophys 4:17 – 34

Taylor SR (1977) Island arc models and the composition of the continental crust. Am Geophys Union Maurice Ewing series I:325 – 335

Taylor SR (1979) Chemical composition and evolution of the continental crust: The rare earth element evidence. In: McElhinny MW (ed) The Earth: its origin, structure and evolution. Academic, London, pp 353 – 376

Taylor SR, Hallberg JA (1977) Rare-earth elements in the Marda calc-alkaline suite: an Archaean geochemical analogue of Andean-type volcanism. Geochim Cosmochim Acta 41:1125 – 1129

Taylor SR, McLennan SM (1981a) The composition and evolution of the continental crust: rare earth element evidence from sedimentary rocks. Philos Trans R Soc Lond A 301:381 – 399

Taylor SR, McLennan SM (1981b) The rare earth element evidence in Precambrian sedimentary rocks: implications for crustal evolution. In: Kröner A (ed) Precambrian Plate Tectonics. Elsevier, Amsterdam, pp 527 – 548

Touret J (1971a) Le faciès granulite en Norvège méridionale. I Les associations minéralogiques. Lithos 4:239 – 249

Touret J (1971b) Le faciès granulite en Norvège méridionale. II Les inclusions fluides. Lithos 4:423 – 436

Wang KY (1982) REE content and its tectonic implications of the ancient metamorphic rocks from Taipingzhai, Qianxi County, Hebei Province. Sci Geol Sin N°2:144 – 151 (in Chinese with English abstract)

Weaver BL (1980) Rare-earth element geochemistry of Madras granulites. Contrib Mineral Petrol 71:271 – 279

Weaver BL, Tarney J (1980) Rare earth geochemistry of Lewisian granulite-facies gneisses, northwest Scotland: implications for the petrogenesis of the Archaean lower continental crust. Earth Planet Sci Lett 51:279 – 296

Weaver BL, Tarney J (1981) Lewisian gneiss geochemistry and Archaean crustal development models. Earth Planet Sci Lett 55:171 – 180

Withford DJ, Nicholls IA, Taylor SR (1979) Spatial variations in the geochemistry of Quarternary lavas across the Sunda arc in Java and Bali. Contrib Mineral Petrol 70:341 – 356

Wildeman TR, Condie KC (1973) Rare earths in Archean graywackes from Wyoming and from the Fig Tree Group, South Africa. Geochim Cosmochim Acta 37:439 – 453

Wildeman TR, Haskin LA (1973) Rare earths in Precambrian sediments. Geochim Cosmochim Acta 37:419 – 438

Zhang RY, Cong BL, Ying YP, Li JL (1981) Ferrifayalite-bearing eulysite from Archaean granulites in Qianan County, Hebei, North China. Tschermaks Min Petr Mitt 28:167 – 187

Zhang RY, Cong BL (1982) Mineralogy and T-P conditions of crystallization of early Archaean granulites from Qianxi County, NE China. Sci Sin 25:96 – 112

Zhang YX, Yen HQ, Wang KD, Li FY (1980) Investigation of komatiite in the Qianxi Group, E. Hebei. J Changchun Geol Inst 1:1 – 8 (in Chinese)

The Most Ancient Rocks in the USSR Territory by U-Pb Data on Accessory Zircons

E. V. BIBIKOVA[1]

Contents

1 Introduction 235
2 The Baltic Shield 237
3 The Ukrainian Shield 239
4 The Aldan Shield and Its Margins 243
5 Crystalline Massifs of the USSR North-East 246
6 General Zircon Chemistry 248
7 Conclusions 249
References 249

Abstract

The early Precambrian formations in the USSR occur in the Baltic, the Ukrainian and the Aldan shields and were studied geochronologically using the U-Pb isotopic method on accessory zircons. It is only within the Ukrainian shield where the isotopic data confirm the existence of Lower Archaean formations. Supracrustal and associated granitoid formations of the Upper Archaean (2.9 – 2.6 Ga) were dated within all Precambrian shields of the USSR. The discovery of high-grade Archaean rocks as old as 3.4 Ga within the Omolon massif of the USSR Far East and their similarity to some Aldan formations gives hope to confirm, by isotope geochronology, the existence of an Early Archaean core within the Aldan shield.

1 Introduction

Studies of the most ancient parts of Precambrian shields, whose formation had been completed by 3.0 Ga ago, show that these terrains can be subdivided into two types: high-grade metamorphic zones, composed of gneisses and migmatites (so-called grey gneisses) and granite-greenstone areas. Geochemical data point to a mantle-derived origin for both types. Rocks of later tectonomagmatic epochs, particularly those of the period between 2.9 – 2.6 Ga ago, are also predominantly direct derivatives of the upper mantle and have geochemical features close to those of the early Archaean. The identification of the earliest crustal remnants

---

1 Vernadsky Institute of Geochemistry and Analytical Chemistry, USSR Academy of Sciences, Kosygina 19, 117975 Moscow V-334, USSR

Archaean Geochemistry (ed. by A. Kröner et al.)
© Springer-Verlag Berlin Heidelberg 1984

among the more widely distributed rocks of later tectonomagmatic cycles therefore needs to be made almost exclusively on the basis of isotope geochronology.

The polymetamorphic nature of most ancient rocks may interfere with isotopic dating. Not all isotopic methods may therefore be useful for these purposes. For example, in several cases the suitability of the Pb-Pb and Rb-Sr whole rock isotopic methods for dating granulite facies rocks is restricted; the removal of U, K, and Rb during the course of metamorphism, along with preservation of radiogenic Pb and Sr, could cause overestimation of the ages (Baadsgaard 1976). The Pb-Pb and Rb-Sr isotopic systems also appear to be disturbed frequently in volcanic rocks of greenstone belts, despite the low grades of metamorphism.

One of the most suitable methods for dating very ancient rocks has turned out to be the U-Pb method on accessory zircons since the mineral zircon is widely distributed in many types of rocks. Uranium, isomorphically replacing zirconium in the structure of the mineral, is firmly held in it during superimposed metamorphism which, as a rule, manifests itself only by removal of part of the radiogenic lead. This prevents overestimation of age values.

The morphological features of accessory zircons enable correlation of isotopic ages with definite events of rock formation and transformation. Thus, accessory zircons formed during granulite metamorphism can be reliably identified by their morphological features (equant, multifaceted forms). This permits dating of both the time of metamorphism manifested in a granulitic terrane and the age of the protolith by using surviving, premetamorphic grains of zircon. Such studies may also prove a primary magmatic nature for zircons in volcanic rocks of greenstone belts. Therefore, the epochs of palaeovolcanism in the Earth's history can be established by dating such authigenous grains of zircon.

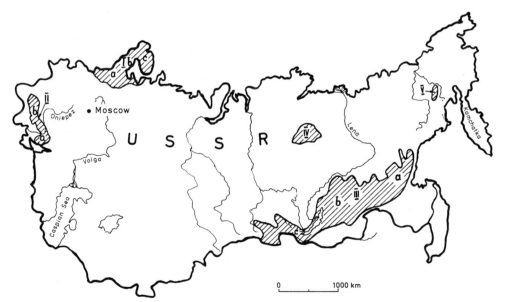

Fig. 1. Location of the Precambrian shields in the territory of the USSR. **I** Baltic shield, regions: *a* Karelia, *b* Belomoria, *c* Kola; **II** Ukrainian shield, regions: *a* Pré-Dnepro, *b* Pobuzhie; **III** Aldan shield, regions: *a* Aldan, *b* Olekma, *c* Pre-Baikal; **IV** Anabar massif; **V** Omolon massif

There are at present reliable data on accessory zircons for many of the most ancient rocks of the Earth, such as the Amitsoq, Uivak and Morton gneisses (Jacobsen and Wasserburg 1978; Baadsgaard et al. 1979; Goldich et al. 1970) and the Isua and Pilbara volcanics (Pidgeon 1978; Michard-Vitrac et al. 1977). The early Precambrian formations in the USSR crop out within the boundaries of three large Precambrian shields — the Baltic and Ukrainian in the European part, and the Aldan with its margin and smaller massifs (Anabar , Omolon, Okhotsk) in the Asian part (Fig. 1). Archaean rocks belonging to the Earth's early crust were identified within all these shields, but reliable isotopic data confirming such identification are few in number. For the last few years we have undertaken isotopic dating on the most ancient rock types of the Baltic, Ukrainian, and Aldan shields by the U-Pb method on accessory zircons. The analytical techniques in the laboratory are similar to those of Krogh (1973). All ages are recalculated using the new decay constants (Steiger and Jäger 1976). Analytical data may be found in the references cited.

2 The Baltic Shield

Within the eastern part of the Baltic shield, geological data outline the three most ancient regions: the granite-greenstone region of central Karelia and the high-grade regions of the White Sea and the Kola Peninsula (Fig. 1, Ia, b, c).

U-Pb isotopic dating of accessory zircons from different rocks of central Karelia permit deciphering of the succession of geological events in this typical granite-greenstone province (Fig. 1, Ia). The vast area here is occupied by granitoids

Table 1. Age of zircons from granitoids in eastern Karelia

No.	Locality	Content (%)		Isotopic composition of lead				Age (Ma)		
		Pb	U	204	206	207	208	207/206	206/238	207/235
1.	Kostomuksha, plagiogranite	0.0097	0.026	0.066	75.92	14.64	9.37	2710	1880	2320
2.	Kostomuksha, Pl-Mi granite	0.041	0.075	0.027	75.73	14.18	10.07	2720	2540	2650
3.	Kostomuksha, Mi-granite	0.026	0.074	0.054	72.43	13.85	13.67	2700	1670	2190
4.	Nadvoizi granodiorite	0.0105	0.026	0.034	72.05	13.91	14.01	2760	1885	2340
5.	Nadvoizi, diorite	0.0138	0.029	0.051	69.73	13.80	16.42	2770	2105	2410
6.	Mashozero, granodiorite	0.0017	0.0068	0.25	65.05	15.72	18.98	2830	1070	1790
7.	Tikshozero	0.019	0.024	0.07	53.33	11.45	33.15	2800	2730	2770
8.	Kartashi, Mi-granite	0.003	0.0054	0.217	67.055	15.45	17.28	2790	2250	2530

Lead correction: 206/204 − 13.28; 207/204 − 14.67

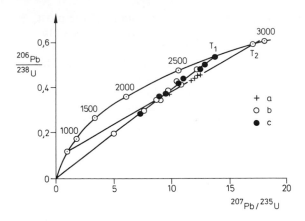

Fig. 2a – c. Concordia plot for zircon fractions from rocks of Karelia. **a** Lopii volcanics; **b** granites in the Karelian basement; **c** granites intruding Lopii volcanics

and granite-gneisses, varying in composition from tonalite, diorite and granodiorite to microcline granite. Part of these belong to the basement of greenstone belts, but some undoubtedly intrude these belts. Numerous age measurements on accessory zircons from these granitoids (both intrusive and those regarded as the basement) made during the last 20 years (Tugarinov and Bibikova 1980) yielded ages of approximately 2740 ± 40 Ma (Table 1, Fig. 2).

We have recently measured the time of formation of the Lopii greenstone belt using zircons separated from felsic volcanics of the Koikar structure in central Karelia (Bibikova and Krylov 1983). Accessory zircons of metadacite are represented by small, prismatic, almost colourless grains with fine internal zonation indicative of a magmatic origin. U-Pb measurements on different zircon fractions were made in the geochronological laboratory of the Geological Survey of Finland. The results of these measurements are given in Table 2 and are plotted as crosses on Fig. 2. They yielded a concordia intercept age of 2935 ± 20 Ma, defining the upper time boundary for the formation of volcanogenic belts in central Karelia. The granite formation and metamorphism which completed the evolution of the granite-greenstone area took place about 2750 Ma ago. Thus, the length of the whole tectonomagmatic cycle is estimated at 200 – 250 Ma. Some data point to the existence of relicts of more ancient greenstone belts (Volcanism 1981), but reliable isotopic data have not confirmed this interpretation.

A decrease in age of granites in a westward direction (Tugarinov and Bibikova 1978) led Rybakov and Lobach-Zhuchenko (1981) to suggest that within the boundaries of the Fenno-Karelian craton the magmatic processes migrated from east to west. These data are in accordance with data of Vidal et al. (1980) and Martin et al. (1983) who dated volcanic rocks of greenstone belts and granitoids of eastern Finland using whole rock Rb-Sr and Sm-Nd isotopic systems. Volcanism in central Karelia took place 2.93 Ga years ago and granitoids intruded 2.85 – 2.75 Ga years ago, while in Finland these events took place 2.65 Ga and 2.62 – 2.5 Ga ago, respectively. The first generation of grey gneisses in eastern Finland is 2.86 Ga old, while there are reliable data of 3.1 Ga for tonalites in central Karelia (Lobikov and Lobach-Zhuchenko 1980) and northern Finland (Kröner et al. 1981; Jahn et al. 1984).

Table 2. U-Pb age of zircons from the Lopii volcanics (Karelia)

No. Fraction (mm)	Content (%)		Isotopic composition of lead				Age (Ma)		
	Pb	U	204	206	207	208	207/206	206/238	207/235
1. 4.2<d<4.6>0.07	0.0159	0.0288	0.013	69.359	16.556	14.072	2882±12	2382±12	2633±6
2. 4.2<d<4.6<0.07	0.0166	0.0310	0.014	68.577	17.581	13.828	2819±2	2299±12	2586±6
3. 4.0<d<4.2	0.0163	0.0299	0.013	68.879	17.152	13.955	2829±3	2345±13	2613±6
4. Magnetic	0.0168	0.0360	0.020	68.784	17.808	13.387	2751±3	2058±11	2462±6
5. HF residue	0.0179	0.0277	0.022	68.606	17.259	14.113	2842±3	2398±12	2646±6

Lead correction: 206/204 − 13.331; 207/204 − 14.541

For the highly metamorphosed deposits of the White Sea and Kola Series we also failed to obtain age values in excess of 2800 − 2900 Ma by the U-Pb method (Tugarinov and Bibikova 1980).

The White Sea Series is composed mainly of gneisses, granite-gneisses and amphibolites with different composition and structure. Our measurements of accessory zircons from all these rocks demonstrated that these rocks were formed 2700 ± 50 Ma ago, and some of them underwent intense metamorphic overprinting 1850 ± 20 Ma ago (Tugarinov and Bibikova 1980).

We also studied the age of highly metamorphosed rocks of the Kola Series, which comprise a single structure within the boundaries of the central Kola Block (Fig. 1, Ic). According to Bondarenko et al. (1968) the rocks of the Kola Series reveal a progressive metamorphic zonation. This is manifested in the gradually increasing intensity of metamorphism from amphibolite to the granulite facies, with a transition zone where mineral assemblages of the granulite facies are observed to progressively overprint those of the amphibolite facies. The oligoclase granites from the area of the Voronya tundra and hypersthene diorites of the central Kola Block (whose abyssal origin is confirmed by the results of oxygen-isotope analyses (Bibikova et al. 1982a) have been regarded as a basement of the Kola Series.

Numerous U-Pb isotopic measurements of accessory zircons from all these rocks, including the so-called basement gneisses of the Kola Series that are metamorphosed to both granulite and amphibolite facies, as well as intrusive granites yielded similar ages of 2.9 − 2.7 Ga (Tugarinov and Bibikova 1980).

Geochronological data for the stratigraphically oldest rocks in the eastern part of the Baltic shield show that the formation of high-grade rocks of the Kola and White Sea Series as well as the evolution of the granite-greenstone assemblages of Karelia took place almost synchronously 2.9 − 2.7 Ga ago. There are no isotopic data indicating the existence of more ancient crust in the eastern Baltic shield.

3 The Ukrainian Shield

The most ancient rocks of the Ukrainian shield occur in the granite-greenstone area of the Middle Dnepro province and the highly metamorphosed region of the

Bug province (Fig. 1, IIa, b). Detailed consideration of the Ukrainian shield evolution during the early Precambrian is given elsewhere in this volume (Shcherback et al. 1984), so we report here only the analysis of U-Pb data on accessory zircons.

The middle Dnepro province can serve as a classical example for a granite-greenstone terrane. It is a large tectonic block, confined by faults. The most widespread rocks here are granites and migmatites of the Dnepro Archaean assemblage, enclosing several tectonized greenstone belts of volcaniclastic and sedimentary deposits known as the Konka Series. The rocks of this Series have been metamorphosed to greenschist facies and are only locally affected by younger granites where the metamorphism increases to the level of amphibolite facies.

We made measurements on accessory zircons from meta-andesites of the Konka Series (Bibikova et al. 1983). Mineralogical studies of these zircons showed a high homogeneity and characteristics of magmatic genesis. On a concordia plot (Table 3, Fig. 3) three strongly discordant zircon fractions give an up-

Table 3. U-Pb age of zircons from volcanics of Konka (Ukrainian shield)

No.	Content (%) Pb	U	Isotopic composition of lead 204	206	207	208	Age (Ma) 207/206	206/238	207/235
I.	0.0103	0.0210	0.144	68.041	16.985	14.830	3030	2065	2590
2.	0.0107	0.0220	0.053	71.206	16.653	11.984	3060	2160	2660
3.	0.0101	0.0208	0.027	72.053	17.053	10.680	3065	2205	2685

Lead correction: 206/204 − 13.28; 207/204 − 14.67

Table 3a. U-Pb age of zircons from tonalites of the Pre-Dnepro region (Ukrainian shield)

No.	Size fraction (mm)	Content (%) Pb	U	Isotopic composition of lead 204	206	207	208	Age (Ma) 207/206	206/238	207/235
1.	+0.125	0.0168	0.0240	0.011	65.641	14.519	19.829	2980	2675	2850
	−0.075	0.0174	0.0272	0.024	69.534	15.339	15.103	2960	2670	2840
	nuclei			0.022	66.577	14.833	18.568	2970		
	envelope			0.014	67.109	14.865	18.012	2970		
2.	+0.100	0.087	0.0194	0.018	73.941	16.246	8.692	2980	2100	2580
	−0.100+									
	+0.065	0.0089	0.0173	0.028	73.471	16.627	9.873	2980	2355	2700
	−0.065	0.0144	0.0233	0.029	73.081	16.006	10.817	2980	2700	2850
3.	m	0.0099	0.0264	0.040	72.086	16.373	11.500	2985	1750	2385
4.	−0.125+									
	+0.075	0.0354	0.0443	0.120	61.613	14.846	23.421	2965	2860	2920
	−0.075+									
	+0.053	0.0291	0.0455	0.007	65.690	14.533	19.769	2980	2560	2800
5.	+0.065	0.0157	0.0254	0.018	70.834	15.860	13.288	2980	2645	2845
	−0.065	0.0086	0.0138	0.011	70.848	15.726	13.415	2980	2670	2850

Lead correction: 206/204 − 13.28; 207/204 − 14.67

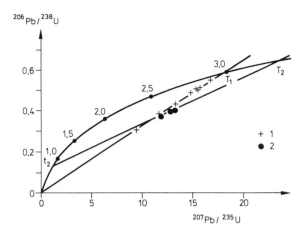

Fig. 3. Concordia plot for zircon fractions from rocks of the Pré-Dnepro region (Ukrainian shield). *1* tonalite; *2* Konka volcanics

per intercept age of 3250 ± 120 Ga. Provided that formation of the zircons was synchronous with lava eruption, the age of 3250 Ga can be regarded as a minimum age for volcanism of the Konka Series. There have been different concepts on the age of granites from the Dnepro region and their correlation with the supracrustal formations. The tendency now is to consider those plutons that are tonalitic in composition as basement to the volcanics of the greenstone belts. We studied the U-Pb isotopic system of accessory zircons extracted mainly from tonalites and granodiorites (possible basement of the belts). The results are given in Table 3 and Fig. 3 and show an excellent alignment of all data points with an upper concordia intercept age of 3000 ± 20 Ma. The data thus clearly reveal a younger age for the tonalites studied compared to the volcanics of the greenstone belts. Geochemically the accessory zircons of these granitoids differ sharply from zircons of primary crust granites in having lower U contents and higher Th/U ratios. The geochemical data are not at variance with an ultrametamorphic origin for these rocks through anatexis.

Within the Orekhovo-Pavlograd zone (a tectonic structure to the east of the Middle Dnepro block), drill holes have recovered rocks of ultramafic-mafic composition that contain accessory zircons with granitic morphology and geochemistry (high U content up to 0.15% and low Th/U ratio). We shall not discuss these data in detail, since they are considered in another paper of this volume (Shcherback et al., this Vol.). The Concordia intercept age of different morphologic and density fractions of these zircons exceeds 3600 Ma.

High-grade metamorphic rocks of the Dniester-Bug Series in the western part of the Ukrainian shield (Fig. 1, IIb) are correlated with the Konka Series of the Middle Dnepro province. The Dniestro-Bug Series contains thick layers of enderbite along with bipyroxene to amphibole-pyroxene crystalline schists and plagiogneisses. Metamorphism of this Series took place under conditions of a moderate-pressure granulitic facies.

The mineralogical and isotopic studies of accessory zircons from the enderbite-gneisses of this region enabled us to establish two differently-aged generations of this mineral (Table 4). About 10% of the sample population is composed of roundish, multifaceted grains of transparent zircon which is interpreted to

On the other hand, within the Olekmo-Stanovoi zone (Fig. 1, IIIb), which used to be regarded as a Proterozoic margin of the Aldan shield, ages of about 2600 Ma were obtained for granulitic zircons in gneisses (Bibikova et al. 1984b). Thus, a considerably earlier epoch of granulitic metamorphism occurred within the boundaries of this zone. This suggests that the age relationship between the rocks of the Aldan and Olekma-Stanovoi zones is inverse, i.e. it is in the Olekma-Stanovoi zone where the most ancient rocks of Aldan shield can be found.

In a south western direction the Aldan shield continues as separate Precambrian blocks. We made a detailed Pb-U isotopic study of accessory zircons from high-grade metamorphic rocks of the Sharyzhalgai Series (North Forebaikal; Fig. 1, IIIc), which correlate with the lowermost part of the Aldan sequence (Tugarinov and Vojtkevich 1970).

Within the Sharyzhalgai Series there are prominent complexes which have been metamorphosed under conditions of the granulite and amphibolite facies. The rocks of these complexes are predominantly represented by mafic schists and gneisses, charnockites and enderbites with subordinate amounts of carbonate rocks. Charnockites are present, as a rule, as the vein material of agmatites, in which the blocks consist of schists and amphibolites (Bibikova et al. 1981).

Mineralogic and isotopic studies of accessory zircons from different rocks of the Sharyzhalgai Series revealed two main generations of this material (Fig. 5). The first generation is represented by sub-idiomorphic grains of pre-granulitic zircon. In charnockites and enderbites the second generation is widespread in addition to the first generation and is represented by equant and short prismatic grains that are transparent with bright lustre; these most probably formed under granulitic conditions. A multiple approach to the analysis of these heterogeneous zircons (fission track investigations, selective decomposition of different generations of the mineral) permitted the establishment of the isotopic age of different zircon generations (Table 5, Fig. 6) (Bibikova et al. 1981).

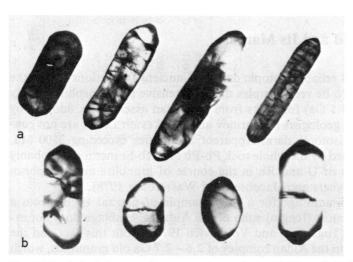

Fig. 5. Morphological types of accessory zircons from enderbites of the Sharyzhalgai block. **a** pre-metamorphic type, T ~ 2700 Ma; **b** granulitic type, T ~ 1900 Ma

Table 5. Analytical data for zircon fractions from rocks of the North Forebaikal and Sayany Provinces

No.	Fraction size mag.	Concentration (%)		Isotopic composition of lead				Age (Ma)		
		Pb	U	204	206	207	208	207/206	206/238	207/235
U-Pb age of accessory zircons from rocks of the Sharyzhalgay Series										
1	− 150, nm	0.0146	0.0361	0.006	79.278	10.584	10.132	2130	1980	2050
1B	+ 150, nm	0.0123	0.0298	0.011	76.700	10.147	14.143	2103	2015	2060
1B	solution			0.011	84.487	10.915	7.587	2130		
1B	residue			0.005	70.667	9.046	20.283	2060		
2	− 125, nm	0.0400	0.0882	0.002	83.603	13.460	2.930	2465	2355	2410
2B	− 150+125	0.0373	0.0841	0.006	83.563	13.456	2.979	2457	2305	2390
2C	− 125, m	0.0488	0.1110	0.004	82.536	13.244	4.215	2454	2270	2370
2	solution		0.1053	0.002	83.832	13.570	2.959	2470		
2	residue		0.0238	0.003	77.763	11.338	10.895	2290		
3	− 150, nm	0.0104	0.0245	0.015	80.038	12.086	7.860	2330	2140	2240
3B	+ 150, nm	0.0149	0.0388	0.014	81.287	11.843	6.587	2267	1990	2130
5	− 150, nm	0.0135	0.0343	0.024	78.585	10.452	10.939	2085	1970	2225
6	− 90+ 65, m	0.0123	0.0318	0.096	70.037	9.578	20.290	1935	1725	1830
7	− 90+ 65	0.0173	0.0358	0.008	75.907	12.603	11.482	2505	2285	2405
7B	+ 125 dark	0.0244	0.0567	0.004	76.555	12.306	11.135	2455	2085	2280
7C	+ 125 light	0.0133	0.0271	0.060	71.693	12.404	15.842	2485	2185	2345
U-Pb age of accessory zircons from Onot plagiogneisses										
1	− 125+100	0.0183	0.0410	0.029	75.916	17.817	6.236	3060	2130	2640
2	+ 65 −100	0.0218	0.0505	0.019	76.051	17.634	6.297	3035	2080	2590
3	− 65	0.0192	0.0488	0.019	75.881	16.997	7.102	2990	1920	2485

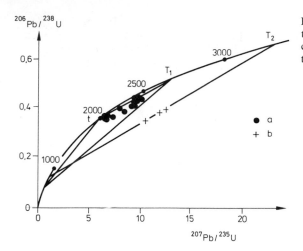

Fig. 6. Concordia plot for zircons fractions from the Pre-Baikal region. **a** zircons from Sharyzhalgai; **b** from Onot tonalites

The accumulation of the original, predominantly volcanogenic deposits was followed by early regional metamorphism that took place about 2600 – 2700 Ma ago (Fig. 6). The orthogenic nature of the Sharyzhalgai rocks and their compositional similarity to volcanics of greenstone belts is revealed by oxygen isotope data [$\delta^{18}O$ for basic crystalline schists lies between $+ 5$ and $+ 6‰$ and for enderbites $+ 5$ and $+ 7.5‰$ (Bibikova et al. 1982a)]. The granulitic stages of metamorphism, including anatexis and formation of charnockites and enderbites, took place within the Sharyzhalgai block boundaries about 1900 Ma ago, as indicated by the lower concordia intercept in Fig. 6.

To the west the intensity of metamorphism decreases, and in the Sayany Province grabens of weakly metamorphosed rocks crop out; these rocks can be correlated with other greenstone belts. Because of the lower degree of metamorphism and available K-Ar ages they were regarded as Proterozoic formations. However, we dated these different fractions of accessory zircons separated from plagioclase gneisses of the Onot graben in the Sayany Province. These measurements yielded a concordia intercept age of 3250 ± 50 Ma (Table 5, Fig. 6) (Bibikova et al. 1982b). If the interpretation of local geologists is correct about the intrusive contact between plagiogneisses and supracrustal rocks of the graben, then the greenstone belts in the Sayany Province must be older than 3.2 Ga and can be regarded as some of the most ancient rocks in the USSR. However, this possibility needs to be examined by more detailed geological and geochronological studies.

5 Crystalline Massifs of the USSR North-East

In addition to the Aldan shield there are a number of other, smaller massifs within the Asian part of the USSR, such as the Anabar, Omolon, and Okhotsk massifs (Fig. 1). They have not yet been studied in detail geochronologically.

Anabar. Within the Anabar crystalline massif there are predominantly high-grade metamorphic formations which may be correlated with the granulites of the Aldan complex. Initial measurements (Bibikova et al. 1984b) of isotopic ages for accessory zircons from high-alumina gneisses of this region yielded concordia intercept age values of 2900 Ma.

Omolon. We have studied in detail the rocks from the Archaean block of the Omolon massif, which occupies the very far eastern part of the Asian continent (Fig. 1, V). Within the Omolon Archaean block there occur predominantly tonalites of variable composition which make up about 70% of the assemblage. Bipyroxene crystalline schists and high-alumina gneisses alternate with them. These rocks were metamorphosed under granulite facies conditions.

A mineralogical study of accessory zircons from these tonalites revealed two generations of this mineral which are significantly different as to morphological features, the content of U and Th and isotopic age. As much as 95% of the zircons are deeply-coloured, sub-idiomorphic prismatic grains which are probably of primary magmatic nature. The second generation is represented by smaller, equant, transparent crystals with bright diamond lustre. They were undoubtedly formed during granulite metamorphism.

The results of the U-Pb isotopic analyses indicate that the zircons of the second generation were formed 2750 Ma ago, while formation of the first zircon generation took place considerably earlier (Table 6, Fig. 7). On a concordia plot 5 size and magnetic fractions determine a discordia line that intersects concordia at 3400 ± 150 and 1100 ± 100 Ma, respectively. Accessory zircons with granulitic habit, extracted from bipyroxene crystalline schists, were formed only 1900 Ma ago. The difference in time between the 2750 Ma old granulitic zircon in the tonalites and the 1900 Ma old crystalline schists possibly results from differences in the nature of granulitic metamorphism during these two stages. Dry conditions of granulite metamorphism of the early stage, without an influx of a fluid phase, could not result in zircon formation in the mafic rocks, since all zirconium occurs as isomorphic substitution inside the mafic materials. The metamorphism of the next stage was accompanied by an influx of alkalies and silica and could have led to a release of zirconium and neocrystallization of zircon. The possibility of formation of at least part of the mafic rocks as dykes cannot be excluded; however, the available data exclude the possibility that the mafic rocks were the protoliths for tonalite formation (an idea supported by many geologists).

Table 6. U-Pb age of zircons from plagiogneisses of the Omolon massif

No.	Size fraction colour	Content (%)		Isotopic composition of lead				Age (Ma)		
		Pb	U	204	206	207	208	207/206	206/238	207/235
1.	a +0.100	0.0283	0.0492	0.007	73.88	18.30	7.81	3160	2530	2900
	b −0.100+0.075	0.0387	0.0650	0.016	72.68	18.25	9.05	3175	2670	2970
	c −0.075	0.0318	0.0548	0.032	71.99	18.33	9.65	3180	2490	2885
	d grey	0.0361	0.0688	0.005	72.28	18.29	6.49	3135	2385	2810
	e brown	0.0459	0.0853	0.008	76.24	18.38	5.37	3125	2460	2840
2.	isometric grain	0.0127	0.0229	0.027	64.40	12.69	22.88	2760	2180	2485

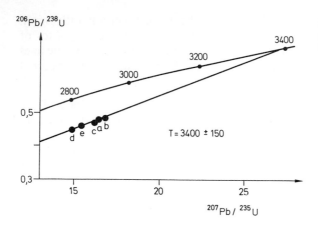

Fig. 7. Concordia plot for premetamorphic zircon fractions from the tonalites of the Omolon massif

The orthogenic nature of the tonalites is also confirmed by the results of an oxygen-isotope study. The $\delta^{18}O$ values range between $+7$ and $+8‰$ and are correlated with SiO_2 content (Bibikova et al. 1982a).

The geochronological data show that within the Omolon massif the most ancient orthogneisses have a tonalitic composition and an age of about 3400 Ma. They underwent granulite metamorphism 2750 Ma ago and again about 1900 Ma ago.

The similarity between the Archaean block of the Omolon massif and the Olekma zone of the Aldan shield suggests that we should focus our studies on the Olekma zone for a search of the most ancient formations of the Aldan shield.

6 General Zircon Chemistry

Geochemical information obtained during isotopic dating of accessory zircons from Archaean rocks suggests to us a specific geochemistry of this mineral that correlates with rocks of different genesis. On a diagram of U versus ($^{208}Pb/^{206}Pb$) rad (Fig. 8), zircons from greenstone belt volcanics (Bibikova and Krylov 1983; Bibikova et al. 1983; Michard-Vitrac et al. 1977; Pidgeon 1978) have a rather low content of U and a distinctive value of the Th/U ratio. Considerably higher contents of U along with lower values of the Th/U ratio define zircons from tonalitic to granodioritic gneisses of magmatic origin (Baadsgaard et al. 1979; Bibikova et al. 1982b; Bibikova et al. 1981; Tugarinov and Bibikova 1978; Jacobsen and Wasserburg 1978; Goldich et al. 1970; Bibikova et al. 1982c). The secondary zircons formed during granulite metamorphism or ultrametamorphism generally have low contents of U while their Th/U ratios are considerably higher than in the two first groups.

Thus, the geochemical features of accessory zircons are apparently related to those processes that led to formation or transformation of the enclosing rocks. They may therefore help to make a correct interpretation of the resulting age.

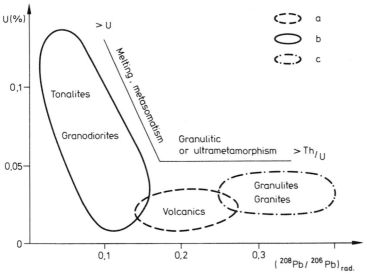

Fig. 8. Plot of U vs $^{208}Pb/^{206}Pb$ rad in accessory zircons from the most ancient rocks of the world. **a** volcanics (Isua, Pilbara, Lopii, Konka); **b** tonalites and tonalitic gneisses (Amitsoq, Uivak, Morton, Omolon, Pobuzhie, Onot); **c** ultrametamorphic granites and granulitic zircons (Ukrainia, Karelia, Omolon)

7 Conclusions

Geochronological studies of the Precambrian shields in the USSR by the U-Pb isotopic method on accessory zircons yield information on early crustal formations with ages of up to 3600 Ma within the central part of the Ukrainian shield. However, the isotopic data also show that the overwhelming majority of Archaean formations in the Baltic shield were formed more recently than 3000 Ma ago.

The discovery of Archaean rocks as old as 3.4 Ga within the Omolon massif of the USSR Far East and their similarity with the sequences of the Olekma zone of the Aldan Shield give us hope of discovering there, and confirming by isotopic geochronology, the most ancient core of the Aldan shield.

References

Baadsgaard H (1976) Further U-Pb dates on zircons from the early Precambrian rocks of the Godthaabsfjord, West Greenland. Earth Planet Sci Lett 33:261 – 267

Baadsgaard H, Collerson KD, Bridgwater D (1979) The Archaean gneiss complex of Northern Labrador. Can J Earth Sci 16:951 – 963

Bibikova EV, Sumin LV, Kirnozova TI, Gracheva TI (1981) A sequence of geological events within the Sharuzhalgay block. Geochemistry 11:1652 – 1664 (in Russian)

Bibikova EV, Grinenko VA, Kiselevsky MA, Shukolyukov YuA (1982a) Geochronological and oxygen-isotope study of the Precambrian granulites of the USSR (1982). Geochemistry 12:1718 – 1782 (in Russian)

Bibikova EV, Khiltova VYa, Gracheva TV, Makarov VA (1982b) Age of the Prisayanje greenstone belt. Dokl Ac Sci USSR 267 (5):1171–1174 (in Russian)
Bibikova V, Makarov VA, Gracheva TV, Lesnaya IM (1982c) Isotopic age of Pobuzhje enderbites. Dokl Ac Sci USSR 263 (1):159–162 (in Russian)
Bibikova EV, Krylov IN (1983) Isotopic age of acid volconites of Karelia. Dokl Ac Sci USSR 268:1231–1234
Bibikova EV, Kirnozova TI, Gracheva TV (1983) Isotopic age of ancient granitoids in PreDnepro region. Geochemistry 7:997–1004 (in Russian)
Bibikova EV, Belov AN, Gracheva TV, Rosen OM (1984a) The upper limit for the age of Anabar granulites. Dokl Ac Sci USSR, to be published (in Russian)
Bibikova EV, Shuldiner VI, Gracheva TV, Panchenko IV, Makarov VA (1984b) Isotopic age of granulites on the West of Stanovoy region. Dokl Ac Sci USSR, to be published (in Russian)
Bondarenko LP, Dagelaysky VB (1968) Geology and metamorphism of the Archaean in the Central part of Kola peninsula. Science, Leningrad (in Russian)
Goldich SS, Hedge CG, Stern TW (1970) Age of Morton and Montevideo gneisses and related rocks, southwestern Minnesota. Bull Geol Soc Am 81:3671–3696
Jacobsen SB, Wasserburg GJ (1978) Interpretation of Nd, Sr, Pb isotope data from Archaean migmatites of Lofoten-Vesteralen, Norway. Earth Planet Sci Lett 41:245–253
Jahn BM, Vidal P, Kröner A (1984) Multi-chronometric ages and origin of Archaean tonalitic gneisses in Finnish Lapland: a case for long crustal residence time. Contrib Mineral Petrol, in press
Krogh TE (1973) A low contamination method for hydrothermal decomposition of zircon and extraction of U and Pb for isotopic age determinations. Geochim Cosmochim Acta 37:485–494
Kröner A, Puustinen K, Hickman M (1981) Geochronology of an Archaean tonalitic gneiss dome in northern Finland and its relation with an unusual overlying volcanic conglomerate and komatiitic greenstone. Contrib Mineral Petrol 76:33–41
Krylov IN, Shafeev AA (1969) The peculiarities of structure of the rocks of Sharyzhalgay series in SW PreBaikalij. In: Geology of PreBaikalye. Guide-book AZOPRO, Irkutsk
Lobikov AF, Lobach-Zhuchenko SB (1980) Isotopic age of granites in Palaya Lamba greenstone belt of Karelia. Dokl Ac Sci USSR 250:729–733 (in Russian)
Martin H, Chauvel C, Jahn BM, Vidal Ph (1983) Rb-Sr and Sm-Nd ages and isotopic geochemistry of Archaean granodioritic gneisses from eastern Finland. Precambrian Res 20:79–91
Michard-Vitrac A, Lancelôt J, Allègre CJ, Moorbath S (1977) U-Pb ages of a single zircon from the early Precambrian rocks West Greenland and Minnesota River-Valley. Earth Planet Sci Lett 35:449–453
Pidgeon RT (1978) 3450 m.y. old volcanics in the Archaean Layered greenstone succession of the Pilbara block, Western Australia. Earth Planet Sci Lett 37:421–428
Rybakov SI, Lobach-Zhuchenko SB (1981) Greenstone belts of the Fenno-Karelian craton. In: Geological, geochemical and geophysical investigation in the eastern part of the Baltic Shield, The Committee for Scientific and Technical Cooperation, Helsinki
Shcherback NP, Bartnitskii EN, Bibikova EV, Boiko VL (1984) Evolution of Ukrainian shield in early Precambrian. This Vol., 251–261
Steiger RH, Jäger E (1976) Subcommission on geochronology: convention of the use of decay constants in geo- and cosmochronology. Earth Planet Sci Lett 36:359–362
Tugarinov AI, Voitkevich GV (1970) Precambrian geochronology of the continents. Nedra, Moskow (in Russian)
Tugarinov AI, Bibikova EV (1978) Some examples of dating metamorphic rocks by U-Pb method. Open File Rep 78-701, USGS, pp 437–439
Tugarinov AI, Bibikova EV (1980) Geochronology of the Baltic Shield on zirconometry data. Science, Moscow (in Russian)
Vidal P, Blais S, Jahn BM, Capdevila R, Tilton GR (1980) U-Pb and Rb-Sr systematics of the Suomussalmi Archaean greenstone belt, eastern Finland. Geochim Cosmochim Acta 44:2033–2044
Volcanism of Archaean greenstone belts in Karelia (1981) Science, Leningrad (in Russian)

Age and Evolution of the Early Precambrian Continental Crust of the Ukrainian Shield

N. P. Shcherbak[1], E. N. Bartnitsky[1], E. V. Bibikova[2] and V. L. Boiko[1]

Abstract

There are three stages in the evolution of the early Precambrian in the Ukrainian shield.

The early Archaean stage is characterized by highly metamorphosed, mainly volcanogenic formations of basic and ultrabasic composition, intruded and reworked by tonalites and plagiogranites. Their age is 3700 ± 200 Ma.

In the composition of the Upper Archaean greenstone stage sediments play an important role along with volcanogenic rocks. The age of the metavolcanics is 3250 ± 140 Ma, that of synorogenic granites is 3000 ± 20 Ma and that of posterogenic granites is 2700 ± 100 Ma.

Early Proterozoic formations of the third stage are represented mainly by sedimentary rocks. They occur transgressively on the Archaean basement. The age of the third stage granites is 2000 ± 100 Ma.

After this stage, the Ukrainian shield became a stable platform.

Recent studies on the geochemistry, petrology, geochronology, and deep structure of the Ukrainian shield by geophysical methods and drilling made it possible to model the early Precambrian crustal evolution of the region in question. Formations older than 2600 Ma are referred to the early Precambrian. This age boundary (the southern part of the East European Platform) is in the zone of the Ingulo-Ingulets and Middle near-Dnieper regions. The boundary is marked by a transgressive deposition (unconformity) of the Krivoy Rog Series (Lower Proterozoic) on the greenschist complexes of the Upper Archaean.

On the basis of tectonic data from Precambrian regions of the Ukrainian shield one can distinguish six blocks or belts (see Fig. 1). As a rule these are separated from each other by deep fractures. Each of the belts has its specific geologic structure, such as stratigraphic sequences of volcanic and sedimentary complexes, conditions of metamorphism and others. Early Precambrian formations are abundant within the following three blocks: Middle near-Dnieper, Dniester-Bug, and Ros-Tikich. Within the other three blocks early Precambrian rocks are of subordinate significance.

---

1 Institute of Geochemistry and Physics of Minerals, Ukrainian Academy of Sciences, Kiev, 252680, USSR
2 Vernadsky Institute of Geochemistry and Analytical Chemistry, USSR Academy of Sciences, Moscow, 117334, USSR

Archaean Geochemistry (ed. by A. Kröner et al.)
© Springer-Verlag Berlin Heidelberg 1984

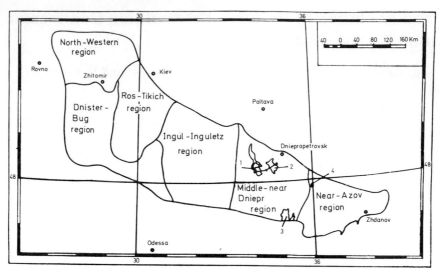

Fig. 1. Geological regions of the Ukrainian shield. *1* Verkhovtsevo area; *2* Sura area; *3* Belozerka area; *4* Novopavlovsk area

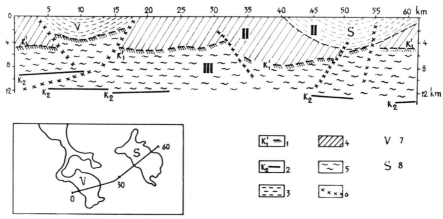

Fig. 2. Geologic-geophysical section. *1* geophysical horizon of layer **II**; geophysical horizon of layer **III**; *3* Konka-Verkhovtsevo series; *4* Dnieper granitoid complex; *5* Auly series; *6* fractures; *7* Verkhovtsevo area; *8* Sura area. Location of section see Fig. 1

Of all early Precambrian formations those of the Middle near-Dnieper belt have been studied in most detail. A geologic-geophysical section of this belt is shown in Fig. 2. There are three distinct layers separated by two seismic boundaries, K_1^1 and K_2 (Sollogub 1982). The upper layer consists of metamorphic rocks of the Konka-Verkhovtsevo series and granitoids of the Dnieper complex belonging to the Upper Archaean (2.6 – 3.0 Ga). The lower layer is generally represented by metamorphosed volcanogenic formations of the Lower Archaean (>3.0 Ga). Below the K_2 seismic boundary seismic velocities indicate a layer of mafic rocks unchanged by granitization processes.

The oldest formations of the Ukrainian shield, known as the Auly Series (Table 2), may be up to 3.8 Ga old and belong to the Lower Archaean section that occurs within the contact zone of the Middle near-Dnieper and near-Azov belts. According to geophysical data the zone is bounded by deep fractures. Within this zone orthogneisses and granitoids of the Upper Archaean are most abundant.

This area consists mainly of orthogneisses and granitoids of Upper Archaean age (see Fig. 3). It also contains metagabbroids, metamorphosed anorthosites and meta-ultrabasites that are intruded by tonalites. Bodies of basic and ultrabasic rocks up to 2 km long and 500 m thick have been intensely changed by superimposed secondary processes. High temperature amphibolite and low temperature granulite facies metamorphism have led to profound recrystallization, and granitization phenomena are also characteristic. The Auly Series section and its full composition have not yet been studied in detail but the (tectonic) thickness is estimated to be about 5000 m.

Within the Ukrainian shield Upper Archaean formations are more abundant than those of the Lower Archaean. The former have been identified within four belts or megablocks. Their composition is quite variable, and they are characterized by different grades of metamorphism (from greenstone to high temperature granulite facies).

The Middle near-Dnieper region, comprising the Konka-Verkhovtsevo series and Dnieper complex granitoids, is one of the typical Upper Archaean granite-greenstone terrains. Metamorphosed sedimentary and volcanogenic formations of the series are characterized by irregular distribution both vertically and laterally; their quantitative relative proportions are also variable. But the general trend here is an upward volumetric decrease of metavolcanics and an increase of metasedimentary rocks in individual belts.

A clear idea about the relationship between metavolcanic and metasedimentary rocks of the greenstone belts is given by a section through the Konka-Verkhovtsevo series in the Verkhovtsevo and Konka-Belozerka regions, the distance between them being about 100 km (see Fig. 4). Here the series is divided into the lower Konka suite and the upper Belozerka suite, and each of them, in turn, is subdivided into three subsuites. This permits a detailed stratigraphic comparison

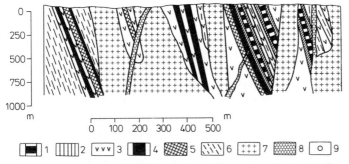

Fig. 3. Schematic geologic section. *1* pyroxenites; *2* anorthosites; *3* gabrros; *4* iron quartzites; *5* oreless quartzites; *6* garnet biotite, biotite-amphibole, amphibole gneisses; *7* biotite plagiogranites; *8* microcline granites; *9* sampling locality for rocks of Auly Series see Table 1 and Fig. 6

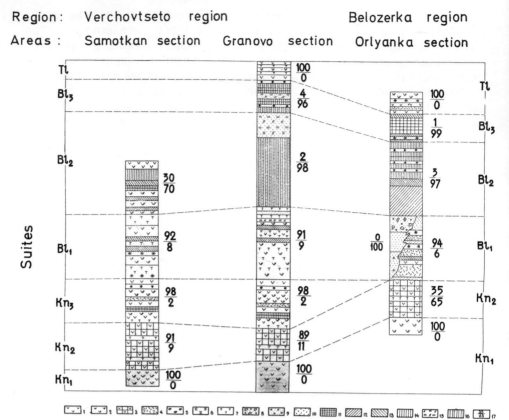

Fig. 4. Stratigraphic columns of the Konka-Verkhovtsevo series. *1* amphibolites, spilites; *2* foliated amphibolites, spilite schists; *3* amphibolite interlayered with iron hornfelds; *4* andesite porphyrites; *5* quartz keratophyres; *6* porphyroids; *7* talc-carbonate rocks, picrites, actinolites, tremolites, prochlorite schists; *8* breccia conglomerates; *9* arcose meta-sandstones and meta-aleurolites; *10* quartz meta-sandstones and meta-aleurolites; *11* iron quartzites; *12* quartz-sericite schists; *13* quartz-chlorite schists; *14* alternating quartz-sericite, quartz chlorite-sericite, quartz sericite-chlorite, and quartz chlorite schists; *15* meta-sandstones with schist intercalations; *16* schists with intercalations of meta-sandstones; *17* percentage of lithology (numerator − volcanogenic rocks, denominator − metasedimentary rocks); suites: *Kn* Konka suite; *Bl* Beloserka suite; *Tl* Teplovo suite

of spatially separated sections. This comparison shows that individual facies are not persistent across the entire region. Thus, one stratigraphic layer within a particular megablock may reflect intensive volcanic activity while some other area of the same stratigraphic level is characterized by sedimentary rocks. The thickness of the metamorphosed volcano-sedimentary series of the Upper Archaean reaches 6 − 7 km.

Petrochemical features of the metavolcanic rocks occurring within the Archaean greenstone belts are evident from the AFM diagram of Fig. 5. The data illustrate the variable and differentiated chemical composition of these formations. Metagabbros and anorthosites of Lower Archaean age reveal tholeiitic affinities while the meta-ultrabasites along the komatiitic series.

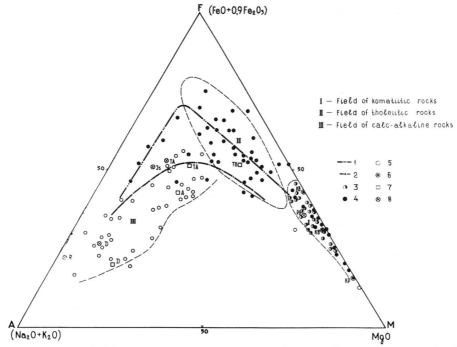

Fig. 5. Petrochemical diagram. *1* trend for tholeiitic suite of the Hawaiian Islands; *2* separation line for tholeiite and calc-alkaline compositions; *3* ultrabasic rocks of the Auly Series; *4* points for metavolcanics of basic composition referring to the Konka-Verkhovtsevo Series; *5* points for metavolcanics of intermediate and acidic composition for the Konka-Verkhovtsevo Series; *6* mean composition of peridotite (*KP*) and basaltic komatiite (*KB*) according to Viljoen and Viljoen (1969); *7* mean composition of tholeiitic basalts (*TB*), tholeiitic andesites (*TA*), andesites (*A*) and dacites (*D*) according to Jakes and White (1972); *8* mean composition of tholeiitic andesites (*TA*), islandites (*Is*), dacites (*D*), rhyolites (*R*) and picrite basalts (*PB*) according to Irvine and Baragar (1971)

The granitoid complexes of Archaean and Proterozoic age have been described by many authors (Polovinkina and Polevaya 1973; Semencnko et al. 1977; Shcherbak et al. 1981). The granitoids differ by their initial $^{87}Sr/^{86}Sr$ ratios that vary between 0.701 and 0.708. More detailed data on isotopic ages of metamorphic rocks and granitoids of the Ukrainian shield are given in a recent review (Shcherbak et al. 1981).

Some new zircon isotope data on Archaean rock types are now available and provide ages for the oldest ultrabasites that are associated with Lower Archaean tonalites and meta-anorthosites discussed above.

Meta-ultrabasites, tonalites and meta-anorthosites each contain a definite variety of zircons. Dark brown, nearly black, non-transparent, irregularly shaped grains of zircon (2 mm in diameter) are characteristic of the ultrabasites; elongated, light brown, low transparent grains measuring 0.1 – 0.2 mm in diameter are identified in tonalites; isometric, transparent pink-red grains with brilliant lustre measuring 0.3 – 0.4 mm in diameter are associated with the meta-anorthosites. The relationships between these varieties of zircon are revealed by micro-

Table 1. U-Pb Analytical data for zircons from core samples of the Auly Series

Sample no.	Rock	Zircon fraction, specific gravity (d), fraction size	Concentration (ppm)		Atom (%)					Age (Ma)		
			Pb	U	^{204}Pb	^{206}Pb	^{207}Pb	^{208}Pb	$^{207}Pb/^{206}Pb$	$^{206}Pb/^{238}U$	$^{207}Pb/^{235}U$
1	Pyroxenite	composite	590	840	0.013	74.060	20.346	5.580	3320	3043	3214
1A	Pyroxenite	(d) −4.6 − +4.2	650	980	0.005	74.691	20.261	5.036	3310	2929	3160
1B	Pyroxenite	(d) +4.6	250	370	0.005	73.001	17.443	9.555	3110	2918	3032
1C	Pyroxenite	(d) −4.2	770	1300	0.010	76.624	18.923	4.440	3155	2727	2981
2A	Pyroxene-chlorite-amphibole-phlogopite rock	−190 μ	140	190	0.027	68.148	17.378	14.447	3190	2950	3095
2B	Pyroxene-chlorite-amphibole-phlogopite rock	+125 μ	240	370	0.015	72.627	18.107	9.251	3165	2808	3022
2C	Pyroxene-chlorite-amphibole-phlogopite rock	−125 μ	350	550	0.018	75.368	18.981	5.632	3180	2849	3048
2D	Pyroxene-chlorite-amphibole-phlogopite rock	+190 μ	500	730	0.026	74.778	20.524	4.666	3310	3001	3190

Decay constants: ^{238}U $1.55125 \cdot 10^{-10}$ yr^{-1}, ^{235}U $9.8485 \cdot 10^{-10}$ yr^{-1}
Atomic abundances: $^{238}U/^{235}U = 137.88$
Correction for order Pb: $^{204}Pb : ^{206}Pb : ^{207}Pb = 1 : 11.15 : 13.00$

scopic studies. The core of the second and third generations contain dark brown zircons characteristic of pryroxenites; the third variety is often observed as envelopes around the first and second varieties. The first and second varieties of zircon are characterized by uranium contents of 0.06 − 0.01% and hafnium contents of 1.4 − 1.1%.

Within ultrabasites and tonalites the zircon is often associated with muscovite.

Zircon from pyroxenite and from pyroxene-amphibole-phlogopite rock collected from the same bore hole (Fig. 3) have been dated by the U-Pb method. The resulting isotopic ages are listed in Table 1. Zircons were separated into fractions according to their specific gravity and grain size. For these fractions U-Pb isotopic analyses were performed according to the method of Manhes et al. (1978). U and Pb were determined on zircon separates by isotope-dilution mass spectrometry using a TSN-206-type mass spectrometer Cameca. Accuracy of the concentration determinations is estimated at 1.0%. Isotopic measurements of Pb were made by the silica gel technique. The accuracy of the measurements is ±0.15%. All age values calculated on the basis of different isotopic ratios are discordant.

U-Pb data on zircon fractions from the altered ultrabasic rock (Table 1, samples IIA − IID) are shown on the concordia plot of Fig. 6 (circles). The intersection of discordia with concordia indicates ages of 3605^{+180}_{-150} and 2290^{+150}_{-130} Ma (2 σ errors), respectively. The age of 2290^{+150}_{-130} Ma coincides with that obtained by the Rb-Sr whole rock isochron method on metavolcancis of the Konka-Verkhovtsevo Series and is interpreted as the age of a secondary thermal process caused by granite and basalt dyke intrusion (Shcherbak et al. 1981).

The points for U-Pb isotopic ratios on specific gravity fractions of zircons from pyroxenite (samples I − IC) scatter considerably about their calculated regression line (dots in Fig. 6). The upper intercept of the regression line for these four points correspond to an age of 3680^{+210}_{-150} Ma. Maybe these zircon fractions are characterized by considerable heterogeneity.

The concordia upper intercept age of the Konka-Verkhovtsevo metavolcanic Series, based on three highly discordant points, is 3250^{+290}_{-140} Ma (Shcherbak et al.

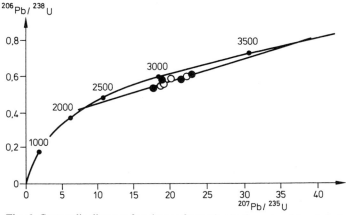

Fig. 6. Concordia diagram for zircons from ultrabasic rocks of the Auly Series. *Circles* are altered ultrabasic rocks, *dots* are pyroxenites

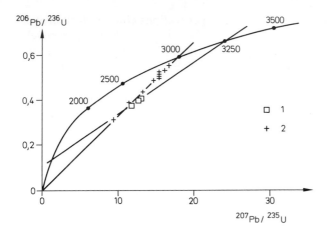

Fig. 7. Concordia diagram for zircons from metavolcanics of the Konka-Verkhovtsevo Series (*1*) and granitoids of the Dnieper complex (*2*)

1982). The corresponding graph (open squares in Fig. 7) is characterized by considerable discordance caused by both superimposed processes and recent disturbances within the uranium-lead system. The same samples of metavolcanics were also dated by the Rb-Sr whole rock method and yielded an estimated age of 2300 Ma (Shcherbak et al. 1981). The lower intercept of the discordia line with concordia (800^{+470}_{-530} Ma) does not seem to reflect a geological event.

As far as zircon dating is concerned granitoids of the Ukrainian shield have been studied in most detail (Shcherbak et al. 1981). The results of isotopic dating of major Archaean granitoid complexes are shown as crosses in Fig. 7.

Archaean granitoids are abundant within the Middle near-Dnieper belt and define a linear array on the concordia diagram with an upper intercept age of 3000 ± 20 Ma (Bibikova et al. 1983).

We must give special consideration to the dating of enderbite (hypersthene-bearing) and charnockite (pyroxene-bearing K-Na granites), which are abundant in the Dniester-Bug region. On the basis of geological data the enderbites have been proved to be older than the charnockites. Extensive areas of enderbitic formations (were subjected to some later metasomatic reworking which has manifested itself in newly formed K-feldspar. Zircons genetically related to the enderbites and charnockites are strikingly different (Lesnaya 1981). Accessory monazite is also genetically associated with charnockites. Enderbite and charnockite zircon ages have also been determined by the U-Pb method (see Fig. 8).

Different zircon fractions (Bibikova et al. 1982) from enderbites (generation I) define a discordia line whose upper intercept with concordia corresponds to an age of 3020^{+20}_{-30} Ma while the lower intercept defines an age of 710^{+300}_{-320} Ma (circles in Fig. 8).

The discordia line defined by data points for zircons from the charnockites intersects concordia at 2110 ± 20 Ma (dots in Fig. 8).

It is evident from these data that enderbites and charnockites do not define a single enderbite-charnockite complex. It should be noted that the enderbites are synchronous with the early stage of granulite metamorphism of the Dniester-Bug series. The charnockites, on the other hand, formed synchronously with the second granulite phase during the Lower Proterozoic.

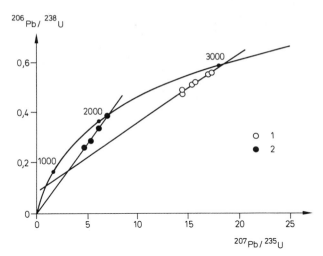

Fig. 8. Concordia diagram for zircons from enderbites (*1*) and charnockites (*2*)

Between the first and the second stages of metamorphism an event of dyke magmatism took place. During the Proterozoic dykes of basic composition were affected by a single episode of metamorphism in conditions of granulite facies.

Thus, the geological history of the Ukrainian shield during the early Precambrian can be divided into three major stages: Lower Archaean, Upper Archaean, and Lower Proterozoic (see Table 2). The early Archaean stage is characterized by the formation of predominantly volcanogenic formations of basic and ultrabasic composition. Maybe these formations mark the beginning of the Auly Series section, the upper age boundary of which has not yet been established.

The Upper Archaean stage is characterized by the accumulation of volcanogenic sedimentary series within the Middle near-Dnieper and Dniester-Bug megablocks. These series consist largely of differentiated sedimentary formations. Discontinuity of sections and abrupt change of facies is a feature of these series; this may be related to the limited size of the original sedimentary basins. The rocks of the different megablocks are characterized by variable grades of metamorphism. Thus, the rocks of epidote-amphibolite facies are characterized by intensive granitoid magmatism. Within the area of granulite facies enderbites are widespread. Granitoids of the Dnieper complex in the Middle near-Dnieper and enderbites in the Dniester-Bug megablocks formed synchronously at about 3.0 Ga ago. They define the end of Archaean crustal consolidation within all megablocks of the Ukrainian shield.

Early Proterozoic series of the third stage are abundant within the Ukrainian shield except in the Middle near-Dnieper megablock. They overlie Archaean formations with angular and stratigraphic discontinuity and are represented by the Teterev, Krivoy Rog, Bug, Ingul-Ingulets, and Central near-Azov Series. These series are all characterized by volcanogenic-sedimentary formations.

The rocks of this stage have been subjected to metamorphism (epidote-amphibolite to granulite facies). The age boundary of 2000 ± 100 Ma is characterized by intensive ultrametamorphism resulting in the formation of granitoids of the Kirovograd-Zhitomir complex and their analogues. The age boundary also

Table 2. Chronology of the early Precambrian continental crust of the Ukrainian shield

Strati-graphical index	Typical series of metamorphic rocks and granitoid complexes	Geochronological boundaries (Ma)	Duration of forma-tion (Ma)	Predominant petrological composition of series and complexes	Geophysical boundaries, their indexes and transverse wave velocity (V), km/s	Predominant facies of meta-morphism	Section level	Notes
Pt₁	Krivoy Rog Series	2000 ± 100	600	Sedimentary	I	Greenschist	At the outcrops	2000 Ma granitization
	Kirovograd-Zhitomir complex	Unconformity		K-granites	$V = 6.0$	The lower boundary is unknown		
		2600 ± 100			K_1			
Ar₂	Konka-Verkhov-tsevo series		600	Sedimentary volcanogenic	II $V = 6.0 - 6.3$	Greenschist, amphibolite and granu-lite (ender-bite)	At the outcrops	2600 Ma granitization 3000 Ma granitoid magmatism
	Dnieper complex			Na and K-Na granites	K_1^1			3200 Ma volcanism
		3200 ± 100						
Ar₁	Auly Series	3700 ± 200	500	Basic, ultra-basic rocks and tonalites	III $V = 6.2 - 6.5$ K_2	Amphibolite and granu-lite	Relicts within layer II	Magmatism 3600 Ma or older

marks the final consolidation of the greater part of the Ukrainian shield into a stable platform.

References

Bibikova EV, Kirnozova TI, Gracheva TV (1983) Isotopic age of the ancient granitic rocks of the Near-Dnieper area. Geokhimia 7:997 – 1004 (in Russian)

Bibikova EV, Lesnaya IM, Gracheva TV, Makarov VA (1982) The isotopic age of enderbites from the Pobuzhye. Dokl Ac Sci SSSR 263 (1):159 – 163 (in Russian)

Irvine TN, Baragar WRA (1971) A guide to the chemical classification of common volcanic rocks. Can J Earth Sci 8:523 – 548

Jakes P, White AGH (1972) Major and trace element abundances in volcanic rocks of orogenic areas. Geol Soc Am Bull 83:29 – 40

Lesnaya IM (1981) The oldest enderbites from the Pobuzhye. Dokl Ac Sci Ukr SSR 2:28 – 31 (in Russian)

Manhes G, Allègre CJ, Dupré B, Hamelin B (1979) Lead-lead systematics, the age of the earth and the chemical evolution of our planet in a new representation space. Earth Planet Sci Lett 44:91 – 104

Polovinkina JIr, Polevaya NI (1973) The Ukrainian crystalline shield. In: Geochronology of the USSR, I. The Precambrian. Nedra, pp 83 – 104 (in Russian)

Semenenko NP, Boiko VL, Orsa VI, Ladieva VD, Yaroshchuk EYu, Bartnitsky EN, Poletaeva LN (1977) Dating of Precambrian metamorphic and magmatic processes of the Middle near-Dnieper area of the Ukrainian shield. Geol J 37:(2)3 – 23 (in Russian)

Sollogub VV (1982) The Earth's crust in the Ukraine. Geophys J 4(4):3 – 25 (in Russian)

Shcherbak NP, Bartnitsky EN, Lugovaya IP (1981) Isotopic geology of the Ukraine. Kiev, Naukova Dumka (in Russian)

Shcherbak NP, Bibikova EV, Zhukov GV, Makarov VA (1982) Isotope dating of palaeovolcanites from the Konka-Verkhovtsevo series of the Middle near-Dnieper area (the Ukrainian shield). Dokl Ac Sci Ukr SSR 11:25 – 28 (in Russian)

Viljoen MG, Viljoen RP (1969) The geology and geochemistry of the lower ultramafic unit of the Onverwacht Group and a proposed new class of igneous rocks. Geol Soc S Afr Spec Publ 2:55 – 86

Significance of Early Archaean Mafic-Ultramafic Xenolith Patterns

A. Y. GLIKSON[1]

Contents

1 Introduction 263
2 Regional Xenolith Patterns 263
3 Xenoliths and the Structure of Granite-Greenstone Terrains 274
4 Summary 279
References 280

Abstract

The distribution patterns of mafic-ultramafic xenoliths within Archaean ortho-gneiss terrains allow subdivision, and thereby recognition of the internal geometry, of batholiths. Principal regional to mesoscopic-scale characteristics of xenolith swarms are outlined from the Pilbara Block (Western Australia) and southern India. Transitions along strike and across strike between stratigraphically low greenstone sequences and xenolith chains establish their contemporaneity. A complete scale gradation exists between greenstone synclines and outcrop-scale xenoliths, defining early greenstone belts as "mega-xenoliths". Xenolith distribution patterns in arcuate dome-syncline gneiss-greenstone terrains define subsidiary gneiss domes within the batholiths. The arcuate terrains represent least-deformed cratonic "islands" within otherwise penetratively foliated gneiss-greenstone crust. The oval gneiss structures evolved by magmatic diapirism followed by late-stage solid-state uprise. Tectonized boundary zones of batholiths contain foliated gneiss-greenstone intercalations derived by deformation of xenolith-bearing intrusive contacts, a process involving inter-thrusting and refolding of plutonic and supracrustal lithologies. The exposure of high grade metamorphic sectors has been related to uplift of deep seated batholithic sectors along reactivated boundaries. The transition from granite-greenstone terrains into gneiss-granulite suites involves a decrease in the abundance of supracrustal enclaves and an increased strain rate. Late Archaean greenstone sequences may locally overlap older gneiss terrains and their entrained xenolith systems unconformably. The contiguity of xenolith patterns suggests their derivation as relics of regional mafic-ultramafic volcanic layers and places limits on horizontal movements between individual crustal units. Combined isotopic-palaeomagnetic studies of Archaean xenoliths may be able to provide limits on theories of early crustal development.

1 Division of Petrology and Geochemistry, Bureau of Mineral Resources, Geology and Geophysics, P.O. Box 378, Canberra, ACT 2601, Australia

Archaean Geochemistry (ed. by A. Kröner et al.)
© Springer-Verlag Berlin Heidelberg 1984

1 Introduction

Some of the oldest rock units identified in Archaean shields occur as supracrustal enclaves within orthogneiss-dominated suites, forming either discrete greenstone belts or discontinuous xenolith swarms. The latter occur in both low-grade granite-greenstone terrains and high-grade gneiss-granulite terrains. Little-deformed intrusive contact zones of diapiric tonalite-trondhjemite plutons contain evidence for the magmatic history of xenoliths, including their progressive injection, disintegration and assimilation by granitic magma. However, in strongly deformed sectors of batholiths these relations are masked where the gneiss and the supracrustals are penetratively co-foliated. Major batholith-greenstone boundary zones often constitute loci of complex tectonic reactivation involving inter-leaving of gneiss and foliated greenstone slices, masking primary granite-greenstone relations.

Examples of early supracrustal enclaves include the ca. 3.8 Ga old Isua outlier and contemporaneous Akilia xenolith association in southwest Greenland (Allart 1976; McGregor and Mason 1977), the pre-3.6 Ga old Uliak swarm in the Nain Complex, Labrador (Collerson and Bridgwater 1979), Dwalile supracrustal remnants in the ca. 3.5 Ga old Ancient Gneiss Complex of Swaziland (Anhaeusser and Robb 1981), pre-3.5 Ga old Sebakwian swarms in the Selukwe and Mashaba areas of Zimbabwe (Stowe 1973), pre-3.4 Ga old. Sargur Group enclaves in the amphibolite to granulite facies gneisses of southern Karnataka (Janardhan et al. 1978) and greenstone xenoliths and intercalations within Pilbara batholiths in Western Australia (Hickman 1975, 1983; Bettenay et al. 1981). This paper examines the regional distribution patterns of some of these xenolith/enclave systems with emphasis on examples from the Pilbara Block and southern Karnataka. The principles underlying these patterns and implications of their geometry to the structure and evolution of the granite-greenstone systems are discussed. It is suggested that the unravelling of xenolith distribution is central to an understanding of the relations between the plutonic and supracrustal components of Archaean terrains and thereby of early crustal evolution.

2 Regional Xenolith Patterns

Because of their incorporation in tonalitic and trondhjemitic gneisses, which in many terrains are more deeply weathered than potassic granites and greenstone belt lithologies (Glikson 1978; Robb 1979), the distribution of xenoliths is generally not well charted. However, in well-exposed gneiss terrains, and where greenstone enclaves are large enough to form resistant outcrops, the xenoliths are well documented. From examples below it is evident that, far from being distributed at random, the enclaves bear systematic relations to associated greenstone synclinoria and to foliation within the enveloping gneiss. Structurally conformable relations between these elements are usually dominant, although locally foliation may intersect xenoliths at low angles. Such relationships apply on all scales, from regional to detailed outcrop patterns.

Some of the most instructive xenolith patterns have been reported from the Selukwe-Gwenoro Dam area, southern Zimbabwe (Stowe 1973), where complete across-strike and along-strike transitions take place between the Selukwe and Ghoko schist belt and mafic-ultramafic xenolith swarms. Where xenoliths abound migmatitic and strongly deformed gneiss dominate, whereas where xenoliths decrease in abundance less foliated tonalitic domes are defined, constituting magmatic cores within the gneiss. An abundance of migmatite along xenolith-gneiss contacts, including progressive disintegration and assimilation of the mafic-ultramafic material, attests of the dominance of high-temperature magmatic processes. Possibly the enhanced migmatitization is related to dehydration of xenoliths and consequent depression of the tonalite solidus. The contact metamorphic grade of the supracrustal rocks increases from greenschist facies in intact schist belts to amphibolite facies along intrusive contacts and xenoliths surrounding the tonalitic cores. Supracrustal xenoliths may be retrogressed along their margins. An essential structural conformity is observed between gneiss and greenstone elements of the Gwenoro Dam complex, including the Somabula tonalite, Natale pluton, and Eastern Granodiorite plutons, regarded as the larger intrusive cores of the gneisses. The oldest gneiss phases are typically uniform, fine-grained sodic varieties, whereas younger felsic magma fractions are more fractionated, coarser-grained rocks (Stowe 1973). Progressive anatexis and digestion of mafic-ultramafic volcanic material was achieved by lit-par-lit injection of near-liquidus tonalitic magma. Detailed petrological and geochemical studies of this key occurrence are required to elucidate the assimilation processes.

It has been suggested that mafic-ultramafic greenstone belt type materials constituted sources of tonalitic-trondhjemitic partial melts which constitute the bulk of the Archaean batholiths (Arth and Hanson 1975; Glikson and Sheraton 1972; Glikson 1979). If so, the nature of xenolith screens may furnish a key for processes whereby sialic segments segregated from mafic-ultramafic crust. The classic outcrops of tonalite-intruded Onverwacht Group in the southwestern part of the Barberton Mountain Land, Transvaal, furnish some of the clearest observations in this regard (Viljoen and Viljoen 1969). In this area continuous transitions occur along strike from the main greenstone synclinoria into xenolith screens within the gneisses, for example a greenstone tongue between the Stolzburg and Theespruit tonalitic plutons (Robb 1981). This author distinguished three sets of relations in migmatites asscociated with mafic-ultramafic screens in this area:

1. Agmatites exhibiting clear intrusive relations between tonalite-throndhjemite and xenoliths, as indicated by lit-par-lit injection, discordant intrusive tongues, mechanical desegregation and progressive assimilation features. The nature of xenolith fragmentation and digestion depends directly on the original rock type, e.g. felsic tuff units are more heavily veined than mafic rocks, becoming progressively more difficult to distinguish from the intrusive gneiss.
2. Remobilized composite units composed of partly desegregated and assimilated mafic-ultramafic material intercalated with gneiss and sheared to obliteration of the original relations between these components.

3. Small deformed basic dykes showing intrusive relations with the gneiss. The dykes appear to be minor since they are common neither in least-deformed supracrustal sequences nor in tonalite gneiss domes away from the main xenolith concentrations.

In some instances deformation of agmatites ensued in remobilisation of less viscous felsic components. The occurrence of angular xenolith fragments and assimilated mafic schlieren in close proximity to each other suggests a differential viscous flow, where parts of the agmatite froze in situ, while other parts were subjected to continuous flow-deformation and progressive magmatic digestion. The bulk of the intrusive gneisses are of Na-rich composition, although K-rich bands and zones are also present.

Mafic-ultramafic screens often outline a ghost stratigraphy contiguous with the greenstone belt with which they merge. This feature was pointed out by Viljoen and Viljoen (1969) in connection with the Rhodesdale batholith, Zimbabwe, where basal units of the Sebakwian Group, including ultramafic rocks, tholeiitic basalts and fuchsite schist, are identified within gneiss along the periphery of the pluton. Preservation of original relationships between the supracrustal and the plutonic units depends on (1) the primary geometry of their contact and (2) subsequent deformation history of the boundary zone. Where the strike of the greenstones is at high angles to the intrusive contact, originally embayed intrusive contacts are commonly well retained, probably owing to the mechanical strength of such interlocked boundaries. An example of such contact occurs along the northwestern flank of the Mount Edgar batholith, Pilbara Block, Western Australia (Figs. 1a – c). In this area transgressive contacts, detached greenstone xenoliths and thermal metamorphic aureoles indicate primary intrusive relations between trondhjemitic gneiss and greenstones of the Talga Talga Subgroup. Intrusive relations are also observed along contacts of other Pilbara batholiths, for example the northern flank of the Shaw batholith (Bettenay et al. 1981). By contrast, where the boundary of batholiths and the strike of greenstones are parallel, primary contact relations are generally obscured due to the susceptibility of these lithological discontinuities to late magmatic and post-consolidation reactivation and to multiple solid state deformation.

The significance of the xenolith patterns extends beyond contact zones and into the interior of composite batholiths. Anhaeusser and Robb (1980) and Robb (1981) used these patterns to delineate a number of plutons and cells southwest of the Barberton Mountain Land. Hickman (1975, 1981) showed that the Mount Edgar batholith can be divided into at least five oval gneiss domes defined by foliations and intervening amphibolite screens which, in places, merge with greenstones of the Talga-Talga Subgroup (Figs. 1a – b). The Shaw batholith contains a high proportion of interleaved mafic-ultramafic material, much of it highly deformed, rendering identification of primary gneiss units difficult. A detailed study of the western part of this batholith (Bickle et al. 1980; Bettenay et al. 1981) unravelled multiple post-consolidation deformation affecting tectonically interleaved gneiss-greenstone units. At least three phases of deformation have been identified in this area, including synmetamorphic horizontal thrusting and recumbent folding (Bickle et al. 1981). Fold axes and foliations in the gneiss are

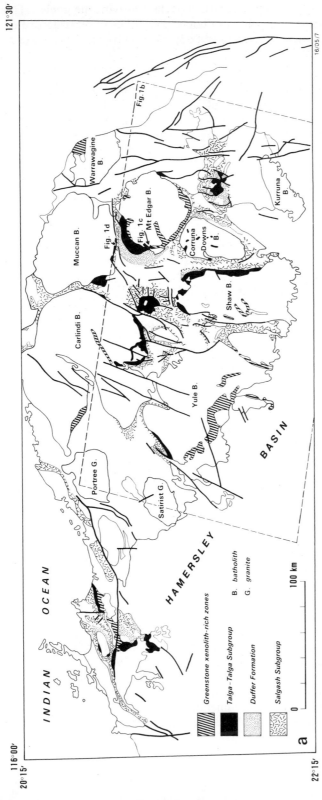

Fig. 1a–d. a Geological sketch map of the Pilbara Block, Western Australia, showing locations of **b**, **c**, and **d**. (After Hickman 1983). **b** Computer enhanced LANDSAT imagery of the eastern and central parts of the Pilbara Block, Western Australia. For location and correlation with geology refer to **a**. *Solid arrows* point to transitions between greenstone tongues and xenolith chains; *large open arrows* point to sheared gneiss-greenstone contact zones; *small open arrows* point to contacts where intrusive relations are well retained (Courtesy of Dr. J. Hantington CSIRO Division of Mineral Physics). **c** A transition from a greenstone tongue of the Talga Talga Subgroup (Mount Ada Basalt) into a deformed xenolith swarm east of Marble Bar. For location refer to **a**. *T* Talga Talga Subgroup; *D* Duffer Formation; *N* gneiss (Mount Edgar batholith); *NX* greenstone xenoliths in gneiss. **d** A deformed gneiss-greenstone contact zone along the southern rim of the Muccan batholith, northeast of Marble Bar. For location refer to **a**. *V* basalt; *U* ultramafics; *d* dolerite in Duffer Formation; *C* chert; *S* sediments; *X* xenoliths in gneiss; *N* gneiss
(Figs. 1b–d see next pages)

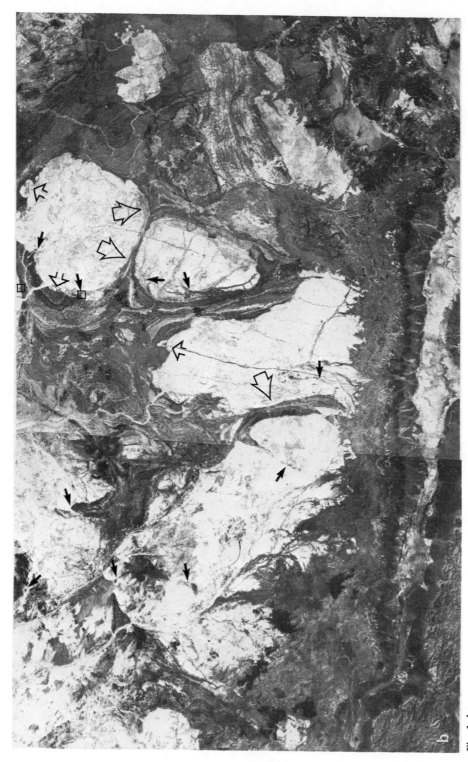

Fig. 1, b

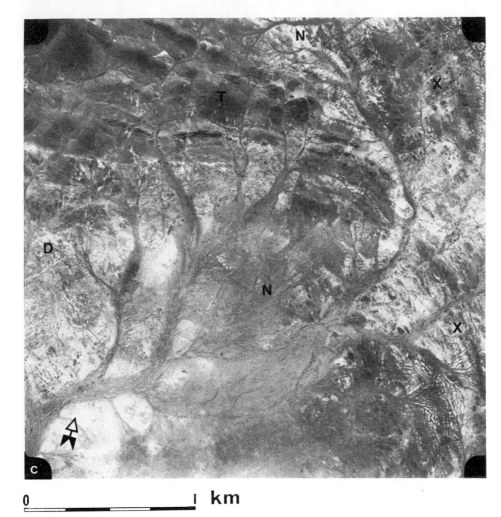

0 _____ı km

Fig. 1, c

oriented at low to intermediate angles to the main gneiss-greenstone belt
boundary. The upper amphibolite facies metamorphism of some intercalated
supracrustal material suggests considerable uplift relative to the juxtaposed
Tambourah greenstone belt. Although intrusive gneiss-greenstone relations were
recorded (Bickle et al. 1981), Bettenay et al. (1981) argue against a view of the
greenstone intercalations as sheared xenoliths, pointing out the absence of assimilation features. These authors favour an interpretation in which at least the
older gneiss phases are remnants of a hypothetical sialic basement to the greenstones. This model is subject to two types of problems:

1. the abundance of mafic-ultramafic interlayers within the gneiss requires the
   existence of penetrative thrust fault systems on the finest (outcrop) scale, in-
   terleaving lithological units which were originally located apart;

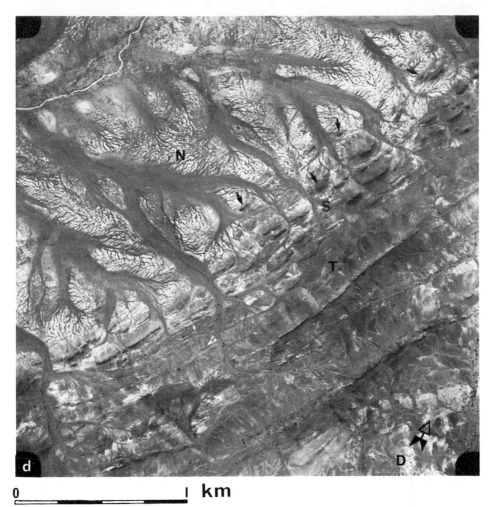

0 | km

Fig. 1, d

2. according to this model the fine interlayering of gneiss and greenstone slices must have involved extensive remelting of an older gneiss. However, anatexis of granodiorite and tonalite compositions must produce abundant K-rich neosome granitic fractions. While the latter are locally widespread, the bulk of the gneisses are of granodiorite-trondhjemite composition and thus unlikely to be products of sialic anatexis. An alternative view of the interleaved gneiss-greenstone suite, supported here, regards it as a deformed and metamorphosed derivative from xenolith/enclave-bearing intrusive contract zones similar to those retained along the northern and eastern parts of the Shaw batholith. Similar superposition of penetrative deformation on intrusive boundaries is observed along the southern margins of the Mount Edgar and the Muccan batholiths (Figs. 1a, b, d). These zones, which range in width

from several hundred metres to over 1 km, consist of gneiss, amphibolite and chlorite schist intercalated on a cm to metre scale. Local remelting resulted in subconcordant apophyses protruding into the schists.

The nature of well preserved gneiss-greenstone contact zones and xenolith screens in the Pilbara generally suggests a lesser degree of magmatic digestion than in the Zimbabwe and Barberton examples cited above. Thermal metamorphism of contact aureoles and xenoliths is mostly up to lower amphibolite facies but rarely attains pyroxene hornfels facies. Mechanical disintegration accompanied by low grade thermal effects dominate, while assimilation and hybridization, although locally observed, are less widespread. This is consistent with a lower temperature liquidus of Pilbara granodioritic-trondhjemitic compositions, as compared to the tonalitic compositions of South African gneisses. Further, since late-stage emplacement of Pilbara batholiths was associated with solid-state deformation of upper and marginal solidified sectors, retrogression and deformation have served to obscure original igneous features.

The South Indian Peninsular gneiss complex in the vicinity of Karnataka greenstone belts contains hosts of mafic-ultramafic enclaves (e.g. Radhakrishna 1975) (Fig. 2a). Some of the best examples occur around the Holenarsipur greenstone belt (Hussain 1980; Naqvi 1971) (Fig. 2c) and in the Sargur area (Janardhan et al. 1978) (Fig. 2d). Deformed ultramafic xenoliths and minor pelitic and anorthositic xenoliths of amphibolite facies abound east and west of the Holenarsipur belt. Some xenoliths are several km long, constituting mini-greenstone belts oriented parallel to the main belts. Thus, an entire hierarchy of sizes exists between the Holenarsipur belt and cm-scale xenoliths. Examples of xenolith types within the Peninsular Gneiss are shown in Fig. 2e. Quarry exposures and continuous sections along irrigation canals allow detailed observations of both intrusive and tectonic relationships between gneiss, greenstone and xenoliths. In places the gneisses are finely interleaved with greenstone intercalations. The density and size distribution of enclaves in the Hamavati River area are as high as to question any spatially distinct definition of greenstone belts. Thus, the main Holenarsipur, Nuggihalli and Sandur belts can be regarded as the largest in a range of supracrustal enclaves which pervade the tonalitic gneisses on every scale. This relation can be observed on LANDSAT imagery, on map scale and in individual outcrops. The LANDSAT imagery of southern Karnataka reveals megascale zonation of felsic and mafic bands whose width is on the scale of tens of km, showing through alluvial cover in little exposed areas (Fig. 2b). The similarity between this pattern und outcrop-scale banding (Fig. 2e) is striking. This terrain grades into charnockites of the Eastern Ghats metamorphic belt. Around Sargur hosts of high-grade enclaves of pyroxene granulite, hornblendite, ultramafics, quartzite, marble and pelitic schist pervade granulite facies orthogneiss, ultramafics being volumetrically the most important constituent (Janardhan et al. 1978) (Fig. 2d). The size and continuity of supracrustal enclaves diminish upon transition into the granulite facies terrain, arcuate belt patterns are almost absent, consistent planar structures dominate, and original lithologies become difficult to identify. Nevertheless, syncline keels and supracrustal xenoliths are retained, demonstrating that the granulite facies metamorphism was superposed

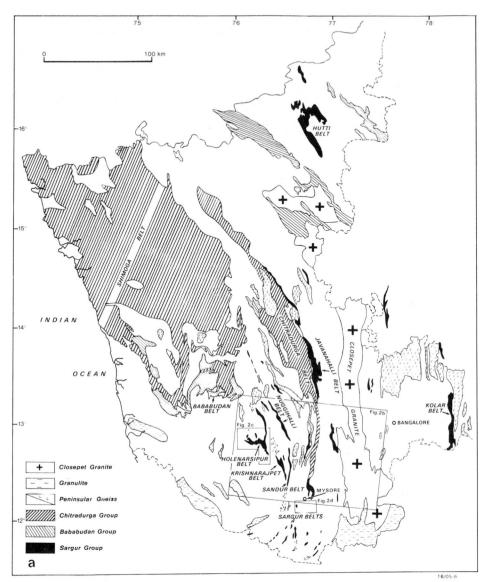

a

16/05·6

Fig. 2a – e. a Geological sketch map of Karnataka State, southern India, indicating locations of **b – d**.
b Part of a computer-enhanced LANDSAT image of southern Karnataka. For location and correlation with geology refer to **a**. Note the light-dark meridional banding of gneiss and amphibolite-rich
zones northwest and northeast of Mysore. (Courtesy of Dr. J. Huntington, CSIRO Division of Mineral Physics). **c** Geological sketch map of part of the Holenarsipur greenstone belt, Karnataka. (After
Naqvi 1981). **d** Amphibolite to granulite facies supracrustal xenoliths in the Sargur area, southern
Karnataka (after Janardhan et al. 1978). *Solid areas* ultramafic-dominated metamorphics; *oblique
hatching* pyroxene granulites; horizontal hatching-pelitic schist; *dots* quartzite; *brick pattern* marble;
dotted lines dykes. **e** Amphibolite xenoliths within the Peninsular Gneiss Complex, southern India.
a – c Gavigudda quarry, 9 km SE of Bangalore; **d** folded interlayered gneiss-amphibolite near Brena-
van dam, Mysore. Scale for **c** and **d**: *hammer* is 30 cm long; scale for *a* and *b*: *pocket knife* (*arrowed*)
is 8 cm long. Note the progression from irregularly shaped, deformed and veined amphibolite xenoliths (**a, b**) to interfoliated, penetratively deformed mafic-felsic banded rocks (**c, d**). Note also the
similarity between the outcrop-scale banding and megascale banding on the LANDSAT imagery (**b**).
Xenolith-gneiss relations in **a** and **b** suggest they could be deformed dykes
(Figs. 2, b – e see next pages)

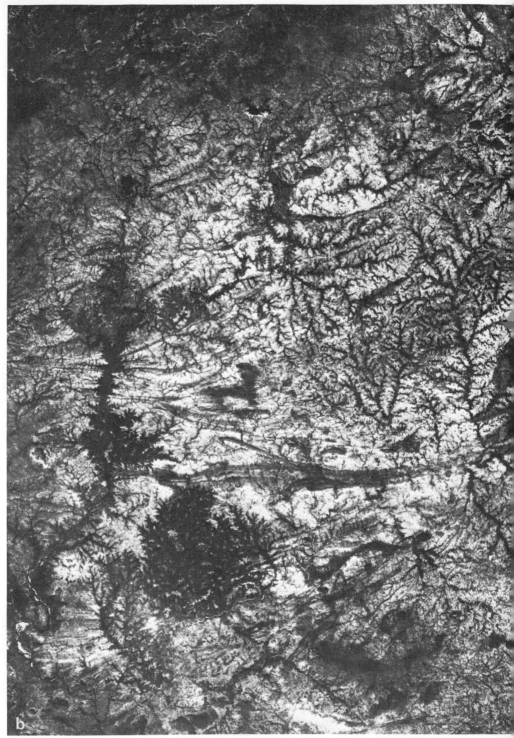

Fig. 2, b

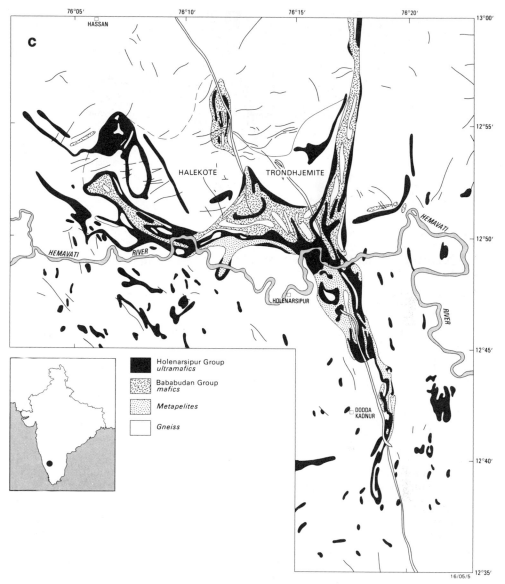

Fig. 2, c

on older granite-greenstone suite. The decreasing abundance of metamorphosed supracrustal material in the high-grade terrain is consistent with its interpretation as a deeper crustal zone and with the synclinal structure of the supracrustal belts. Given the northward structural tilt of the Indian Shield (Pichamuthu 1968), variations in abundance, size, and texture of the xenoliths with metamorphic grade allow their use as markers in the study of vertical crustal zonation.

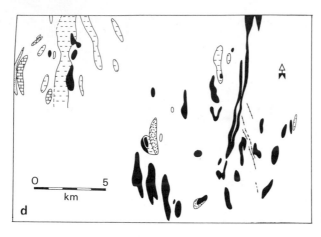

0   5
km

d

Fig. 2, d

3 Xenoliths and the Structure of Granite-Greenstone Terrains

Two contrasted concepts have been advanced regarding the structure of granite-greenstone terrains and the relations between their components. One school views supracrustal belts as discrete depositories formed in subsiding or down-faulted zones above an older sial basement (Baragar and McGlynn 1976; Binns et al. 1976; Archibald et al. 1978, 1981; Bettenay et al. 1981, *see also Groves and Batt, this Vol., eds.*). Proponents of this approach refer to older cratonic blocks as potential basement for the greenstones, e.g. the ca. 3.8 Ga old Sand River gneisses, Limpopo belt (Barton 1981), or the 3.6 – 3.0 Ga old western gneiss terrain, Yilgarn Block (Gee et al. 1981). An alternative concept hinges on (1) the commonly intrusive granite-greenstone relations and (2) the geochemical and isotopic characteristics of the tonalite-trondhjemite gneisses which suggest their derivation by anatexis of greenstone-type precursors. In this model sialic nuclei formed by partial melting at the root zones of subsiding mafic-ultramafic crustal sectors, either in open ensimatic regimes or in divergent Red Sea type rift zones straddling older cratons (Glikson 1972, 1976, 1979; Naqvi 1981; Anhaeusser 1973, 1981). The differences between concepts of granite-greenstone relations corresponding to the two models are illustrated in Fig. 3.

The xenolith relations described above provide constraints for these alternative models. Principal relevant observations are:

1. Where contacts between gneisses and main greenstone belts are least deformed, intrusive relations are observed, including magmatic discordance, apophyses, xenoliths, and thermal aureoles. Late Archaean sequences may overlie older gneisses unconformably, e.g. Belingwe belt (Bickle et al. 1975), Jones Creek, Yilgarn (Durney 1972), Finland (Blais et al. 1978).
2. Xenoliths commonly decrease in abundance from greenstone belt contacts toward inner parts of batholiths. However, they also form intra-batholith screens representing synclinal keels which are often seen to merge with the main supracrustal zones (Fig. 1a, b, c; 2a, b, c).

Fig. 2, e

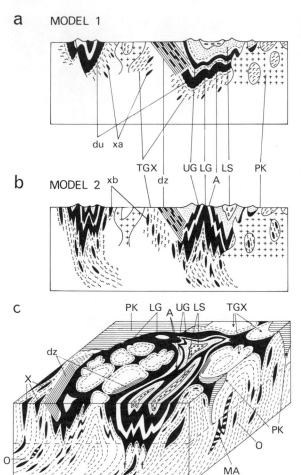

a   MODEL 1

b   MODEL 2

c

Fig. 3. Models illustrating alternative interpretations of gneiss-greenstone cross-sections and xenolith patterns. **a** *Model 1* concept of gneiss-greenstone basement-cover relation, involving deformed unconformities (*du*). **b** *Model 2* a concept of gneiss-greenstone relations involving primary and deformed intrusive contacts. **c** A portrayal of model 2 in block diagram, showing the transition from granite-greenstone into gneiss-granulite terrains with increasing crustal depth. *LG* lower greenstones; *UG* upper greenstones; *A* acid volcanics and/or sediments between lower and upper greenstones; *TGX* tonalite-trondhjemite-granodiorite gneisses with ultramafic-mafic xenoliths; *xa* xenolith distribution according to model 1; *xb* xenolith distribution according to model 2; *PK* post-kinematic granites; *dz* deformed zone; *du* deformed unconformity; *LS* late sedimentary sequences. *O* orthogneiss (xenolith-bearing); *MA* mafic and anorthositic intrusions; *f* fault

3. Arcuate xenolith screens serve as structural markers which define the internal geometry of the batholiths, outlining subsidiary gneiss domes, e.g. Selukwe, Barberton, Pilbara (Figs. 1a, b).

4. The density and structure of xenolith patterns defines greenstone belts as the largest in a series of enclaves, rather than discrete primary depositories within the gneiss terrain − a distinction of fundamental genetic importance.

5. There is rarely evidence for the primary configuration of early Archaean greenstone depositories. The widespread distribution of the xenoliths sets minimum limits on the original extent of the volcanic basin or crust from which they were derived. By contrast, some late Archaean greenstone sequences and late clastic sedimentary units have formed within inter-batholithic troughs or linear down-faulted basins, e.g. Moodies Group, Barberton (Eriksson 1980), Gorge Creek Group, Pilbara (Eriksson 1981).

The geometry of xenolith patterns furnishes critical evidence regarding granite-greenstone relations. Since the xenoliths merge or correlate with early greenstone units of the main synclinal belts, as indicated by continuous transitions, xenolith screens can be used as regional reference markers. Early greenstones and their derived xenoliths may be interpreted as the relics of simatic crustal regimes intruded by Na-rich felsic melts generated by partial melting of subsiding infracrustal zones (Glikson 1972, 1979). On the other hand, late Archaean greenstone sequences, postdating older gneiss-greenstone terrains, probably formed above heterogeneous crust. Since late greenstone sequences commonly strike parallel or subparallel in relation to older structural grains, concordant relations are the rule, and distinctons between older xenoliths and younger greenstone units may be difficult. This is exemplified by the relations between the Bulawayan Group (ca. 2.64 Ga) and the Mashaba gneiss containing Sebakwian Group xenoliths (pre-3.5 Ga) in Zimbabwe. Thus, in the Midland belt Bulawayan and Sebakwian rocks strike parallel and are separated by a long-acting unconformity. In the Javanahalli area, India (Fig. 2a), the Sargur Group is juxtaposed with younger parallel striking Dharwar Group volcanic-sedimentary rocks of the Chitradurga belt (Naqvi 1981). The parallel juxtaposition of temporally distinct supracrustal sequences results in stratigraphic correlation problems, particularly where isotopic age resetting and tectonic reactivation took place.

The large-scale structure of gneiss-greenstone systems varies considerably. Early Archaean terrains tend to have arcuate dome-syncline patterns, i.e. the Isua enclave (southwest Greenland), Zimbabwean, Kaapvaal, and Pilbara terrains. However, some early Archaean greenstone belts such as the Sargur Group are mostly, although not exclusively, linear. Late Archaean terrains have either arcuate or linear structural grain. The latter are defined by fold axes, elongated axes of plutons and fault systems, i.e. Yilgarn (Hallberg and Glikson 1981), Superior Province (Goodwin 1978) and Dharwar belts in India (Pichamuthu 1968). In detail, however, the large-scale linear grain contains smaller-scale arcuate dome-syncline structures, for example, in the Abitibi belt (Superior Province, Canada) (Goodwin 1978), and parts of the Eastern Goldfields and Murchison provinces (Yilgarn) (Hallberg and Glikson 1981). The late Archaean Slave Province of Canada shows mainly domal patterns (Henderson 1981). Older gneiss-greenstone structures have commonly controlled younger supracrustal structures, e.g. the superposition of Bababudan Group volcanics above Sargur Group volcanics in the Holenarsipur belt of India (Fig. 2c). Another example is the concordance between late-stage clastic sediments, including granite pebble-bearing conglomerates, and older volcanic synclines in spite of major intervening tectonic events, emplacement of batholiths, uplift and erosion. Thus, the Moodies Group in the Barberton Mountain Land, the Gorge Creek Group in the Pilbara, and the Chitradurga Group in India, all of which are essentially structurally conformable with underlying greenstone successions, have clearly postdated major plutonic and uplift episodes (Viljoen and Viljoen 1969; Hickman 1983; Naqvi 1981).

Factors controlling the arcuate vs. linear structure of gneiss-greenstone terrains include (1) the crustal level and the stress regime under which plutons were emplaced and (2) subsequent post-consolidation deformation history. Lateral transitions from arcuate to linear structural terrains are observed, for example, from the bifurcating northern Holenarsipur belt to the south (Hussain 1980)

(Fig. 2b, c). However, in most instances, the two types of terrain are separated by faults, by younger granites or by unexposed zones. The good preservation of primary volcanic and sedimentary structures and textures in arcuate greenstone enclaves, as contrasted with penetrative deformation in foliated gneiss-xenolith suites, suggests that the former may have originally occupied high crustal levels. In this model the least-deformed domal segments constitute little-deformed rigid islands in an otherwise foliated gneiss-greenstone crust. These nuclei contrast with the dominantly planar fabric of high-grade deep-seated gneiss-granulite terrains, e.g. the Ancient Gneiss complex of Swaziland (Hunter 1974), the Amitsoq and Uivak gneisses of Greenland and Labrador (Bridgwater et al. 1976) and the Eastern Ghats belt, southern India (Pichamuthu 1968). The low angle dips of gneissosity common in these terrains (Bridgwater et al. 1974; Park 1981) conceivably reflect viscous drag associated with differential horizontal movements at deep crustal levels, possibly associated with lateral detachments along the Moho.

The linear grain of greenstone synclinoria, batholiths and fault systems in late Archaean terrains such as the Yilgarn (Hallberg and Glikson 1981) and Superior Provinces (Goodwin 1978) are considered to be primary rather than superposed trends. In the Yilgarn the dominant north-northwest strike partly results from the elongate shape of domal intrusions such as the Widgiemooltha, Pioneer and Spargoville domes and similar bodies in the Edjudina area northeast of Kalgoorlie (Glikson 1978). These shapes may reflect an early Archaean tectonic grain that controlled anatectic processes. Possibly the secular structural variations reflect the declining geothermal gradients throughout the Archaean. The confinement of late Archaean basins between older cratonic blocks could have been a factor in the development of oriented intercratonic stress regimes at high crustal levels. Further, lower heat flow and a more brittle behaviour of the crust in the late Archaean could be a reason for the more pronounced faulting. Examples of younger deformation superposed on gneiss-greenstone systems are known from intersecting mobile zones. e.g. the Limpopo belt (Coward et al. 1976) and the Grenville belt (Wynne-Edwards 1976).

Interpretations of the three-dimensional structure of gneiss-greenstone terrains and concepts on their evolution are inherently intertwined. Since seismic and gravity data can only yield generalized density-averaged profiles (e.g. Drummond et al. 1981), an understanding of detailed transitions depends on surface observations in structurally tilted terrains where a range of crustal depths is exposed. Inherent is the question of the contemporaneity of low and high grade terrains, even where they are juxtaposed with each other (Glikson 1983). Continuous transitions have been observed in the Indian Shield in southern Karnataka (Allen et al. 1982; Raith et al. 1982; Janardhan et al. 1982; *see also Hansen et al., this Vol. and Condie and Allen, this Vol.*) (Fig. 2a, b) and in the Rice Lake − Pikwitonei belt area, Manitoba (Weber and Scoates 1978). Isotopic evidence suggests a broad contemporaneity of the spatially separated Barberton gneiss-greenstone terrain and the Ancient Gneiss complex in Swaziland (Barton 1981). Where continuous transitions occur they are best traced by means of the supracrustal xenolith swarms whose abundance generally decreases with higher metamorphic grade. This agrees with gravity evidence for a pinching-out downward of greenstone synclines at depths up to about 10 km (Constable 1978; cited in Archibald et al. 1981). The lower crust beneath gneiss-greenstone terrains is interpreted from seismic refraction studies as intermediate granulite of ca. 2.9 g/cm^3 density (Mathur 1976; Drummond et al. 1981). This value is similar to average felsic to intermediate charnockites including supracrustal xenoliths. The conceived vertical crustal zonation is illustrated in Fig. 3. As transitions between gneiss-greenstone and gneiss-granulite suites are commonly obscured or are interrupted by faults, closely spaced seismic reflection studies are needed to throw light on the structural and metamorphic zonation.

4 Summary

Archaean xenolith patterns provide important information bearing on theories of early crustal development, including the possibility of plate tectonics processes (Windley 1981; Kröner 1981), as follows:

1. Xenolith swarms and early volcanic units of greenstone belts can be traced into, or correlated with, each other, constituting contemporaneous units within individual terrains.
2. A complete scale hierarchy exists between early mafic-ultramafic successions of greenstone belts, intermediate-size enclaves, outcrop-size and inch-size xenoliths. Thus, the older greenstone units of the main belts can be regarded as mega-xenoliths.
3. In arcuate dome-syncline gneiss-greenstone terrains xenolith screens represent synclinal keels which allow identification of subsidiary domes of gneiss within the batholiths.
4. Highly deformed marginal batholith zones commonly contain foliated gneiss-greenstone intercalations which parallel the contacts, representing deformation and interthrusting of originally intrusive boundary zones.
5. Domal-arcuate gneiss-greenstone terrains signify the least deformed Archaean cratonic "islands" within, and at high levels of, an otherwise penetratively deformed crust.
6. Transitions from high-level granite-greenstone systems into deep crustal gneiss-granulite zones can be traced using supracrustal enclaves and xenoliths. In general the abundance of greenstone xenoliths decreases and their parallel deformation increases with crustal depth.
7. Late Archaean volcanic and sedimentary units of greenstone belts may overlap older sialic basement unconformably and therefore do not, in these instances, bear the same relation to xenolith swarms as early greenstone units.

Acknowledgements. I wish to thank M. A. Etheridge, A. M. Goodwin, A. Kröner, J. W. Sheraton, S. Sun, and R. J. Tingey for their comments. This paper is published with the permission of the Director, Bureau of Mineral Resources, Geology and Geophysics.

References

Allaart JH (1976) The pre-3760 m. y. old supracrustal rocks of the Isua area, Central West Greenland and the associated occurrence of quartz banded ironstone. In: Windley BF (ed) Early history of the earth, John Wiley and Sons, London, pp 177 – 190

Allen P, Condie KC, Narayana BL (1982) The Archaean low to high grade transition near Krishnagiri, Southern India (abstract). India-U.S. Precambrian Workshop, National Geophys Res Inst Hyderabad, p 17

Anhaeusser CR (1973) The evolution of the early Precambrian crust of Southern Africa. Phil Trans Roy Soc London A273:359 – 388

Anhaeusser CR (1981) Geotectonic evolution of the Archaean successions in the Barberton Mountain Land, South Africa. In: Kröner A (ed) Precambrian plate tectonics. Elsevier, Amsterdam, pp 137 – 156

Anhaeusser CR, Robb LJ (1980) Regional and detailed field and geochemical studies of Archaean trondhjemitic gneisses, migmatitis and greenstone xenoliths in the southern part of the Barberton Mountainland, South Africa. Precambrian Res 11:373 – 397

Anhaeusser CR, Ribb LJ (1981) Magmatic cycles and the evolution of the Archaean crust in the eastern Transvaal and Swaziland. Geol Soc Aust Spec Publ 7:457 – 468

Archibald NJ, Bettenay LF, Binns RA, Groves DI, Gunthorpe RJ (1978) The evolution of Archaean greenstone terrains, East Goldfields Province, Western Australia. Precambrian Res 6:103 – 131

Archibald NJ, Bettenay LF, Bickle MJ, Groves DI (1981) Evolution of Archaean crust in the Eastern Goldfields province of the Yilgarn Block, Western Australia. Geol Soc Aust Spec Publ 7:491 – 504

Arth JG, Hanson GN (1975) Geochemistry and origin of the early Precambrian crust of northeastern Minnesota. Geochim Cosmochim Acta 39:325 – 362

Baragar WRA, McGlynn JC (1976) Early Archaean basement in Canadian Shield: A review of the evidence. Geol Surv Can Paper 76:14

Barton JM (1981) The patterns of Archaean crustal evolution in southern Africa as deduced from the evolution of the Limpopo mobile belt and the Barberton granite-greenstone terrain. Geol Soc Aust Spec Publ 7:21 – 32

Bettenay LF, Bickle MJ, Boulter CA, Groves DI, Morant P, Blake TS, James BA (1981) Evolution of the Shaw batholith – an Archaean granitoid-gneiss dome in the Eastern Pilbara, Western Australia. Geol Soc Aust Spec Publ 7:361 – 372

Bickle MJ, Martin A, Nisbet EG (1975) Basaltic and peridotitic komatiites and stromatolites above a basal unconformity in the Belingwe greenstone belt, Rhodesia. Earth Planet Sci Lett 27:155 – 162

Bickle MJ, Bettenay LF, Boulter CA, Groves DI, Morant P (1980) Horizontal tectonic interaction of an Archaean gneiss belt and greenstone terrain involving horizontal tectonics: evidence from the Pilbara Block, Western Australia. Geology 8:525 – 529

Bickle MJ, Morant P, Bettenay LF, Boulter CA, Groves DI (1982) Archaean tectonics in the 3500 MA Pilbara Block: Structural and metamorphic tests of the batholith concept (abstract). Geol Assoc Canada Miner Assoc Canada Joint Ann Meeting, Winnipeg, Abstracts Vol. p 38

Binns RA, Gunthorpe RJ, Groves DI (1976) Metamorphic patterns and development of greenstone belt in the Eastern Yilgarn block, Western Australia. In: Windley BF (ed) Early history of the earth. Wiley and Sons, London pp 331 – 350

Blais S, Auvray B, Capdevila R, Jahn BM, Hameurt J, Bertrand JM (1978) The Archaean greenstone belts of Karelia and the komatiitic and tholeiitic series. In: Windley BF, Naqvi SM (eds) Archaean geochemistry. Elsevier, Amsterdam, pp 87 – 108

Bridgwater D, McGregor VR, Myers JS (1974) A horizontal tectonic regime in the Archaean of Greenland and its implications for early crustal thickening. Precambrian Res 1:158 – 165

Bridgwater D, Keto L, McGregor VR, Myers JS (1976) Archaean gneiss complex in Greenland. In: Escker A, Stuart Watt W (eds) Geol Surv Greenland, pp 18 – 75

Collerson KD, Bridgwater D (1979) Metamorphic development of early Archaean tonalitic and trondhjemitic gneisses: Saglek area, Labrador. In: Barker F (ed) Trondhjemites, dacites and related rocks. Elsevier, Amsterdam, pp 205 – 273

Condie KC, Allen P (1984) Origin of Archaean charnockites from southern India. This Vol., 182 – 203

Coward MP, Lintern BC, Wright L (1976) The precleavage deformation of the sediments and gneisses of the northern part of the Limpopo belt. In: Windley BF (ed) Early history of the earth. John Wiley and Sons, London, pp 323 – 330

Drummond BJ, Smith RW, Horwitz RE, Horwitz RG (1981) Crustal structure in the Pilbara and northern Yilgarn Blocks from deep seismic sounding. Geol Soc Aust Spec Publ 7:33 – 42

Durney DW (1972) A major unconformity in the Archaean, Jones Creek, Western Australia. J Geol Soc Aust 19:251 – 259

Eriksson KA (1980) Transitional sedimentation styles in the Moodies and Fig Tree Groups, Barberton Mountain Land, South Africa: evidence favouring an Archaean continental margin. Precambrian Res 12:141 – 160

Eriksson KA (1981) Archaean platform to trough sedimentation, east Pilbara Block, Australia. Geol Soc Aust Spec Publ 7:235 – 244

Gee RD, Baxter JL, Wilde SA, Williams IR (1981) Crustal development in the Archaean Yilgarn Block, Western Australia. Geol Soc Aust Spec Publ 7:43 – 56

Glikson AY (1972) Early Precambrian evidence of a primitive ocean crust and island nuclei of sodic granite. Geol Soc Am Bull 83:3323 – 3344

Glikson AY (1976) Stratigraphy and Evolution of primary and secondary greenstones: significance of data from southern hemisphere shields. In: Windley BF (ed) Early history of the earth. Wyllie and Sons, London, pp 257 – 278

Glikson AY (1978) Archaean granite series and the early crust, Kalgoorlie System, Western Australia. In: Windley BF, Naqvi SM (eds) Archaean geochemistry. Elsevier, Amsterdam, pp 151 – 174

Glikson AY (1979) Early Precambrian tonalite-trondhjemite sialic nuclei. Earth Sci Reviews 15:1 – 73

Glikson AY (1983) Exposed cross sections through the continental crust – a discussion. Earth Planet Sci Lett 64:168 – 170

Goodwin AM (1978) The nature of Archaean crust in the Canadian shield. In: Tarling DH (ed) Evolution of the earth crust. Academic Press, London, pp 175 – 218

Groves DI, Batt WD (1984) Spatial and temporal variations of Archaean metallogenic associations in terms of evolution of granitoid-greenstone terrains with particular emphasis on the Western Australian shield. This Vol., 73 – 98

Hallberg JA, Glikson AY (1981) Archaean granite-greenstone terrains of Western Australia. In: Hunter DR (ed) Precambrian of the Southern Hemisphere. Elsevier, Amsterdam, pp 33 – 96

Hansen EC, Newton RC, Janardhin AS (1984) Pressures, temperatures and metamorphic fluids across an unbroken amphibolite facies to granulite facies transition in Southern Karnataka, India. This Vol., 161 – 181

Hickman AH (1975) Precambrian structural geology of part of the Pilbara region. Geol Surv Western Austr Ann Rep 1974:68 – 73

Hickman AH (1983) Geology of the Pilbara Block and its environs. Geol Surv Western Austr Bull 127

Hunter DR (1974) Crustal development in the Kaapvaal craton: part 1 – The Archaean, Precambrian Res 1:259 – 294

Hussain SM (1980) Geological, Geophysical and Geochemical studies over the Holenarsipur schist belt, Karnataka. Unpubl Ph D thesis, Osmania Univ Hyderabad, India

Janardhan AS, Srikantappa C, Ramachandra HM (1978) The Sargur schist complex – an Archaean high grade terrain in Southern India. In: Windley BF, Naqvi SM (eds) Archaean geochemistry. Elsevier, Amsterdam, pp 127 – 150

Janardhan AS, Newton RC, Hansen EC (1982) The transformation of amphibolite facies gneiss to charnockite in southern Karnataka and northern Tamil Nadu (abstract). India-US Precambrian Workshop, Nat Geophys Res Inst Hyderabad: 14

Kröner A (1981) Precambrian plate tectonics. In: Kröner A (ed) Precambrian Plate Tectonics, Elsevier, Amsterdam, pp 57 – 83

Mathur SP (1976) Relation of Bouguer anomalies to crustal structure in Southwestern and Central Australia. Bur Miner Resour J Aust Geol Geophys 1:277 – 286

McGregor VR, Mason B (1977) Petrogenesis and geochemistry of metabasaltic and metasedimentary enclaves in the Amiseq gneisses, West Greenland. Am Mineral 62:887 – 904

Naqvi SM (1981) The oldest supracrustals of the Dharwar craton, India. J Geol Soc India 22:458 – 469

Park RC (1981) Origin of horizontal structures i.e. high grade Archaean terrains. Geol Soc Aust Spec Publ 7:481 – 490

Pichamuthu CS (1968) The Precambrian of India. In: Rankama K (ed) The Precambrian, Vol. 3, Interscience Publishers, New York, pp 1 – 96

Radhakrishna BP (1975) The two greenstone groups in the Dharwar craton. Indian Mineral 16:12 – 15

Raith M, Raase P, Lal RK, Ackermand D (1982) Regional geothermometry in the granulite facies terrain of South India (abstract). India-US Precambrian Workshop. Nat Geophys Res Inst Hyderabad, p 31

Robb LJ (1979) The distribution of granitophile elements in Archaean granites of the eastern Transvaal, and their bearing on geomorphological and geological features of the area. Econ Geol Res Unit, Univ Witwatersrand Inform Circ 129

Robb LJ (1981) Detailed studies of select migmatite outcrops in the region southwest of the Barberton greenstone belt and their significance concerning the nature of the early crust in this region. Geol Soc Aust Spec Publ 7:337 – 350

Stowe CW (1973) The older tonalite gneiss complex in the Selukwe area, Rhodesia. Geol Soc S Africa Spec Publ 3:85 – 96

Viljoen MJ, Viljoen RP (1969) A reappraisal of granite-greenstone terrains of shield areas based on the Barberton model. Geol Soc S Africa Spec Publ 2:245 – 274

Weber W, Scoates RFJ (1978) Archaean and Proterozoic metamorphism in the northwestern Superior Province and along the Churchill-Superior boundary, Manitoba. Geol Surv Canada Paper 78: 10, pp 5 – 16

Windley BF (1981) Precambrian rocks in the light of the plate tectonics concept. In: Kröner A (ed) Precambrian plate tectonics. Elsevier, Amsterdam, pp 1 – 16

Wynne-Edwards HR (1976) Proterozoic ensialic orogenesis: The millipede model of ductile plate tectonics. Am J Sci 276:927 – 953

Subject Index

Abitibi belt, Canada 35, 55, 81–90, 277
accessory zircons 235–249
accretion, energy of 12, 13
  heterogeneous 26, 30
  history 1–20
  inhomogeneous 29
  sequence 9–12
  temperature 11
  time 13
accretionary lapilli 84
Aldan shield 235–237, 243–247, 249
alteration, synvolcanic 87
Ancient Gneiss Complex, Swaziland 55,
  263, 278
anorthositic protocrust 34
Archaean batholith(s) 262, 264
  bimodal mafic-felsic suite 48, 51, 57, 61,
    62, 64, 65, 67, 81, 188
  charnockite(s), see charnockite
  continental crust, composition 47–68,
    229, 230
  granulite, see granulite
  greywacke 56–61, 65
  heat flow 67, 229
  metallogenetic associations 73–96
  sandstones 56, 57
  seawater 116, 117, 121, 126, 132, 134
  sediments 34, 47–68, 217
  shales 50–54, 63–67
  TTG rocks (tonalite-trondhjemite-
    granodiorite) 205, 219, 221, 227, 276
Auly Series, Ukrainian shield 253, 255, 256,
  259, 260

back-arc basin 55, 116
Baltic shield 235–239, 249
Barberton greenstone belt 115–134
  (Mountain Land) 29, 31, 32, 35, 36, 39,
    55, 77–79, 83, 84, 86, 88, 116, 119–121,
    125–127, 133, 146, 264, 265, 270, 276, 277
basaltic magma, wet 197
basanite 19
basement recycling 57
batch melting curves 192, 194–197
BIF (banded iron formation) 78, 79, 81–83,
  87, 89, 92–94, 206, 207, 209, 215, 220, 221,
  228, 253, 254

bimodal suite, see Archaean bimodal mafic-
  felsic suite
boninite 39

C1 abundance 3, 8, 9, 147, 149, 154
calc-alkaline fractionation trend 188
caliche 4
Cape Smith 29, 33
carbonic metamorphism 177–179, 228
chalcophile elements 18
charnockite(s) 161, 162, 164, 166–168, 173,
  175, 176, 179, 182–201, 244, 246, 258, 259,
  279
  (granulite) protolith(s) 197, 198, 213, 215,
    216, 223, 225, 236
charnockitization 199
chemical alteration patterns 115–134
chert 121, 266
chilled margin 122
chromitite 149, 152, 155
CO_2-streaming, purging 162, 177, 199
condensation behaviour 2
continental collision 179
  (flood) basalt 221, 222, 224, 227
core formation 9, 10, 12, 26, 28–30, 32, 33
  mantle differentiation 26, 30–34
cosmochemical constraints 4
crustal composition 48–50
  contamination 35, 36
  extension, thinning 84, 85, 89, 90, 93, 95
  growth 49, 68
  recycling 26
  underplating, see underplating
cumulate rocks 107, 198, 199

delamination 40
density, minimum for basaltic compositions 108
depleted Archaean tholeiite (TH1) 192, 197,
  198
Dharwar Group, India 141, 142, 277
discriminant function analysis 205, 215,
  216, 219
disproportionation of Fe 10, 18
Dnieper region, belt, complex 252, 253,
  258–260

early atmosphere   20
eclogite melting   39–41, 67
enderbite   241–244, 246, 258, 259
enriched Archaean tholeiite (TH2)   192, 197,
    198
Eu-anomaly   193, 197, 198, 205, 217,
    219–221, 226, 227, 230
    in sedimentary rocks   49–55, 62, 65, 199,
    220, 229
eucrite parent body (EPB)   7
eulysite   209, 219–221
Fig Tree Group   51–53, 55–58, 60, 118, 131
Finland   29, 39, 177, 221, 225, 229, 238, 274
fluid inclusions   163–165, 167–170, 175, 192
fluid-phase (metamorphism)   182, 193, 195,
    197, 225, 247
fore-arc basin   55
fractionation sequence   12
fuchsite schist   265

garnet lherzolite   220
geobarometry   146, 155, 161–164, 167, 170,
    173–176, 178, 184, 210
geochemical model studies   198
geothermometry   146, 155, 162–164, 170,
    178, 184, 209
gold mineralization   74, 86–89, 92, 95
Gorge Creek Group   52–55, 60, 277
Gorgona   29, 33
granite extraction   199, 200
    greenstone terrain(s) 73–96, 177, 183, 219,
    235, 237, 239, 240, 262, 263, 273–279
granitoid batholith   78, 262, 263
    diapirism, domes   78, 80, 81, 95
granulite metamorphism, facies, terrane   75,
    161–179, 182–201, 204–230, 236, 239, 241,
    243, 244, 246–249, 253, 258, 259, 270, 271,
    274, 276, 278, 279
greenstone belt, sequence   32, 33, 35, 39, 40,
    54, 55, 57, 61, 73, 77, 85, 92, 94–96,
    115–134, 139, 140, 146, 153, 163, 222, 224,
    236, 240, 241, 246, 248, 251, 254, 262, 265,
    268, 270, 274, 276, 277, 279
greenstones, platform-phase   73, 77–80, 85,
    86, 89, 92–95
    rift-phase   73, 74, 77, 78, 80–83, 87–90,
    92, 94, 95

hafnium content, in zircon   257
heat flow, see Archaean heat flow
Hebei Province, China   204–230
heterogeneity of I_{Nd} value(s)   212
high-grade (gneiss) terranes   74–76, 139,
    142,145,162–164,167,176,178, 191, 219,
    262, 263, 273, 278
homogeneous accretion model   10, 20
horizontal thrusting   162, 265
hot water plumes   86

igneous fractionation trend   188
inhomogeneous accretion model 1, 2, 9,
    12–14, 17, 18, 20, 29
intracratonic rifting   55, 140, 142
intracrustal melting   48, 229–230
island arc basalts, tholeiites   39, 41, 50, 63,
    224
isochron rotation   213
isotope homogenizaion   213, 223
isotopic equilibrium   19

Kaapvaal craton   85, 86, 88, 116, 178, 277
Kabbal (durga), India   161, 165, 166, 168,
    169, 177, 183, 184, 191
Kalgoorlie sequence   52–54, 60
Kambalda   34, 52, 60, 62, 63, 81, 91
Karelia   237–239, 249
Karnataka, southern India   138–156,
    161–179, 263, 270–274, 278
Keweenawan   41
khondalite   156
Kilbourne Hole   4
Komati Formation   116–119, 123, 126–129,
    132, 134
komatiite   4, 25, 29, 31–33, 36, 38–40, 73,
    74, 78, 79, 81, 85–90, 92–95, 99–113,
    115–117, 119, 127–134, 138–140,
    146–149, 153, 155, 209, 215, 217, 218, 221,
    254, 255
    magma   115, 119, 126
komatiitic magmas, experimental
    studies   108

Landsat imagery   266, 267, 270, 271
layered anorthosite   139
    ultramafic complex   138–156
least squares mixing method   129, 130
Limpopo belt   139, 156, 163, 178, 278
low pressure fractional crystallization   99,
    100
Luliang orogeny   205, 211

mafic-ultramafic crust   264
    (greenstone) xenolith (patterns)   262–279
magma mixing   110–113
    ocean   13, 14, 17
magmatic diapirism   262
    overplating, see overplating
    underplating, see underplating
mantle, chemical stratification   30–32, 36
    chemistry   1–20
    composition   4–6, 25–44, 67
    convection   13, 31, 33, 40, 42
    diapir   164, 178
    enrichment process   26, 41
    heterogeneity   25–27, 32, 34–36, 42
    metasomatism   16, 26, 27, 40, 41
    nodules, xenoliths   27, 33, 41

primitive   1, 3, 4, 6–9, 13–15, 27, 28, 67
sink of volatiles   20
source   99, 100
Mars   17, 18
metal segregation   9, 10, 30
metamorphic dehydration   162
fluids   161–179
gradient   162
metasomatized mantle   226
meteorite bombardment   32
microthermometry   170
Moodies Group   51–53, 55–57, 60, 118, 276, 277
MORB   6, 16, 17, 37, 38, 50, 63, 192, 196–198, 222, 224
multiphase differentiation   226
Munro Township   29, 39, 99–113, 127, 134
Murchison C 2 meteorite   2

Nd isotopes   3, 83, 211–215, 238
Newton Township, Canada   35
noble metals   32
Norseman-Wiluna belt   80–84, 88, 90–92, 94

ocean floor (ridge) basalt, tholeiite   115, 120, 126, 127, 156, 192, 197, 222, 224
island basal   37, 224
Omolon massif, USSR   235–237, 246–249
Onverwacht Group   55, 116–118, 122, 128, 131, 134, 264
outgassing   12
overplating   178–179
oxygen fugacity   2, 7, 10, 12, 154
in the core   10
isotopes   115–134, 239, 246, 248

Peninsular Gneiss   141, 142, 161, 166, 177, 270, 271, 275
pigeonite   110
Pilbara Block (craton)   29, 35, 36, 39, 51–53, 55, 63, 64, 77–80, 82–91, 94, 237, 249, 262, 263, 265, 266, 270, 276, 277
pillow lava   115–134
plagiogranite   251, 253, see also tonalites and trondhjemites
planetary accretion   2
planetesimals   2, 11
plate tectonics   14, 179, 228, 279
porphyry-style mineralization   86, 87, 89
primitive mantle, see mantle, primitive
chemical composition   27–35, 41, 42, 67
Pultusk bronzite chondrite   18
pyrolite   38, 39
pyroxene texture, acicular   99, see also spinifex texture

Qianxi Group   204–210, 212, 213, 217, 219, 223

REE geochemistry in Archaean
granulites   217–230
in sedimentary rocks   47–68
mobility   212, 213, 222–223
refractory elements   3, 4, 6
Rhodesian craton, Zimbabwe   29, 35, 178, 201, see also Zimbabwe (craton)

San Carlos spinel-lherzolite   3, 16, 19
Sargur Group, supracrustal series, schist
belt   138–143, 145–148, 151, 153, 155, 156, 263, 270, 277
seawater alteration   115, 122, 125, 126
shergotty parent body (SPB)   7, 17, 18
Si deficiency   4, 8
siderophile elements   9, 10
Sino-Korean Paraplatform   205
Slave Province   52, 56, 84, 277
solar condensation temperature   28
solar nebula   2, 10, 11
spinel-lherzolite   3, 4, 6, 14–17, 19, 220
spinifex texture   103, 105, 107, 110–112, 127, 128, 155
subducted ocean crust   34, 38, 40
subduction   13, 14, 17
Superior Province   56, 77, 99, 277, 278
supracrustal enclaves, xenoliths   263, 264, 270, 278, 279
surface contamination   20

tectonic mixing   188, 230
thin-skinned thrusting   205, 228, 230
tonalite(s) and trondhjemite(s)   40, 41, 51, 52, 57, 61, 62, 64, 166, 182, 184, 188–191, 193, 195, 197–200, 205, 209, 238, 241, 246–249, 251, 253, 255, 257, 263–265, 269, 270, 273, 274, 276
fractionation   198
solidus   264
trace elements, in Archaean sedimentary
rocks   50–54
two component accretion model   9, 20

Ukrainian shield   235–237, 239–243, 249, 251–261
underplating   26, 27, 40, 179, 205, 228
U-Pb zircon data   211, 235–249, 255–260

volatile elements   7–12, 14, 17–20, 28
volatiles, in Earth's mantle and crust   19, 20
volatilization studies   2

Warrawoona Group   55
Western Australia   73–98, 263, 265, 266
West Greenland (Isua sequence)   29, 35, 51, 52, 54, 55, 60, 63, 64, 77, 249, 263, 277

xenolith fragment(ation)   264

Yellowknife Supergroup   55, 60
Yilgarn Block (craton)   29, 51–53, 55, 60,
  62–64, 66, 77, 78, 81–83, 85, 87–94, 133,
  274, 276, 278

Zimbabwe (craton)   77, 84, 88, 90, 94,
  263–265, 270, 277, *see also* Rhodesian
  craton zircon chemistry   248–249

Crustal Evolution of Southern Africa

3.8 Billion Years of Earth History

By A. J. Tankard, M. P. A. Jackson, K. A. Eriksson,
D. K. Hobday, D. R. Hunter, W. E. L. Minter

With a contribution by S. C. Eriksson

1982. 182 figures. XV, 523 pages
ISBN 3-540-90608-8

Contents: Tectonic Framework. – Archean Crustal Evolution: Granite Greenstone Terrane: Kaapvaal Province. Granulite-Gneiss Terrane: Limpopo Province. – Early Proterozoic Supracrustal Development: The Golden Proterozoic. The Transvaal Epeiric Sea. The Bushveld Complex: A Unique Layered Intrusion; The Vredefort Dome: Astrobleme or Gravity-Driven Diapir? The Earliest Red Beds. – Proterozoic Orogenic Activity: Namaqua-Natal Granulite-Gneiss Terranes. The Pan African Geosynclines. – The Gondwana Era: The Cape Trough: An Aborted Rift. The Intracratonic Karoo Basin. – After Gondwana: Fragmentation and Mesozoic Paleogeography. Kimberlites and Associated Alkaline Magmatism. Changing Climates and Sea Levels: The Cenozoic Record. – References. – Index.

The southern African subcontinent is a key segment of the Earth's crust, providing a record of geological processes that occurred over the entire span of terrestrial history. This volume, of interest to a broad spectrum of geologists studying every time period, is a contemporary analysis of the crustal evolution of this unique region. Site of the discovery of the earliest evidence of life to date, this area is known for its remarkable geological formations and great mineral wealth, particularly gold, diamonds, and strategic minerals. Six experts have collaborated to make this book a modern, process-oriented approach, focusing on the dynamics of crustal structure. "In the book", B. F. Windley writes in the foreword, "the reader will find a detailed review of factual data, together with a balanced account interpretive models." With outstanding maps and illustrations prepared especially for this volume. **The Crustal Evolution of Southern Africa** is an authoritative, multidisciplinary approach to a region of great geological importance.

Springer-Verlag
Berlin
Heidelberg
New York
Tokyo

L. J. Salop

Geological Evolution of the Earth During the Precambrian

Translated from the Russian by V. P. Grudina
1983. 78 figures. XII, 459 pages. ISBN 3-540-11709-1

Contents: General Problems in Division of the Precambrian. – The Katarchean: Rock Records. Geologic Interpretation of Rock Record. – The Paleoprotozoic (Archeoprotozoic): Rock Records. Geologic Interpretation of Rock Record. – The Mesoprotozoic: Rock Records. Geologic Interpretation of Rock Record. – The Neoprotozoic: Rock Records. Geologic Interpretation of Rock Record. – The Epiprotozoic: Rock Records. Geologic Interpretation of Rock Record. – The Eocambrian (Vendian sensu stricto): Rock Records. Geologic Interpretation of Rock Record. – Geologic Synthesis. – References. – Subject Index. – Index of Local Stratigraphic Units and of Some Intrusive Formation.

Geologic evolution from approximately $3.7 \cdot 10^9$ y up to the beginning of the Cambrian period is the subject of this study. The author, head of the Precambrian Geology Department of the All-Union Research Geology Institute in Leningrad, proposes herein a new, detailed division of the Earth's early history. His geologic conclusions are based on rock record data collected the world over, as well as on new methods for studying older formations and the refinement of existing methods, in particular the division, dating and correlation of "silent" metamorphic strata.

The scope of the book embraces such problems as

– the evolution of sedimentogenesis, of tectonic structures and of larger elements of the Earth's crust
– the periodicity of tectogenesis
– the relation between tectonic processes and magmatism
– the origin of the oldest astroblemes and their appearance on the Earth's surface, as well as possible causes for changes in organic evolution.

The author concludes with a consideration of a directed and irreversible geologic evolution of the planet.

With the great significance new data on the Precambrian bear for the theoretical and philosophical foundations of science, showing as they do previously unknown general regularities and unexpected evolutionary relations of different geologic phenomena and processes, this book will prove required reading for the Precambrian specialist and advanced student and will also interest the researcher in other geologic branches.

Springer-Verlag
Berlin
Heidelberg
New York
Tokyo